Wiring Regulations in Brief

About the author

Ray Tricker (MSc, IEng, FIET, FCMI, FIQA, FIRSE) is the Principal Consultant of Herne European Consultancy Ltd – a company specialising in ISO 9000 Management Systems- is also an established Butterworth-Heinemann author (18 titles). He served with the Royal Corps of Signals (for a total of 37 years) during which time he held various managerial posts culminating in being appointed as the Chief Engineer of NATO's Communication Security Agency (ACE COMSEC).

Most of Ray's work since joining Herne has centred on the European Railways. He has held a number of posts with the Union International des Chemins de fer (UIC) (e.g. Quality Manager of the European Train Control System (ETCS)) and with the European Union (EU) Commission (e.g. T500 Review Team Leader, European Rail Traffic Management System (ERTMS) Users Group Project Co-ordinator, HEROE Project Co-ordinator) and currently (as well as writing books on diverse subjects as Optoelectronics, Medical Devices, ISO 9001:2000 and Building Regulations for Elsevier under their Butterworth-Heinemann and Newnes imprints!) he is busy assisting small businesses from around the world (usually on a no-cost basis) produce their own auditable Quality and/or Integrated Management Systems to meet the requirements of ISO 9001:2000, ISO 14001 and OHSAS standards etc. He is also a UKAS Assessor for the assessment of certification bodies for the harmonisation of the trans-European high-speed railway network.

To Claire

Wiring Regulations in Brief

A Complete guide to the requirements of the
16th Edition of the IEE Wiring Regulations,
BS 7671 and Part P of the Building

Ray Tricker

ELSEVIER

AMSTERDAM • BOSTON • HEIDELBERG • LONDON
NEW YORK • OXFORD • PARIS • SAN DIEGO
SAN FRANCISCO • SINGAPORE • SYDNEY • TOKYO

Butterworth-Heinemann is an imprint of Elsevier

Butterworth-Heinemann is an imprint of Elsevier
Linacre House, Jordan Hill, Oxford OX2 8DP
30 Corporate Drive, Suite 400, Burlington, MA 01803

First published 2007

British Library Cataloguing in Publication Data
A catalogue record for this book is available from the British Library

Library of Congress Cataloging-in-Publication Data
A catalog record for this book is available from the Library of Congress

ISBN 13: 978-0-7506-6851-4
ISBN 10: 0-7506-6851-2

For information on all Butterworth-Heinemann publications visit
our web site at http://books.elsevier.com

Typeset by Charon Tec Ltd (A Macmillan Company), Chennai, India
www.charontec.com

Printed and bound in the UK, by MPG Books Ltd.
07 08 09 10   10 9 8 7 6 5 4 3 2 1

Contents

Preface

The Industrial Revolution during the 1800s was responsible for causing poor living and working conditions in ever-expanding, densely populated urban areas. Outbreaks of cholera and other serious diseases (through poor sanitation, damp conditions and lack of ventilation) forced the government to take action. Building control took on the greater role of health and safety through the first Public Health Act in 1875 and this eventually led to the first set of national building standards (i.e. The Building Regulations).

As is the case with most official documents, as soon as they were published, they were almost out of date and consequently needed revising. So it wasn't too much of a surprise to learn that the committee responsible for writing the Public Health Act of 1875 had overlooked the increased use of electric power for street lighting and for domestic purposes. Electricity was beginning to become increasingly popular but, as there were no rules and regulations governing their installation at that time, the companies or person responsible simply dug up the roads and laid the cables as and where they felt like it!

From a health and safety point of view the government of the day expressed extreme concern at this exceedingly dangerous situation and so in 1882 *The Electric Lighting Clauses Act* (modelled on the previous 1847 *Gas Act*) was passed by Parliament. This legislation was implemented by '*Rules and Regulations for the prevention of Fire Risks Arising from Electric Lighting*' and it is this document that is the forerunner of today's IEE Wiring Regulations. Since then, this document has seen a succession of amendments, new editions and new titles and is now the well-known and respected 16th Edition of the IEE Wiring Regulations (i.e. BS 7671: 2000 *Requirements for Electrical Installations*).

The current legislation for all building control, on the other hand is the Building Act 1984 which is implemented by the Building Regulations 2000. These Building Regulations are a set of minimum requirements designed to secure the health, safety and welfare of people in and around buildings and to conserve fuel and energy in England and Wales. They are basic performance standards which are supported by a series of documents that correspond to the different areas covered by the regulations. These are called 'Approved Documents' and they contain practical and technical guidance on ways in which the requirements of the Building Act 1984 can be met.

Since the introduction of the Public Health Act in 1875 there has always, therefore, been a direct link between electrical installations and building control and parts of all of the Approved Documents have an affect on these sorts of installation. With the publication of Approved Document P for 'Electrical Safety' in 2005, however, the design, installation, inspection and testing of electrical installations has now become inextricably linked to building control and the purpose of this book is to attempt to draw all of the various requirements together.

Over the past 120 years there have been literally hundreds of books written on the subject of electrical installations, but the aim of *Wiring Regulations in Brief* is not just to become another book on the library shelf to be occasionally looked at. The intention is that it will provide professional engineers, students and (to a lesser degree) the unqualified DIY fraternity with an easy-to-read reference source to the official requirements of BS 7671: 2001 and The Building Regs for electrical safety and electrical installations.

Although BS 7671: 2001 is well structured and has separate sections for all the main topics (e.g. safety protection, selection and erection of equipment and so on) it is not the easiest of standards to get to grips with for a particular situation. Occasionally it can be very confusing and requires the reader to constantly flick backwards and forwards through the book to find what it is all about.

For example, Regulation (471-08-02) states that: '*installations which are part of a TN system, shall meet the requirements for earth fault loop impedance and of circuit protective conductor impedance as specified by **Regulations 413-02-08 and 413-02-10 to 413-02-16** (see also **Regulation 413-02-02**). Where the disconnection times specified by **Regulation 413-02-08** cannot be met by the use of an overcurrent protective device, **Regulation 413-02-04** shall be applied.*' This looks pretty confusing at first glance but then you will find that if you looked at the first reference (i.e. 413-02-08) which concerns maximum disconnection times, you will find that this in turn cross-refers to 413-02-04, 413-02-12 and 413-02-13, which in turn cross-refers to 413-02-02, 413-02-06 to 413-02-26 (inclusive) and 547-02-01. You will then begin to see the extent of the problem.

The intention of *Wiring Regulations in Brief*, therefore, is to peel away some of this confusion and 'officialise' and provide the readers with an on-the-job reference source that they can quickly use without having to delve backwards and forwards though the standard.

Please note, however, that this is **only** the author's impression of the most important aspects of the Wiring Regulations and their association with the Building Regulations. It should, therefore, only be treated as an aide mémoire to the Regulations and electricians should **always** consult BS 7671 to satisfy compliance.

Wiring Regulations in Brief is structured as follows:

Chapter 1 Introduction	The background to BS 7671, what it contains and a description of its unique numbering system, objectives and legal status. The effect that the Wiring Regulations have on other Regulations and how this British Standard can be implemented.

Chapter 2 Domestic buildings	The requirements of the Building Act 1984, the Building Regulations: 2000 and their Approved Documents (which provide guidance for conformance) and how these Building Control Regulations interrelate with the Wiring Regulations. A résumé of the responsibilities for electrical installations. The types of inspections and tests that have to be completed and the requirements for records. The contents of Approved Document P for electrical safety and other relevant Approved Documents (such as those for Fire Safety, Access and Facilities for Disabled People, Conservation of Fuel and Power, Resistance to the Passage of Sound etc.) together with a listing of all the most important requirements that directly concern electrical installations.
Chapter 3 Mandatory requirements	In this chapter you will find a list all of the mandatory and the more essential requirements for the design, installation and maintenance of safe electrical installations that are called up in the Wiring Regulations and the Building Regulations. As this chapter is primarily intended for reference purposes, explanations have been kept to an absolute minimum as more detail is available elsewhere in this book and/or other publications.

 Note: Whilst the requirements from the Wiring Regulations are normally prefaced by the 'shall' (meaning that this section **is** a mandatory requirement), you will notice that the Building Regulations use the words 'should' (i.e. recommended), 'may' (i.e. permitted) or 'can' (i.e. possible). The reason for this is that Approved Documents reproduce the actual *requirements* contained in the Building Regulations relevant to a particular subject area. This is then followed by *practical and technical guidance* (together with examples) showing how the requirements could be met in some of the more common building situations. There may, however, be alternative ways of complying with the requirements to those shown in the Approved Documents and you are, therefore, under no obligation to adopt any particular solution in an Approved Document – and you may prefer to meet the requirement(s) in some other way, but you **must** meet the requirement!.

Chapter 4 Earthing	This chapter reminds the reader about the different types of earthing systems and earthing arrangements. It then lists the main requirements for safety protection (direct and indirect contact), protective conductors and protective equipment before briefly touching on the test requirements for earthing.

Chapter 5 External influences	Currently Chapter 32 of the Regulations concerning external influences is still under development and is, therefore, at too early a stage for adoption as the basis for a national standard. It is anticipated that full details will be included in future revisions of this standard, but in advance of the eventual publication of Chapter 32, a list of external influences and their characteristics have been included as an appendix (i.e. Appendix 5) to the Regulations. Meanwhile, the BS EN 60721 series of standards on environmental conditions has been available for some time and, in the absence of any 'official' regulations from BSI/IEC, Chapter 5 provides guidance on external influences. Also included in this chapter are extracts from the current Regulations that have an impact on the environment.
Chapter 6 Electrical equipment, components, accessories and supplies	The amount of different types of equipment, components, accessories and supplies for electrical installations currently available is enormous and any attempt to cover every type, model and/or manufacture would prove an impossible task for a book such as this. The intention of this chapter, therefore, is to provide a catalogue of all the different types identified and referred to in the Wiring Regulations (e.g. luminaries, RCDs, plugs and sockets etc.) and then make a list of the specific requirements that are sprinkled throughout the Regulations. For your convenience this catalogue has been compiled in alphabetical order.
Chapter 7 Cables and conductors	Within the Wiring Regulations there is frequent reference to different types of cables (e.g. single core, multicore, fixed, flexible etc.) conductors (such as live supply, protective, bonding etc.) and conduits, cable ducting, cable trunking and so on. Unfortunately, as is the case for equipment and components, the requirements for these items are liberally sprinkled throughout the Standard. The aim of Chapter 7, therefore, is to provide a catalogue of all the different types identified and referred to in the Wiring Regulations in three main headings (e.g. cables, conductors and conduits etc.) and then make a list of their essential requirements.

Chapter 8 Special installations and locations	Whilst the Regulations apply to all electrical installations in buildings, there are also some indoor and out-of-doors special installations and locations that are subject to special requirements owing to the extra dangers they pose. Chapter 8 considers the requirements for these special locations and installations.
Chapter 9 Inspection and testing	To meet the requirements for electrical safety, it is essential for any electrician engaged in inspection, testing and certification of electrical installations to have a **full** working knowledge of the IEE Wiring Regulations. The electrician must also have above-average experience and knowledge of the type of installation under test in order to carry out **any** inspection and testing. Chapter 9 provides a consolidated list of how electrical installations shall be inspected and tested as well as a brief insight into some of the test equipment that may be used.
Chapter 10 Installation, maintenance and repair	This final chapter of the book provides some guidance on the requirements for installation, maintenance and repair of electrical installations. It lists the Regulations' requirements for these activities with respect to electrical installations and (in an Appendix) provides an example stage audit checklist for designers and engineers to use.

These chapters are then supported by the following appendices:

Appendix A	Symbols used in electrical installations
Appendix B	List of symbols
Appendix C	SI units for existing technology
Appendix D	Acronyms and abbreviations
Appendix E	British Standards currently used with the Wiring Regulations (by standard and by title)
Appendix F	List of useful contacts

plus a full index.

It is hoped that the following symbols will help you get the most out of this book:

Need to be careful (e.g. very necessary requirement, a potential minefield or legal/statutory requirement)

A good idea or a useful reminder

and for your assistance, I have also highlighted all the really essential and/or mandatory requirements of a particular section as shown in the following example:

 Protection against electric shock shall be provided.   410-01-01

 Note: If any reader has any thoughts about the contents of this book (such as areas where perhaps they feel I have not given sufficient coverage, too much coverage, omissions and/or mistakes etc.) then please let know by e-mailing me at ray@herne.org,uk and I will make suitable amendments in the next edition of this book.

In the meantime Enjoy!

Ray Tricker

Acknowledgements

I would like to thank the Institution of Engineering and Technology (IET) for giving me permission to reproduce the following Tables and Figures:

Tables 4.3, 4.4, 4.5, 4.6, 5.4, 5.18, 6.6, 6.7, 6.8, 6.9, 6.12, 6.14, 6.15, 7.2, 7.3, 7.4, 7.6, 7.7, 8.1, 8.2, 8.3, 8.4, 8.5, 8.6, 8.7, 8.8, 8.9, 8.10, 9.5, 9.7, 9.8, 9.9, 9.14, 9.16, 9.17, 9.18 and 9.21, and Figures 8.5, 8.7, 9.11, 9.14, 9.16, 9.17, and 9.27 are taken from The IEE Wiring Regulations: BS7671: 2001 incorporating Amendments 1 & 2: 2004 (The IEE, London, UK in agreement with BSI, 2004) ISBN 0863413730.

Figures 9.4, 9.5, 9.6, and 9.8 are taken from The IEE On-Site Guide (BS 7671:2001 16th Edition Wiring Regulations including Amendments 1 & 2: 2004) (IEE Publication, 2004) ISBN 0863413749.

I would also like to thank the following organisations for providing me with assistance in the preparation of this book and for giving me permission to use copyrighted material for illustration purposes in the following Tables and Figures:

BRE Certification Ltd for use of their logo in Figure 2.3.

BSI for use of their logo in Figure 2.3, for giving permission to reproduce Figure 5.8 and for details of specific fuse and circuit breaker ranges extracted from British Standards in tables 6.8, 8.3, 8.4, 8.8, 8.9, 9.8 and 9.17.

CORGI Competent Persons Scheme for use of their logo in Figure 2.3.

ELECSA Limited for use of their logo in Figure 2.3.

NAPIT Certification Limited for use of their logo in Figure 2.3.

NICEIC for use of their logo in Figure 2.3.

OFTEC for use of their logo in Figure 2.3.

TrustMark for use of their logo in Figure 2.4.

TLC for use of their pictures for Figures 9.29, 9.32, 9.34, and 9.33.

Fluke for use of their pictures for Figures 9.30, 9.31, and 9.35.

 Note: Please see Appendix F for full contact details of all these organisations.

In addition I would like to give due recognition to the following tables and figures which are reprinted by kind permission from Elsevier.

Figure 1.3 is taken from Introduction to Health and Safety at Work, Second Edition, Hughes & Ferrett, 2005, ISBN 0750666234.

Tables 4.2 and 6.4, and figures 4.13, 4.14, 4.15, 4.16, 4.19, 4.26, 4.28, 4.29, 9.2, 9.4, and the inside back cover diagram are taken from 16th Edition IEE Wiring Regulations: Explained and Illustrated, Seventh Edition, Scaddan, 2005, ISBN 0750665394.

Tables 4.7, 9.1, and 9.3 and figure 8.8 are taken from Electrical Installation Work, Fifth Edition, Scaddan, 2005, ISBN 0750666196.

Figures 9.3, 9.7, 9.9, and 9.34 are taken from 16th Edition IEE Wiring Regulations: Inspection, Testing and Certification, Fifth Edition, Scaddan, 2005, ISBN 0750665416.

Appendix A is taken from Basic Electrical Installation Work, Fourth Edition, Linsley, 2005, ISBN 0750666242.

I have also used the following from some of my previous publications:

Figures 5.4, 5.5, 5.6, 5.7, 5.9, 5.10, 5.11, 5.12, and 5.13 are taken from Environmental Requirements for Electromechanical and Electronic Equipment, Tricker, 1999, ISBN 0750639024

Figures 2.2, 2.3, 2.5, 2.6, 2.9, 2.10, 2.11, 2.12, 4.24, 9.18, 9.19, 9.21, 9.22, and 9.24 are taken from Building Regulations in Brief, Fourth Edition, Tricker, 2006, ISBN 075068058X

 Note: The BSI logo, Kitemark and the Kitemark symbol are produced with permission of the British Standards Institution and are the registered trademarks of such in the United Kingdom and other in other countries around the World.

Further information

Further reading

Copies of the Wiring Regulations may be obtained from either the IET:
P.O. Box 96
Stevenage
SG1 2SD, UK
Tel: +44 (0)1438 767328
Email: sales@theiet.org
or online at www.iee.org/shop

or BSI:
BSI Customer Services
389 Chiswick High Road
London W4 4AL, UK
Tel: +44 (0)20 8996 9001
Email: orders@bsi-global.com
or online at www.bsi-global.com/bsonline

Further assistance

The IET publishes a range of books and runs courses and services for industry
to support the use and application of BS 7671. Of particular interest are the
IEE's series of Guidance Notes which offers extensive, industry-endorsed
guidance to designers and installers in the effective use of BS 7671.

Seven IEE Guidance Notes are currently available. These are:

1 Selection and Erection of Equipment
2 Isolation and Switching
3 Inspection and Testing
4 Protection against Fire
5 Protection against Electric Shock
6 Protection against Overcurrent
7 Special Locations

Other products

IEE's On-site guide

A convenient, practical guide for electricians which covers domestic installations and smaller industrial and commercial installations up to 100 A, 3-phase.

The most widely used guide in the industry, it removes the need for detailed calculations and outlines a detailed inspection and testing regime for installations. Comprehensive checklists and procedures are provided.

IEE's Commentary on BS 7671: 2001

Written for designers and managers by the IEE's former Principal Engineer, Paul Cook, this title provides clear interpretations of and guidance to the Regulations.

Codes of practice

The *IEE Code of Practice for In-Service Inspection and Testing of Electrical Equipment* offers guidance for the inspection, testing and maintenance of electrical appliances, plus advice on compliance with health and safety legislation.

The *IEE Electrical Maintenance – Code of Practice* offers guidance on electrical aspects of building maintenance, including electrical installation, fire alarms, emergency lighting and more, plus detailed guidance on legal responsibilities.

CD-ROM of IEE Wiring Regulations

The CD-ROM version of BS 7671 (incorporating all amendments) is a fully structured electronic reference tool, hyperlinked and fully cross-referenced. As well as BS 7671, the CD-ROM contains the *On-Site Guide*, all seven *Guidance Notes* and both *IEE Codes of Practice*.

To order any of these publications, contact The IET:
by telephone: +44 (0)1438 767328
by fax: +44 (0)1438 742792
by email: sales@theiet.org
via their website: www.iee.org/shop

 For free downloads and updates you can join their email list by visiting www.iee.org/technical

Other publications

IEE On-Site Guide (BS 7671, 16th Edition Wiring Regulations). The Institution of Electrical Engineers. ISBN 0-85296-987-2, 2002

IEE Guidance Note 1: Selection and erection of equipment. 4th edition. The Institution of Electrical Engineers. ISBN 0-85296-989-9, 2002

IEE Guidance Note 2: Isolation and switching. 4th edition. The Institution of Electrical Engineers. ISBN 0-85296-990-2, 2002

IEE Guidance Note 3: Inspection and testing. 4th edition. The Institution of Electrical Engineers. ISBN 0-85296-991-0, 2002

IEE Guidance Note 4: Protection against fire. 4th edition. The Institution of Electrical Engineers. ISBN 0-85296-992-9, 2003

IEE Guidance Note 5: Protection against electric shock. 4th edition. The Institution of Electrical Engineers. ISBN 0-85296-993-7, 2002

IEE Guidance Note 6: Protection against overcurrent. 4th edition. The Institution of Electrical Engineers. ISBN 0-85296-994-5, 2003

IEE Guidance Note 7: Special locations. 2nd edition (incorporating the 1st and 2nd amendments). The Institution of Electrical Engineers. ISBN 0-85296-995-3, 2003

New wiring colours. Leaflet published by the IEE, 2004. Available for downloading from the IEE website at www.iee.org/cablecolours

ECA comprehensive guide to harmonised cable colours, BS 7671: 2001 Amendment No 2. Electrical Contractors' Association, March 2004

New fixed wiring colours – A practical guide. National Inspection Council for Electrical installation Contracting (NICEIC), Spring 2004

Building Regulations in Brief. 4th edition. By Ray Tricker published by Butterworth-Heinemann. ISBN 0-7506-8058-X (2006)

The Building Regs but not Part P. Article published by the IEE, Spring 2004. Available for downloading from the IEE website at http://www.iee.org/Publish/WireRegs/ IEE_Building-Regs.pdf

Electrical Installers' Guide to the Building Regulations. NICEIC and ECA, August 2004. Available from www.niceic.org.uk and www.eca.co.uk

1

Introduction

1.1 Introduction

The IEE Wiring Regulations is a 321-page document that defines the way in which all electrical installation work must be carried out. It does not matter whether the work is carried out by a professional electrician or an unqualified DIY enthusiast, the installation **must** comply with the Wiring Regulations.

The current edition (i.e. as at 1 January 2006) of the Regulations is 'The Brown Book'.

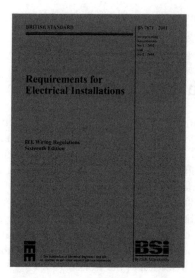

Figure 1.1 BS 7671: 2001 cover of standard publication

Or, to give it its official name:

> *BS 7671: 2001 Requirements for Electrical Installations (Incorporating Amendments No 1: 2002 and No 2: 2004), The IEE Wiring Regulations (Sixteenth Edition)*

This British Standard is published with the full support of the BEC (i.e. the British Electrotechnical Committee, who are the UK national body responsible for formal standardisation within the electrotechnical sector) in partnership with the BSI (i.e. the British Standards Institution – who have ultimate responsibility for all British Standards produced within this sector) and The Institution of Electrical Engineers (IEE) (now renamed as the IET – Institution of Engineering and Technology) who, with over 130,000 members, are Europe's largest grouping of professional engineers involved in power engineering, communications, electronics, computing, software, control, informatics and manufacturing.

The technical authority for this standard is The National Committee for Electrical Installations (JPEL/64), which is a Joint IET/BSI Technical Committee responsible for all the work previously undertaken by the IEE Wiring Regulations Committee and BSI Technical Committee PEU64. Copyright is held jointly by the IET and BSI.

 Note: Please note that all references in this book to the *'Wiring Regulations'* or the *'Regulation(s)'*, where not otherwise specifically identified, refer to BS 7671: 2001 (2004), *Requirements for Electrical Installations*.

BS 7671: 2001 was issued on 1 June 2001 and came into effect on 1 January 2002.

 All installations that were (or are) designed after 1 January 2002 **must** comply with this edition, as amended and expanded.

1.2 Historical background

The first public electricity supply in the UK was at Godalming in Surrey, in November 1881, and mainly provided street lighting. At that time, there were no existing rules and regulations available to control their installations and so the electricity company just dug up the roads and laid the cables in the gutters. This particular electricity supply was discontinued in 1884.

On 12 January 1882, the steam-powered Holborn Viaduct Power Station opened and this facility supplied 110 V d.c. for both private consumption and street lighting. Once more, as there was no one in authority to tell them how to lay the cables, and their positioning was dependent on the electrician responsible for that particular section of the work.

Later on in 1882 *The Electric Lighting Clauses Act* (modelled on the previous 1847 *Gas Act*) was passed by Parliament and this enabled the Board of Trade to authorise the supply of electricity in any area by a local authority, company or person and to grant powers to install this electrical supply (including breaking up the streets) through the use of the 1882 *'Rules and Regulations for the prevention of Fire Risks Arising from Electric Lighting'*.

This document was the forerunner of today's IEE Wiring Regulations and historically, since 1882, there have been a succession of amendments and new editions of the Regulations as shown in Table 1.1 below.

Table 1.1 Succession of amendments and new editions of Wiring Regulations

1882	First edition	Entitled *'Rules and Regulations for the prevention of Fire Risks Arising from Electric Lighting'*
1888	Second edition	
1897	Third edition	Entitled *'General Rules recommended for Wiring for the Supply of Electrical Energy'*
1903	Fourth edition	Entitled *'Wiring Rules'*
1907	Fifth edition	
1911	Sixth edition	
1916	Seventh edition	
1924	Eighth edition	Entitled *'Regulations for the Electrical Equipment of Buildings'*
1927	Ninth edition	
1934	Tenth edition	
1939	Eleventh edition	Revised issue (1943), reprinted with minor amendments (1945), Supplement issued (1946), Revised Section 8 (1948)
1950	Twelfth edition	Supplement issued (1954)
1955	Thirteenth edition	Reprinted, 1958, 1961, 1962 and 1964
1966	Fourteenth edition	Reprinted, 1968, 1969, 1970 (in metric units), 1972, 1973, 1974 and 1976

By now this continual updating was seen as a bit of a problem, particularly to designers and installers who had to ensure that they were always working in compliance to the latest Regulations. With the publication of the fifteenth edition, therefore, it was decided that in future reprints of the same edition would be contained in one of five different coloured covers (i.e. red, green, yellow, blue and brown) and a new edition would be published when the brown covered reprint required updating.

 For this reason, (and following the recent amalgamation of the Institution of Electrical Engineers (IEE) with the Institution of Incorporated Engineers (IIE) to form the new Institution of Engineering and Technology (IET)) it is anticipated that the seventeenth edition will be published within the next few years (rumour indicated 2008!)

Table 1.2 BS 7671: 2001 – publication details

1981	Red cover	Fifteenth edition	Entitled *'Regulations for Electrical Installations'*
1983	Green cover		Reprinted incorporating amendments
1984	Yellow cover		Reprinted incorporating amendments
1986	Blue cover		Reprinted incorporating amendments
1987	Brown cover		Reprinted incorporating amendments
1988	Brown cover		Reprinted with minor corrections
1991	Red cover	Sixteenth	Reprinted with minor corrections in 1992 Reprinted as BS 7671 in 1992 Amendment No 1 issued Dec 1994

(continued)

Table 1.2 (*continued*)

1994	Green cover	Reprinted incorporating Amendment No 1
		Amendment No 2 issued Dec 1997
1997	Yellow cover	Reprinted incorporating Amendment No 2
		Amendment No 3 issued Apr 2000
2001	Blue cover	BS 7671: 2001 issued Jun 2001 (see
		Note below)
		Amendment No 1 issued Feb 2002
		Amendment No 2 issued Mar 2004
2004	Brown cover	Reprinted incorporating Amendments
		No 1 and No 2

Note: BS 7671: 2001 now includes some important changes that were required in order to maintain technical alignment with CENELEC harmonisation documents.

1.3 What does the standard contain?

This standard '*contains the rules for the design and erection of electrical installations so as to provide for safety and proper functioning for the intended use*' and is based on the plan agreed internationally (i.e. through CENELEC) for the '*arrangement of safety rules for electrical installations*'.

The structure of BS 7671: 2001 is given in Table 1.3.

Table 1.3 BS 7671: 2001 – Structure

Part 1	Sets out the scope, object and fundamental principles.
Part 2	Defines certain terms used throughout the Regulations.
Part 3	Identifies the characteristics of an installation that will need to be taken into account in choosing and applying the requirements of the subsequent Parts of the Regulations.
	These characteristics may vary from one part of an installation to another and need to be assessed for each location to be served by the installation.
Part 4	Describes the basic measures that are available for the protection of persons, property and livestock and against the hazards that may arise from the use of electricity.
Part 5	Describes the precautions that need to be taken in the selection and erection of the equipment of an installation.
Part 6	Identifies particular requirements for special installations or locations.
Part 7	Covers inspection and testing.

Any intended departure from the requirements of Parts 1 to 6 requires special consideration by the installation designer of the installation and **must** be documented in the Electrical Installation Certificate specified in Part 7.

The seven parts of the standard are then supported by the following Appendices:

Table 1.4 BS 7671: 2001 – Appendices

Appendix	Title	Description and remarks
1.	British Standards to which reference is made in the Regulations	Reproduced in the Reference section of this book.
2.	Statutory Regulations and associated memoranda	Details of all the Statutory Regulations, legislation and EU Harmonised Directives that electrical installations are required to comply.
3.	Time current characteristics of overcurrent protective devices	Details of time/current characteristics for: • fuses to BS 1361 • semi enclosed fuses to BS 3036 • fuses to BS 88-2.1 and BS 88-6 • circuit breakers to BS EN 80898 and BS EN 61009.
4.	Current-carrying capacity and voltage drop for cables and flexible cords	Schedules of: • installation methods for cables (e.g. cleated, in conduits, on trays, in trenches etc.) • current rating tables (e.g. armoured cables, mineral insulated cables, fire resistant cables, screened cables etc.) • correction factors (for groups of cables, mineral insulated cables, cables installed in trenches, ambient temperature where protection is against short circuits and overload) • copper conductors • aluminium conductors.
5.	Classification of external influences	Lists and schedules of external influences having an influence on electrical installations and which consist of: (for details see Table 1.5).
6.	Model forms for certification and reporting	Reproduced in Section 9.7 (p. 507)
7.	Harmonised cable core colours	Current details of cable core marking and colours that are to be used in all installations (for details see inside cover).

Table 1.5 List of external influences relevant to electrical installations

Environment	Utilisation	Buildings
Ambient (°C)	Capability	Materials
Temperature & humidity	Resistance	Structure
Altitude (metres)	Contact with earth	
Water	Evacuation	
Foreign bodies	Materials	
Corrosion		
Impact		
Vibration		
Other mechanical stresses		
Flora		
Fauna		
Radiation		
Solar		
Seismic		
Lightning		
Movement of air		
Wind		

1.3.1 What about the standard's numbering system?

The numbering system used to identify specific requirements in BS 7671: 2001 is as follows:

- the first digit signifies a Part
- the second digit a Chapter
- the third digit a Section
- and subsequent digits the Regulation number.

Example

The Section number **413** is made up as follows:

- PART 4 – PROTECTION FOR SAFETY
- Chapter **41** (first chapter of Part 4) – Protection against electric shock
- Section **413** (third section of Chapter 41) – Protection against indirect contact.

1.4 What are the objectives of the IEE Wiring Regulations?

Current legal requirements for employee competence in electrical work now call for anyone involved in certain electrical activities – for example, simply choosing the size or type of cable or fuse – to be aware of the regulative requirements associated with such work. The IEE Wiring Regulations (commonly referred to as BS 7671) is the traditionally recognised Approved Code of Practice for those who are involved in (or supervise) electrical work such as electrical maintenance, control or instrumentation.

The stated intention of wiring safety codes is to *"provide technical, performance and material standards that will allow efficient distribution of electrical energy and communication signals, at the same time protecting persons in the building from electric shock and preventing fire and explosion" (IET)*. In other words,

> **To ensure the protection of people and live stock from fire, shock or burns from any installation that complies with their requirments**

The Regulations form the basis of safe working practice throughout the electrical installation industry.

1.5 What is the legal status of the IEE Wiring Regulations?

Although the IEE Wiring Regulations have always been held in high esteem throughout Europe, they had no legal status that would require Continentals who were carrying out installation work in the UK to abide by them. This problem was overcome in October 1992 when the IEE Wiring Regulations became a British Standard, BS 7671 – thus providing them with international status.

 Note: Although the Regulations are *non-statutory regulations* they may, however, be used as evidence in a court of law to claim compliance with a statutory requirement.

1.6 What does it cover?

As shown below, the IEE Wiring Regulations covers both electrical installations and electrical equipment.

1.6.1 Electrical installation

> **Definition**
>
> For the purpose of the Regulations
>
> *Electrical installations* (or *installation*) means *any assembly of associated electrical equipment supplied from a common origin to fulfil a specific purpose and having certain co-ordinated characteristics.*

The Regulations are applicable to electrical installations in and for:

- residential premises
- commercial premises
- public premises
- industrial premises
- agricultural and horticultural premises
- prefabricated buildings
- caravans, caravan parks and similar sites
- construction sites, exhibitions, fairs and other installations in temporary buildings
- highway power supplies and street furniture, and outdoor lighting.

The Regulations include requirements for:

- circuits supplied at non-final voltages (up to and including) 1000 V a.c. or 1500 V d.c.

 Note: Although the preferred frequencies are 50 Hz, 60 Hz and 400 Hz, the use of other frequencies for special purposes is not excluded.

- circuits (but not apparatus and/or equipment internal wiring) that is operating at voltages greater than 1000 V and derived from an installation having a voltage not exceeding 1000 V a.c. (e.g. discharge lighting, electrostatic precipitators)
- wiring systems and cables not specifically covered by an appliance standard
- consumer installations external to buildings
- fixed wiring for communication and information technology, signalling; command and control etc. (but not apparatus and/or equipment internal wiring)
- the addition to (or alteration of) installations and parts of existing installations affected by an addition or alteration.

Although the Regulations are intended as a standard for electrical installations, in certain cases, they may need to be supplemented by the requirements and/ or recommendations of other British Standards or by the requirements of the person ordering the work. Such cases could include (amongst others) the following:

- electric signs and high-voltage luminous discharge tube installations – BS 559
- emergency lighting – BS 5266
- electrical apparatus for explosive gas atmospheres – BS EN 60079 and BS EN 50014
- electrical apparatus for use in the presence of combustible dust – BS EN 50281
- fire detection and alarm systems in buildings – BS 5839

- installations subject to the Telecommunications Act 1984 – BS 6701 Part I
- electric surface heating systems – BS 6351
- electrical installations for open cast mines and quarries – BS 6907.

The Regulations do **not** apply to the following installations:

- '*distributor's equipment*' as defined in the Electricity Safety, Quality and Continuity Regulations 2002
- railway traction equipment, rolling stock and signalling equipment
- motor vehicle equipment (except those to which the requirements of the Regulations concerning caravans are applicable)
- equipment on board ships
- equipment of mobile and fixed offshore installations
- aircraft equipment
- those aspects of mines and quarries specifically covered by Statutory Regulations
- radio interference suppression equipment (except so far as it affects safety of the electrical installation)
- lightning protection of buildings covered by BS 6651
- those aspects of lift installations covered by BS 5655.

 Note: For installations in premises, which are subject to statutory control (e.g. via a licensing or other authority), the requirements of that authority will need to be confirmed and these requirements then complied with in the design and implementation of that installation.

1.6.2 Electrical equipment

Definition

For the purpose of the Regulations

Electrical equipment (or *Equipment*) means *any item used for generation, conversion, transmission, distribution or utilisation of electrical energy, such as machines, transformers, apparatus, measuring instruments, protective devices, wiring systems, accessories, appliances and luminaries.*

The Regulations are only applicable to the actual selection and application of items of electrical equipment **in** an electrical installation.

The Regulations do **not** deal with requirements for the construction of assemblies of electrical equipment, which are required to comply with the appropriate standards.

1.7 What affect does using the Regulation have on other regulations?

The requirements of the IEE Wiring Regulations also have an affect on the implementation of other Statutory Instruments such as:

- The Building Act 1984
- The Disability Act 2003
- The Electricity at Work Regulations 1989
- The Fire Precautions Act 1971
- The Health and Safety at Work Act 1974.

1.7.1 The Building Act 1984

The Building Act 1984 (as implemented by the Building Regulations 2000) is the enabling Act under which all Building Regulations have been made. The Secretary of State (under the power given in the Building Act 1984), in order to:

- secure the health, safety, welfare and convenience of persons in or about buildings and of others who may be affected by buildings or matters connected with buildings
- further the conservation of fuel and power
- prevent waste, undue consumption, misuse or contamination of water

may make regulations with respect to the design and construction of buildings and the provision of services, fittings and equipment in (or in connection with) buildings.

Note: The current regulations governing the Building Regulations 2000 are SI 2000/2531 (as amended) – a .pdf copy of which can be downloaded from: http:// www.opsi.gov.uk/si/si2000/20002531.htm

For many years, the UK has managed to maintain a relatively high standard of electrical safety within buildings (domestic and non-domestic) based on voluntary controls centred on BS 7671. With the growing number of electrical accidents occurring in the 'home', the government has now been forced to implement a legal requirement for safety in all electrical installation work in dwellings.

As from 1 January 2005, therefore **all** new electrical wiring or electrical components for domestic premises (or small commercial premise linked to domestic accommodation) must now to be designed and installed in accordance with the Building Regulations, Part P (which is based on the fundamental principles set out in Chapter 13 of the BS 7671: 2001).

In addition, **all** fixed electrical installations (i.e. wiring and appliance fixed to the building fabric such as socket outlets, switches, consumer units and ceiling fittings) **must** now be designed, installed, inspected, tested and certified to BS 7671.

Part P of the Building Regulations also introduces the requirement for the cable core colours of all a.c. power circuits to align with BS 7671.

 Note: Part P only applies to fixed electrical installations that are intended to operate at low voltage or extra-low voltage which are not controlled by the Electricity Supply Regulations 1988 as amended, or the Electricity at Work Regulations 1989 as amended.

Competent Persons Scheme

Under Part P of the Building Regulations, all domestic installation work **must** now be inspected by Local Authority Building Control officers **unless** the work has been completed out by a *'Competent Person'* who is able to self-certify the work. The IEE supports the Part P Competent Person Scheme.

 For more details about the Building Regulations, visit: http://www.communities.gov.uk/index.asp?id=1130474 or see *Building Regulations in Brief*, 4th edn (ISBN 0-7506-8058-X).

1.7.2 The Disability Discrimination Act 2005

The Disability Discrimination Act 2005 (DDA) is:

 Definition

An Act to make it unlawful to discriminate against disabled persons in connection with employment, the provision of goods, facilities and services or the disposal or management of premises; to make provision about the employment of disabled persons; and to establish a National Disability Council.

From the point of view of the BS 7671: 2001, The Disability Act 1995 as amended in 2005 makes it unlawful:

- for a trade organisation to discriminate against a disabled person
- for a qualifications body to discriminate against a disabled person
- for service providers to make it impossible or unreasonably difficult for disabled persons to make use of that service.

 For more details about the DDA see: http://www.direct.gov.uk/disabledpeople/fs/en%20

1.7.3   The Electricity at Work Regulations 1989

Electricity is recognised as a major hazard for not only can it kill (research has shown that the majority of electric shock fatalities occur at voltages up to

240 V) it can also cause fires and explosions. Even non-fatal shocks can cause severe and permanent injury. Most of the electrical risks can be controlled by using suitable equipment, following safe procedures when carrying out electrical work and/or ensuring that all electrical equipment and installations are properly maintained.

Additional precautions will also be required for harsh and particular conditions (i.e. wet surroundings, cramped spaces, work out of doors or near live parts of equipment). For this reason the Electricity at Work Regulations 1989 is used to impose health and safety requirements for electricity used at work. Overall the Regulations require that:

- all electrical systems shall be constructed and maintained to prevent danger
- all work activities are to be carried out so as not to give rise to danger.

Whilst the majority of the Regulations concern hardware requirements, others are more generalised. For example:

- *installations shall be of proper construction*
- *conductors shall be insulated*
- *means of cutting off the power and for electrical isolation shall be available.*

In brief, the Regulations concern:

Systems, work activities and protective equipment	All systems shall at all times be constructed to prevent, so far as is reasonably practicable, danger.	Regulation 4
Strength and capability of electrical equipment	No electrical equipment is to be used where its strength and capability may be exceeded so as to give rise to danger.	Regulation 5
Adverse or hazardous environments	Electrical equipment sited in adverse or hazardous environments must be suitable for those conditions.	Regulation 6
Insulation, protection and placing of conductors	Permanent safeguarding or suitable positioning of live conductors is required.	Regulation 7
Earthing and other suitable precautions	Equipment must be earthed or other suitable precautions must be taken (e.g. the use of residual current devices, double insulated equipment, reduced voltage equipment etc.).	Regulation 8
Integrity of reference conductors	Nothing is to be placed in an earthed circuit conductor which might, without suitable precautions, give rise to danger by breaking the electrical continuity or by introducing a high impedance.	Regulation 9

(continued)

Connections	All joints and connections in systems must be mechanically and electrical suitable for use.	Regulation 10
Means for protecting from excess current	Suitable protective devices should be installed in each system to ensure all parts of the system **and** users of the system are safeguarded from the effects of fault conditions.	Regulation 11

 Note: Regulations 5 to 11 in effect, therefore, place a duty on the designer, installer and end user to ensure the suitability and protection of all electrical equipment.

Regulation 12	Means of cutting off the supply and for isolation	Where necessary to prevent danger, suitable means shall be available for cutting off the electrical supply to any electrical equipment.
Regulation 13	Precautions for work on equipment made dead	Adequate precautions must be taken to prevent electrical equipment, which has been made dead in order to prevent danger, from becoming live – whilst any work is carried out.
Regulation 14	Work on or near live conductors	No work can be carried out on live electrical equipment unless this can be properly justified.
		which means that risk assessments are required.
		If such work is to be carried out, suitable precautions must be taken to prevent injury.
Regulation 15	Working space, access and lighting	Adequate working space, adequate means of access and adequate lighting shall be provided at all electrical equipment on which or near which work is being done in circumstances that may give rise to danger.
Regulation 16	Competence to prevent danger and injury	No person shall engage in work that requires technical knowledge or experience to prevent danger or injury, unless he has that knowledge or experience, or is under appropriate supervision.

 For more information about the Electricity at Work Regulations 1989, contact: The Health and Safety Executive Local Authorities Enforcement Liaison Committee (HELA) via one of the following links:

- www.hse.gov.uk/lau/lacs/19-3.htm
- Tel: 020 7717 6441, fax: 020 7717 6418

- HSE Infoline – 0845 345 0055 (a 'one-stop' shop, providing rapid access to expert advice and guidance)
- email: LAU.enquiries@hse.gsi.gov.uk

Or to download a .pdf copy of the Electricity at Work Regulations 1989 (Statutory Instrument 1989 No. 635) go to:

- http://www.opsi.gov.uk/si/si1989/Uksi_19890635_en_1.htm

1.7.4 The Fire Precautions (Workplace) Regulations 1997

The Fire Precautions (Workplace) Regulations 1997 (as amended by the Fire Precautions (Workplace) (Amendment) Regulations 1999) stipulates that:

Definition

All offices, shops, railway premises and factories which have more than 20 persons employed in the building (or more than 10 persons employed anywhere other than on the ground floor) require a Fire Certificate.
Any hotel or boarding house providing sleeping accommodation for more than six persons (guests or staff) or where this sleeping accommodation is above the first floor or below the ground floor, require a Fire Certificate.

When a Fire Certificate is issued the owner or occupier is required to provide and maintain:

- the means of escape
- other means for ensuring that the means of escape can be safely and effectively used at all material times
- means of fighting fire
- means of providing warning in case of fire.

These requirements are reflected in the electrical installation.

 For further information about the Fire Precautions (Workplace) (Amendment) Regulations 1999 (Statutory Instrument 1999 No. 1877) visit: http://www. communities.gov.uk/index.asp?id=1123799 or for a copy of the Act, use the following link: http://www.fire.org.uk/si/wp/Statutory%20Instrument%201999% 20No_%201877.htm

1.7.5 The Health and Safety at Work Act 1974

Any company with more than five employees is legally obliged to possess a comprehensive health and safety policy.

Figure 1.2 The Health and Safety at Work Act 1974

Over the years, the IEE Wiring Regulations have been regularly used by HSE in their guidance and/or installation notices and installations which conform to BS 7671: 2001 (as amended) are regarded by HSE as likely to achieve conformity with the relevant parts of the Electricity at Work Regulations 1989. In certain instances where the Regulations have been used they may also be accompanied by Codes of Practice approved under Section 16 of the Health and Safety at Work Act 1974.

 Although some existing installations may have been designed and installed to conform to the standards set by earlier editions of the Wiring Regulations, this does **not** necessarily mean that they will fail to achieve conformity with the relevant parts of the Electricity at Work Regulations 1989.

 For further information about the Health and Safety at Work Act 1974 visit: www.hse.gov.uk/lau/lacs/38-2.htm

1.8 How are the IEE Wiring Regulations implemented?

Although the IEE Wiring Regulations rely (primarily) on British Standards for their implementation (see Reference section for details) they do, however, include the policy decisions made in a number of Statutory Instruments and by the Council of European Communities in the relative EU Harmonised Directives.

1.8.1 Statutory Instruments

In Great Britain the following classes of electrical installations are required to comply with the Statutory Regulations (see Table 1.6).

Table 1.6 Statutory Instruments affecting Electrical Installations

Type of Electrical Installation	Statutory Instrument
Distributors' installations generally (subject to certain exemptions)	Electricity Safety, Quality and Continuity Regulations 2002 • SI 2002 No 2665
Building generally (subject to certain exemptions)	Building Standards (Scotland) Regulations 1990 as amended • SI 1990 No 2179 (S. 187) • SI 1993 No 1457 (S. 191) • SI 1994 No 1266 (S. 65) • SI 1996 No 2251 (S. 183) • SI 1997 No 2157 (S. 150)
Work activity Places of work	The Electricity at Work Regulations 1989 as amended
Non-domestic installations	• SI 1989 No 635 • SI 1997 No 1993 • SI 1999 No 2550
Cinematograph installations	Cinematograph (Safety) Regulations 1955 (as amended under the Cinematograph Act, 1909, and/or Cinematograph Act, 1952) • SI 1982 No 1856
Machinery	The Supply of Machinery (Safety) Regulations 1992 as amended • SI 1992 No 3073 • SI 1994 No 2063
Theatres and other places licensed for public entertainment, music, dancing, etc.	Conditions of licence under: • In England and Wales – The Local Government (Miscellaneous provisions) Act 1982 • In Scotland, The Civic Government (Scotland) Act 1982
High-voltage luminous tube	Conditions of licence under: • In England and Wales – The Local Government (Miscellaneous provisions) Act 1982 • In Scotland, The Civic Government (Scotland) Act 1982

 The full text of **all** Statutory Instruments that have been published since 1987 are now available from The Office of Public Sector Information (OPSI) via their website: http://www.opsi.gov.uk/stat.htm

With effect from July 1999, Statutory Instruments which have also been made by the National Assembly for Wales have been published via the Wales Legislation section: (http://www.opsi.gov.uk/legislation/wales/w-stat.htm).

Whilst the series of Scottish Statutory Instruments have been published via the Scottish Legislation section: (http://www.opsi.gov.uk/legislation/scotland/s-stat.htm).

1.8.2 CENELEC Harmonised Documents

As well as British Standards, The Wiring Regulations also include requirements from CENELEC Harmonisation Documents (HDs) such as:

HD 193 1982	Voltage bands Part 1 and Definitions
HD 308 S2 2001	Identification of cores in cables and flexible cords
HD 384.1 2000	Scope, Object and Fundamental Principles I
HD 384.2 1986	Definitions Part 2 (1997)
HD 384.3 1995	Assessment of general characteristics
HD 384.4.41 1996	Protection against electric shock
HD 384.4.42 1985	Protection against thermal effects (1992), (1994)
HD 384.4.43 1980	Protection against overcurrent Part 4, Chapter 43 (1999)
HD 384.4.443 1999	Protection against overvoltages of atmospheric origin or due to switching
HD 384.4.45 1989	Protection against undervoltage
HD 384.4.47 1999	Isolation and switching
HD 384.4.47 1988	Application of measures for protection against electric shock
HD 384.4.473 1980	Application of measures for protection against overcurrent (1995)
HD 384.4.482 1997	Protection against fire where particular risks of danger exist
HD 384.5.51 1996	Selection and erection of equipment, common rules
HD 394.5.52 1995	Wiring systems (1998)
HD 384.5.523 1991	Wiring systems, current-carrying capacities
HD 384.5.537 1998	Switchgear and controlgear, devices for isolation and switching
HD 384.5.54 1988	Earthing arrangements and protective conductors
HD 384.5.551 1987	Other equipment, low-voltage generating sets
HD 384.5.56 1985	Safety services
HD 384.6.61 1992	Initial verification
HD 384.7.702 1992	Special location – Swimming pools
HD 384.7.703 1991	Special location – Locations containing a hot air sauna heater
HD 384.7.704 1999	Construction and demolition site installations
HD 384.7.705 1991	Special location – Agricultural and horticultural premises
HD 384.7.706 1991	Special location – Restrictive conductive locations
HD 384.7.707 1992	Special location – Caravan parks and caravans
HD 384.7.714 1999	Outdoor lighting installations

 Note: The dates in brackets refer to the year when an amendment was issued.

2

Domestic buildings

"**All electrical installations must be accommodated in ways that meet the requirement of the Building Regulations**"

Building Regulations
Approved Document P (3.1)

Figure 2.1 Mandatory requirements for domestic buildings

2.1 The Building Act 1984

By Act of parliament, the Secretary of State is responsible for ensuring that the health, welfare, safety and convenience of persons living in or working in (or nearby) buildings are secured. This Act is called the Building Act 1984 and one of its prime purposes is to assist in the conservation of fuel and power, prevent waste, undue consumption, and the misuse and contamination of water.

It imposes on owners and occupiers of buildings a set of requirements concerning the design and construction of buildings and the provision of services, fittings and equipment used in (or in connection with) buildings.

2.2 The Building Regulations

The current legislation in England and Wales is the Building Regulations 2000 (Statutory Instrument No. 2531) which is made by the Secretary of State for the Environment under powers delegated by Parliament under the Building Act 1984.

They are a set of minimum requirements and basic performance standards designed to secure the health, safety and welfare of people in and around buildings and to conserve fuel and energy in England and Wales.

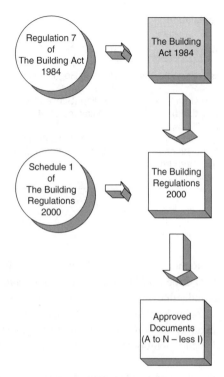

Figure 2.2 Implementing the Building Act

2.3 Approved Documents

The Building Regulations are supported by a series of separate documents which correspond to the different areas covered by the Regulations. These are called 'Approved Documents' and they contain practical and technical guidance on ways in which the requirements of Schedule 1 and Regulation 7 of the Building Act 1984 can be met.

Each Approved Document reproduces the actual *requirements* contained in the Building Regulations relevant to the subject area; this is then followed by *practical and technical guidance* (together with examples) showing how the requirements can be met in some of the more common building situations. There may, however, be alternative ways of complying with the requirements to those shown in the Approved Documents and you are, therefore, under no obligation to adopt any particular solution in an Approved Document – if you prefer to meet the requirement(s) in some other way.

2.3.1 What about the rest of the United Kingdom?

As shown in Table 2.1 The Building Act 1984 does not apply to Scotland or Northern Ireland.

Table 2.1 Building legislation

	Act	Regulations	Implementation
England & Wales	Building Act 1984	Building Regulations 2000	Approved Documents
Scotland	Building (Scotland) Act 2003	Building (Scotland) Regulations 2004	Technical Handbooks
Northern Ireland	Building Regulations (Northern Ireland) Order 1979	Building Regulations (Northern Ireland) 2000	'Deemed to satisfy' by meeting supporting publications

Scotland

Within Scotland, the requirements for buildings are controlled by the *Building (Scotland) Act 2003* and the *Building (Scotland) Regulations 2004* then set the functional standards under this Act. The methods for implementing these requirements are similar to England and Wales, except that the guidance documents (i.e. for achieving compliance) are contained in two *Technical Handbooks*, one for domestic work, and one for non-domestic. Each handbook has a general section 0 and then 6 technical sections.

The main procedural difference between the Scottish system and the others is that a building warrant is **still** required before work can start in Scotland and certain facilities in dwellings are still required in section 3 of the domestic technical handbook.

Northern Ireland

On the other hand, *Building Regulations (Northern Ireland) Order 1979* (as amended by *The Planning and Building Regulations (Amendment) (NI) Order 1990*) is the main legislation for Northern Ireland and the *Building Regulations (Northern Ireland) 2000* then details the requirements for meeting this legislation.

Supporting publications (such as British Standards, BRE publications and/ or technical booklets published by the Department) are then used to ensure that the requirements are implemented (i.e. *deemed to satisfy*).

2.4   Electrical safety

For many years, the UK has managed to maintain relatively high electrical safety standards with the support of voluntary controls based on BS 7671, but with a growing number of electrical accidents occurring in the 'home', the government have now been forced to consider legal requirement for safety in electrical installation work in dwellings and certain other buildings.

Table 2.2 Legislative cross-reference

England & Wales		Scotland		Northern Ireland	
Part A	Structure	Section 1	Structure	Technical Booklet D	Structure
Part B	Fire safety	Section 2	Fire	Technical Booklet E	Fire safety
Part C	Site preparation and resistance to contaminants and water	Section 3	Environment	Technical Booklet C	Preparation of site and resistance to moisture
Part D	Toxic substances	Section 3	Environment	Technical Booklet B	Materials and workmanship
Part E	Resistance to the passage of sound	Section 5	Noise	Technical Booklet G	Sound insulation of dwellings
Part F	Ventilation	Section 3	Environment	Technical Booklet K	Ventilation
Part G	Hygiene	Section 3	Environment	Technical Booklet P	Sanitary appliances and invented hot water storage systems
Part H	Drainage and waste disposal	Section 3	Environment	Technical Booklet J	Solid waste in buildings
				Technical Booklet N	Drainage
Part J	Combustion appliances and fuel storage systems	Section 3	Environment	Technical Booklet L	Heat-producing appliances and liquefied petroleum gas installations
Part K	Protection from falling, collision and impact	Section 4	Safety	Technical Booklet H	Stairs, ramps and protection from impact
Part L	Conservation of fuel and power	Section 6	Energy	Technical Booklet F	Conservation of fuel and power
Part M	Access and facilities for disabled people	Section 4	Safety	Technical Booklet R	Access for facilities and disabled people
Part N	Glazing	Section 6	Energy	Technical Booklet V	Glazing
Part P	Electrical safety	Section 4	Safety		

As from 1 January 2005, therefore, all new electrical wiring or electrical components for domestic premises (or small commercial premise linked to domestic accommodation) have had to be designed and installed in accordance with the Building Regulations, and in particular, Part P which is based on the fundamental principles set out in Chapter 13 of the BS 7671: 2001 (i.e. *'The IEE Wiring Regulations'*). In addition, all fixed electrical installations (i.e. wiring and appliances that are attached to, in or part of a building (such as socket-outlets, switches, consumer units and ceiling fittings on the consumer's side of the electricity supply meter) have now to be designed, installed, inspected, tested and certified to BS 7671.

 Part P also introduced the requirement for new cable core colours for a.c. power circuits and with effect 31 March 2006, **all** new installations or alterations to existing installations **must** use these new (harmonised) colour cables. (Further information concerning cable identification colours for extra-low voltage and d.c. power circuits is available from the IET website at www.iee.org/cablecolours.)

Table 2.3 Identification of conductors in a.c. power and lighting circuits

Conductor	Colour
Protective conductor	Green-and-yellow
Neutral	Blue
Phase of single phase circuit	Brown
Phase 1 of three-phase circuit	Brown
Phase 2 of three-phase circuit	Black
Phase 3 of three-phase circuit	Grey

 For single-phase installations in domestic premises, the new colours are the same as those for flexible cables to appliances (namely green-and-yellow, blue and brown for the protective, neutral and phase conductors respectively).

 Note: Part P applies only to fixed electrical installations that are intended to operate at low voltage or extra-low voltage which are **not** controlled by the Electricity Supply Regulations 1988 as amended, or the Electricity at Work Regulations 1989 as amended.

2.4.1 What is the aim of Approved Document P?

The aim of Part P is to increase the safety of householders by improving the design, installation, inspection and testing of electrical installations in dwellings when they (i.e. the installations) are being newly built, extended or altered.

 It is understood that the government is also intending to introduce a scheme whereby existing domestic installations are checked at regular intervals (as well as when they are sold and/or purchased) to make sure that they comply. This would mean, of course, that if you had an installation, which was not correctly certified, then your house insurance might well **not** be valid!

2.4.2 Who is responsible for electrical safety?

Basically:

- **The owner** needs to determine whether the works being carried out are either minor or notifiable work. If the work is notifiable, then the owner needs to make sure that the person(s) carrying out the work is either registered under one of the self-certified schemes (see Figure 2.3) or is able to certify their work under the local authority Building Control route.
- **The designer** needs to ensure that all electrical work is designed, constructed, inspected and tested in accordance with the BS 7671 and either falls under a Competent Persons Scheme or the local authority Building Control Approval route.
- **The builder/developer** needs to ensure that they have electricians who can self-certify their work or who are qualified/experienced enough to enable them to sign off under the Electrical Installation Certification form.

2.4.3 What are the statutory requirements?

In future all electrical installations will need to:

- be designed and installed to protect against mechanical and thermal damage
- be designed and installed so that they will not present an electrical shock or fire hazard
- be tested and inspected to meet relevant equipment/installation standards
- provide sufficient information so that persons wishing to operate, maintain or alter an electrical installation can do so with reasonable safety
- comply with such requirements placed by the Building Regulations.

What does all this mean?

With a few exceptions any electrical work undertaken in your home which includes the addition of a new electrical circuit, or involves work in your:

- kitchen
- bathroom
- garden area

must, as from 1 January 2005, be reported to the local authority Building Control for inspection. This includes any work undertaken professionally, by you or another family member, or by a friend.

The **ONLY** exception is when the installer has been approved by a Competent Persons organisation such as ELECSA (see Figure 2.3 below).

Authorized competent person self-certification schemes for installers who can do all electrical installation work	Authorized competent person self-certification schemes for installers who can do electrical work only if it is necessary when they are carrying out other work
 **BRE Certification Ltd** www. brecertification. co.uk Phone: 01923 664100	 **CORGI Competent Persons Scheme** www. corgi-group. com Phone: 0870 401 2300
BSI **British Standards Institution** www. bsi-global.com Phone: 01442 230442	 **ELECSA Limited** www.elecsa. org.uk Phone: 0870 749 0080
 **ELECSA Limited** www.elecsa. org.uk Phone: 0870 749 0080	 **NAPIT Certification Limited** www.napit. org.uk Phone: 0870 444 1392
 **NAPIT Certification Limited** www.napit. org.uk Phone: 0870 444 1392	 **NICEIC Certification Services Ltd** www.niceic. org.uk Phone: 01582 531000
 **NICEIC Certification Services Ltd** www.niceic. org.uk Phone: 01582 531000	**Registration Services** **OFTEC** www.oftec.org Phone: 0845 658 5080

Figure 2.3 Authorised competent person self-certification schemes for installers

2.4.4 What types of building does Approved Document P cover?

Part P applies to **all** electrical installations in **(and around)** buildings or parts of buildings comprising:

- dwelling houses and flats
- dwellings and business premises that have a common supply
- land associated with domestic buildings
- fixed lighting and pond pumps in gardens
- shops and public houses with a flat above
- common access areas in blocks of flats such as corridors and stairways
- shared amenities of blocks of flats such as laundries and gymnasiums.

Table 2.4 provides the details of works that are notifiable to local authority and/or must be completed by a company registered as a 'Competent Firm'.

Table 2.4 Notifiable work

Locations where work is being completed	Extensions and modifications to circuits	New circuits
Bathrooms	Yes	Yes
Bedrooms	Yes	Yes
Communal area of flats	Yes	Yes
Conservatories		Yes
Dining rooms		Yes
Garages (integral)		Yes
Garages (detached)		Yes*
Greenhouses	Yes	Yes
Halls		Yes
Hot air saunas	Yes	Yes
Kitchen	Yes	Yes
Kitchen diners	Yes	Yes
Landings		Yes
Lounge		Yes
Paddling pools	Yes	Yes
Remote buildings	Yes	Yes
Sheds	Yes	Yes
Shower rooms	Yes	Yes
Stairways		Yes
Studies		Yes
Swimming pools	Yes	Yes
TV Rooms		Yes
Workshops (remote)	Yes	Yes

*if the installation requires outdoor wiring.

2.4.5 What is a competent firm?

For the purposes of Part P, the Government has defined 'Competent Firms' as electrical contractors:

- who work in conformance with the requirements to BS 7671
- whose standard of electrical work has been assessed by a third party

- who are registered under the NICEIC Approved Contractor scheme and the Electrotechnical Assessment Scheme.

2.4.6 What is a competent person responsible for?

When installation work is undertaken by a competent person, that person is responsible for:

- ensuring compliance with BS 7671 and all relevant Building Regulations
- providing the person ordering the work with a signed Building Regulations compliance certificate (see Appendix 1 to this chapter)
- providing the relevant Building Control Body with an information copy of the certificate
- providing the person ordering the work with a completed Electrical Installation Certificate.

2.4.7 Who is entitled to self-certify an installation?

Part P affects **every** electrical contractor carrying out fixed installation and alteration work in homes. **Only** registered installers are entitled to self-certify the electrical work, however, **and** they must be registered as a competent person under one of the schemes shown in Figure 2.3:

Working with industry and consumer organisations, the government has recently developed the TrustMark initiative for builders and specialist firms that work on the home. Schemes that are capable of delivering '*agreed competence and customer-care standards*' are approved to use the TrustMark brand by a Board consisting of industry and consumer representatives, with government observers. The brand is owned by the DTI, which licences it to the Board.

www.trustmark.org.uk

Figure 2.4 The TrustMark initiative (Logo, reproduced courtesy of TurstMark).

The TrustMark replaces the Quality Mark scheme which closed on 31 December 2004 because too few firms joined. For more information about TrustMark see the website at: www.trustmark.org.uk.

2.4.8 What are the consequences of not obtaining approval?

Failure to comply with the requirements of Part P **is a criminal offence** and local authorities have the power to require the removal or alteration of work that does not comply with the Building Regulations. In addition, a Completion Certificate for works will not be issued – which could cause severe problems in the future as:

- the electrical installation may not be safe
- you will have no record of the work done
- you may have difficulty in selling your home without records of the installation and the relevant safety certificates.

2.4.9 When do I have to inform the local authority Building Control Body?

All proposals to carry out electrical installation work must be notified to the local authority's Building Control Body before work begins, **unless** the proposed installation work is undertaken by a person registered as a competent person under a government-approved Part P Self-Certification Scheme or the work is agreed non-notifiable work such as:

- connecting an electric gate or garage door to an existing isolator (but, be careful, the installation of the circuit up to the isolator is notifiable!)
- fitting and replacing cookers and electric showers (unless a new circuit is required)
- installing equipment (e.g. security lighting, air conditioning equipment and radon fans) that is attached to the outside wall of a house (unless there are exposed outdoor connections and/or the installation is a new circuit, or an extension of a circuit in a kitchen, or special location, or is associated with a special installation)
- installing fixed equipment where the final connection is via a 13 A plug and socket (unless it involves fixed wiring and the installation of a new circuit or the extension of a circuit in a kitchen or special location)
- installing prefabricated, 'modular' systems such as kitchen lighting systems and armoured garden cabling that are linked by plug and socket connectors (provided that products are CE-marked and that any final connections in kitchens and special locations are made to existing connection units or points, e.g. a 13 A socket outlet)
- installing or upgrading main or supplementary equipotential bonding (provided that the work complies with other applicable legislation, such as the Gas Safety (Installation and Use) Regulations)
- installing mechanical protection to existing fixed installations (provided that the circuit's protective measures and current-carrying capacity of conductors are unaffected by increased thermal insulation)
- re-fixing or replacing the enclosures of existing installation components
- replacement, repair and maintenance jobs
- replacing fixed electrical equipment (e.g. socket outlets, control switches and ceiling roses) which do not require the provision of any new fixed cabling

- replacing the cable of a single circuit cable (where damaged, for example, by fire, rodent or impact – provided that the replacement cable has the same current-carrying capacity, follows the same route and does not serve more than one sub-circuit through a distribution board)
- work that is not in a kitchen or special location, which does not involve a special installation and which only consists of:
  - adding lighting points (light fittings and switches) to an existing circuit
  - adding socket outlets and fused spurs to an existing ring or radial circuit (provided that the existing circuit protective device is suitable and supplies adequate protection for the modified circuit)
- work that is not in a special location and only concerns:
  - adding a telephone, extra-low voltage wiring and equipment for communications, information technology, signalling, command, control and other similar purposes
  - adding prefabricated equipment sets (and their associated flexible leads) with integral plug and socket connections.

All of this work can be completed by a DIY enthusiast (family member or friend) but still needs to be installed in accordance with manufacturers' instructions and done in such a way that they do not present a safety hazard. This work does **not** need to be notified to a local authority Building Control Body (unless it is installed in an area of high risk such as a kitchen or a bathroom etc.) **but**, all DIY electrical work (unless completed by a qualified professional – who is responsible for issuing a Minor Electrical Installation Certificate) will **still** need to be checked, certified and tested by a competent electrician.

Any work that involves adding a new circuit to a dwelling will need to be either notified to the Building Control Body (who will then inspect the work) or needs to be carried out by a competent person who is registered under a government-approved Part P Self-Certification Scheme.

Work involving any of the following will also have to be notified:

- consumer unit replacements
- electric floor or ceiling heating systems
- extra-low voltage lighting installations (other than pre-assembled, CE-marked lighting sets)
- garden lighting and/or power installations
- installation of a socket outlet on an external wall
- installation of outdoor lighting and/or power installations in the garden or that involves crossing the garden
- installation of new central heating control wiring
- solar photovoltaic (p.v.) power supply systems
- small-scale generators (such as microCHP units).

 Note: Where a person who is **not** registered to self-certify intends to carry out the electrical installation, then a Building Regulation (i.e. a Building Notice or Full Plans) application will need to be submitted together with the appropriate fee, based on the estimated cost of the electrical installation. The Building Control Body will then arrange to have the electrical installation inspected at first fix stage and tested upon completion.

In any event the electrical work will still need to be certified under BS 7671 by a suitably competent person who will be responsible for the design, installation, inspection and testing of the system (on completion) and have the confidence of completing a certificate to say that the work is satisfactory and complies with current codes of practice.

The main things to remember are:

- is the work notifiable or non-notifiable?
- does the person undertaking the work need to be registered as a competent person?
- what records (if any) need to be kept of the installation?

Figure 2.5 below is a quick guide to the requirements.

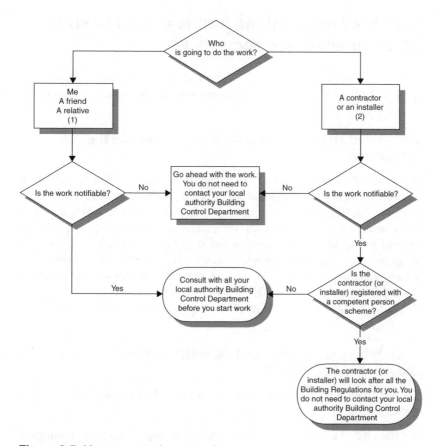

Figure 2.5 How to meet the new rules

Work completed by yourself, a friend or a relative

You do **not** need to tell your local authority's Building Control Department about non-notifiable work such as:

- repairs, replacements and maintenance work
- extra power points or lighting points or other alterations to existing circuits (**unless** they are in a kitchen, bathroom, or outdoors)
- You **do** need to tell them about most other work.

 Note: If you are not sure about this, or you have any questions, ask your local authority's Building Control Department.

Work completed by a contractor or an installer

If the work is of a notifiable nature then the installer(s) must be registered with one of the schemes shown in Figure 2.3.

2.5 What inspections and tests will have to be completed and recorded?

As shown in Table 2.5, there are four types of electrical installation certificates and one Building Regulations compliance certificate that have to be completed.

2.5.1 What should be included in the records of the installation?

All 'original' Certificates should be retained in a safe place and be available to any person inspecting or undertaking further work on the electrical installation in the future. If you later vacate the property, this Certificate will demonstrate to the new owner that the electrical installation complied with the requirements of British Standard 7671 at the time that the Certificate was issued. The Construction (Design and Management) Regulations require that for a project covered by those Regulations, a copy of this Certificate, together with the schedules, is included in the project health and safety documentation.

Figure 2.6 indicates how to chose what type of inspection is required.

2.5.2 Where can I get more information about the requirements of Part P?

Further guidance concerning the requirements of Part P (Electrical Safety) is available from:

- The IET (Institution of Engineering Technology) at http://www.iee.org/publish/wireregs/partp.cfm

Table 2.5 Types of installation

Type of inspection	When is it used?	What should it contain?	Remarks
Minor Electrical Installation Works Certificate	For additions and alterations to an installation that do not extend to the provision of a new circuit	Relevant provisions of Part 7 of BS 7671	An example of a Minor Electrical Installation could be (for example) the addition of a socket outlet or a lighting point to an existing circuit
Electrical Installation Certificate (short form)	For use when one person is responsible for the design, construction, inspection and testing of an installation	A schedule of inspections and a schedule of test results as required by Part 7 (of BS 7671)	
Electrical Installation Certificate (full)	For the initial certification of a new installation or for the alteration and/or addition to an existing installation where new circuits have been introduced	A schedule of inspections and test results as required by Part 7 (of BS 7671). A certificate, including guidance for recipients (standard form from Appendix 6 of BS 7671)	An electrical installation certificate is not to be used for a periodic inspection
Periodic Inspection Report	For the inspection of an existing electrical installation	A schedule of inspections and a schedule of test results as required by Part 7 (of BS 7671)	For safety reasons, electrical installations need to be inspected at appropriate intervals by a competent person
Building Regulations Compliance Certificate (see Appendix 1)	Confirmation that the work carried out complies with the Building Regulations	The basic details of the installation, the location, completion date and name of the installer	A purchaser's solicitor may request this document when you sell your property. Looking further ahead, it may be required as one of the documents that will make up your 'Home Information Pack' when it is introduced possibly in the future.

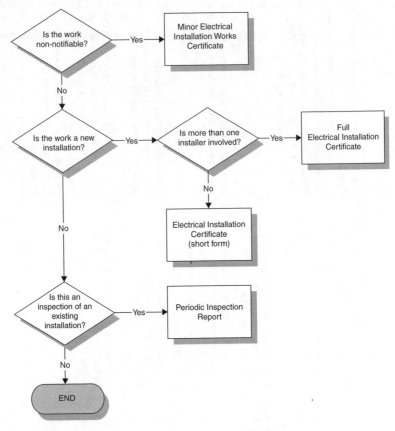

Figure 2.6 Choosing the correct Inspection Certificate

- The NICEIC (National Inspection Council for Electrical Installation Contracting) at www.niceic.org.uk
- The ECA (Electrical Contractors' Association) at www.niceic.org.uk or www.eca.co.uk.

To download .pdf copies of Part P, go to the following: http://www.planningportal. gov.uk/england/professionals/en/4000000001253.html

2.6 The requirement

Although:

- Part B (Fire safety)
- Part E (Resistance to the passage of sound)
- Part J (Combustion appliances and fuel storage systems) and
- Part K (Protection from falling, collision and impact)

all have a number of requirements concerning electrical safety and electrical installations (see below for details), the main requirements are contained in:

- Part P (Electrical safety) together with:
  - Part L (Conservation of fuel and power) and where necessary:
  - Part M (Access and use of buildings).

Figure 2.7 Building Regulations

Electrical work

Reasonable provision shall be made in the design and installation of electrical installations in order to protect persons operating, maintaining or altering the installations from fire or injury.

(Approved Document P1)

 Note: Whilst Part P makes requirements for the safety of fixed electrical installations, this does not cover system functionality (such as electrically powered fire alarm systems, fans and pumps), which are covered in other Parts of the Building Regulations and other legislation.

Conservation of fuel and power

Reasonable provision shall be made for the conservation of fuel and power in buildings by:

a. *limiting heat gains and losses:*
   i. *through thermal elements and other parts of the building fabric; and*
   ii. *from pipes, ducts and vessels used for space heating, space cooling and hot water services;*
b. *providing and commissioning energy-efficient fixed building services with effective controls; and*
c. *providing to the owner sufficient information about the building, the fixed building services and their maintenance requirements so that the building can be operated in such a manner as to use no more fuel and power than is reasonable in the circumstances.*

(Approved Document L1)

 Responsibility for achieving compliance with the requirements of Part L rests with the person carrying out the work. That 'person' may be, for example, a developer, a main (or sub) contractor, or a specialist firm directly engaged by a private client.

 Note: The person responsible for achieving compliance should either themselves provide a certificate, or obtain a certificate from the subcontractor, to show that commissioning has been successfully carried out. The certificate should be made available to the client and the Building Control Body.

Access and facilities for disabled people

In addition to the requirements of the Disability Discrimination Act 1995 precautions need to be taken to ensure that:

- *new non-domestic buildings and/or dwellings (e.g. houses and flats used for student living accommodation etc.)*
- *extensions to existing non-domestic buildings*
- *non-domestic buildings that have been subject to a material change of use (e.g. so that they become a hotel, boarding house, institution, public building or shop)*

are capable of allowing people, regardless of their disability, age or gender to:

- *gain access to buildings*
- *gain access within buildings*
- *be able to use the facilities of the buildings (both as visitors and as people who live or work in them)*
- *use sanitary conveniences in the principal storey of any new dwelling.*

(Approved Document M)

 For detailed guidance on:

- access and facilities for disabled people
- the requirements for conservation of fuel and power and (in particular)
- the requirements for electrical safety in domestic buildings

see our 'sister' publication *Building Regulations in Brief.* (ISBN. 0-7506-8058-X)

Figure 2.8   Building Regulations in Brief.

2.6.1 Meeting the requirement

Electrical installations

Electrical installations must be inspected and tested during, at the end of installation and before they are taken into service to verify that they:

• comply with Part P (and any other relevant Parts) of the Building Regulations	P 1.7
• are safe to use, maintain and alter	
• meet the relevant equipment and installation standards	P 0.1b
• meet the requirements of the Building Regulations.	P 3.1

Any proposal for a new mains supply installation (or where significant alterations are going to be made to an existing mains supply) **must** be agreed with the electricity distributor.   P 1.2

Design

Electrical installations must be designed and installed (suitably enclosed and appropriately separated) so that they:

• comply with the requirements of BS 7671 as amended	P 1.4
• comply with Part P (and any other relevant Parts) of the Building Regulations	P 1.7 and 3.1
• comply with the relevant equipment and installation standards	P 0.1b
• do not present an electric shock or fire hazard to people	P 0.1a
• provide mechanical and thermal protection	P 1.3
• provide adequate protection against mechanical and thermal damage	P 0.1a
• provide adequate protection for persons against the risks of electric shock, burn or fire injuries	P 1.3
• are safe to use, maintain and alter.	P 1.7

 Note: See Appendix A of Part P to the Building Regulations for details of some of the types of electrical services normally found in dwellings, some of the ways they can be connected and the complexity of wiring and protective systems that can be used to supply them.

Extensions, material alterations and material changes of use

 In accordance with Regulation 4(2) the **whole** of an existing installation **does not** have to be upgraded to current standards, but only to the extent necessary for the new work to meet current standards except where upgrading is required by the energy efficiency requirements of the Building Regulations.

Where any electrical installation work is classified as an extension, a material alteration or a material change of use, the work must consider and include:

• confirmation that the mains supply equipment is suitable and can carry the additional loads envisaged	P 2.1b–2.2
• the earthing and bonding systems are satisfactory and meet the requirements	P 2.1a–2.2c
• the necessary additions and alterations to the circuits which feed them	P 2.1a
• the protective measures required to meet the requirements	P 2.1a–2.2b
• the rating and the condition of existing equipment (belonging to both the consumer and the electricity distributor) are sufficient.	P 2.2a

 See Figure 2.9 for details of some of the types of electrical services normally found in dwellings, some of the ways they can be connected and the complexity of wiring and protective systems that can be used to supply them.

 Note: Appendix C to Part P of the Building Regulations offers guidance on some of the older types of installations that might be encountered during alteration work whilst Appendix D provides guidance on the application of the new harmonised European cable identification system.

Electricity distributors' responsibilities

The electricity distributor is responsible for:	
• evaluating and agreeing proposals for new installations or significant alterations to existing ones	P 1.2
• installing the cut-out and meter in a safe location	
• ensuring that it is mechanically protected and can be safely maintained	P 1.5
• taking into consideration the possible risk of flooding.	P 1.5

 Note: See OPDM publication 'Preparing for flooding', which can be downloaded from www.opdm.gov.uk

Distributors are also required to:	
• maintain the supply within defined tolerance limits	P 3.8
• provide an earthing facility for new connections	

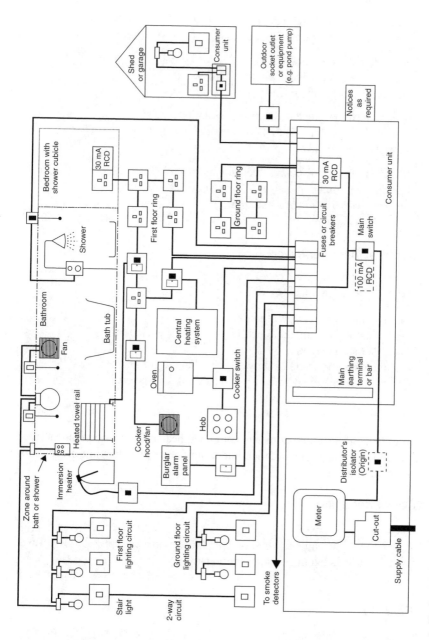

Figure 2.9 Typical fixed installations that might be encountered in new (or upgraded) existing dwellings

- provide certain technical and safety information to consumers to enable them to design their installations.

Distributors (and meter operators) must ensure that their equipment on consumers' premises:

- is suitable for its purpose   P 3.9
- is safe in its particular environment
- clearly shows the polarity of the conductors.

Distributors are prevented by the Regulations from   P 3.12
connecting installations to their networks, which do not
comply with BS 7671.

Distributors may disconnect consumers' installations   P 3.12
which are a source of danger or cause interference with
their networks or other installations.

Electrical installation work

Electrical installation work:

- is to be carried out professionally   P 1.1
- is to comply with the Electricity at Work Regulations
  1989 as amended
- may only be carried out by persons that are competent to   P 3.4a
  prevent danger and injury while doing it, or who are
  appropriately supervised.

 Note: Persons installing domestic combined heat and power equipment **must** advise the local distributor of their intentions before (or at the time of) commissioning the source.

Consumer units

Accessible consumer units should be fitted with a childproof   P 1.6
cover or installed in a lockable cupboard.

Earthing

All electrical installations shall be properly earthed.   P App C

 Note: The most usual type is an electricity distributor's earthing terminal, provided for this purpose near the electricity meter.

All lighting circuits shall include a circuit protective conductor.	P App C
All socket outlets which have a rating of 32 A or less and which may be used to supply portable equipment for use outdoors, shall be protected by a residual current device (RCD).	P App C
Distributors are required to provide an earthing facility for new connections.	P 3.8
It is **not** permitted to use a gas, water or other metal service pipe as a means of earthing for an electrical installation (this does not rule out, however, equipotential bonding conductors being connected to these pipes).	P App C
New or replacement non-metallic light fittings, switches and/or other components must not require earthing (e.g. non-metallic varieties) unless new circuit protective (earthing) conductors are provided.	P App C
Socket outlets that will accept unearthed (two-pin) plugs must **not** use supply equipment that needs to be earthed.	P App C
Where electrical installation work is classified as an extension, a material alteration or a material change of use, the work must consider and include earthing and bonding systems that are satisfactory and meet the requirements.	P 2.1a–2.2c

 See Figure 2.10 for details of some earth and bonding conductors that might be part of an electrical installation.

Equipotential bonding conductors

Main equipotential bonding conductors are required to all water service pipes, gas installation pipes, oil supply pipes and certain other 'earthy' metalwork that may be present on the premises.	P App C

The installation of supplementary equipotential bonding conductors is required for installations and locations

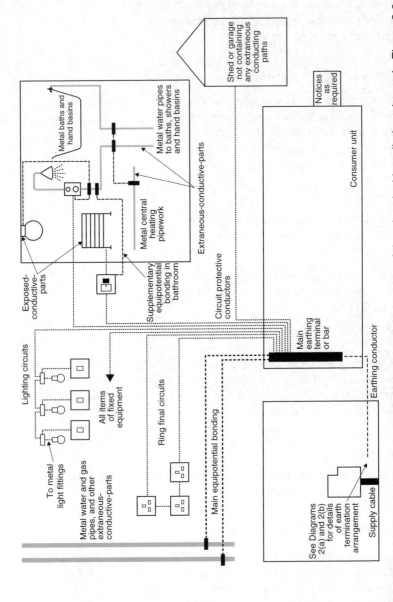

Figure 2.10 Typical earth and bonding conductors that might be part of the electrical installation shown in Figure 2.9

where there is an increased risk of electric shock
(e.g. bathrooms and shower rooms).   P App C

The minimum size of supplementary equipotential   P App C
bonding conductors (without mechanical protection)
is 4 mm².

Types of wiring or wiring system

Cables concealed in floors (and walls in certain   P App C
circumstances) are required to have an earthed metal
covering, be enclosed in steel conduit, or have additional
mechanical protection.

Cables to an outside building (e.g. garage or shed), if run   P App C
underground, should be routed and positioned so as to give
protection against electric shock and fire as a result of
mechanical damage to a cable.

Heat-resisting flexible cables are required for the final   P App C
connections to certain equipment (see makers' instructions).

 PVC insulated and sheathed cables are likely to be suitable for much of the
wiring in a typical dwelling.

Electrical components

All electrical work should be inspected (during
installation as well as on completion) to verify that the
components have:

- been made in compliance with appropriate British   P 1.11a i
  Standards or harmonised European Standards
- been selected and installed in accordance with BS 7671   P 1.11a ii
- been evaluated against external influences (such as   P 1.11a ii
  the presence of moisture)
- been tested as per sections 71–74 of BS 7671: 2001   P 1.9
- been tested using appropriate and accurate instruments   P 1.12
- been tested to check satisfactory performance with   P 1.11b
  respect to continuity of conductors, insulation
  resistance, separation of circuits, polarity, earthing

and bonding arrangements, earth fault loop impedance
and functionality of all protective devices including
residual current devices

- not been visibly damaged (or are defective) so as to   P 1.11a ii
  be unsafe
- had their test results recorded using the models in   P 1.12
  Appendix 6 of BS 7671.

 Note: Inspection and testing of DIY work should **also** meet the above requirements.

Socket outlets

Where necessary, socket outlets should comply with the requirements of Part M.	P 1.6
Socket outlets should be wall-mounted.	M 4.30a and b
Socket outlets that will accept unearthed (two-pin) plugs must **not** be used to supply equipment that needs to be earthed.	P App C
Socket outlets which have a rating of 32 A or less and which may be used to supply portable equipment for use outdoors **shall** be protected by an RCD.	P App C
Socket outlets should be located no nearer than 350 mm from room corners.	M 4.30
Switched socket outlets should indicate whether they are 'on'.	M 4.30k
Front plates should contrast visually with their backgrounds.	M 4.30 m
Mains and circuit isolator switches should clearly indicate whether they are 'on' or 'off'.	M 4.30l
Older types of socket outlet that are designed to accommodate non-fused plugs must **not** be connected to a ring circuit.	P App C
The colours red and green should **not** be used in combination as indicators of 'on' and 'off' for switches and controls.	M 4.28

Wall sockets

Wall sockets shall meet the following requirements:

Table 2.6 Building Regulations requirements for wall sockets

Type of wall	Requirement	Section
Timber framed	Power points may be set in the linings provided there is a similar thickness of cladding behind the socket box	E p14
	Power points should not be placed back to back across the wall	E p14
Solid masonry	Deep sockets and chases should **not** be used in separating walls	E 2.32
	The position of sockets on opposite sides of the separating wall should be staggered	E 2.32f
Cavity masonry	The position of sockets on opposite sides of the separating wall should be staggered	E 2.65e
	Deep sockets and chases should **not** be used in a separating wall	E 2.65d2
	Deep sockets and chases in a separating wall should **not** be placed back to back	E 2.65d2
Framed walls with absorbent material	Sockets should: • be positioned on opposite sides of a separating wall	E 2.146b
	• not be connected back to back	E 2.146b2
	• be staggered a minimum of 150 mm edge to edge	E 2.146b2

Wall-mounted switches and socket outlets

Where necessary, wall-mounted switches and socket outlets should comply with the requirements of Part M.	P 1.6
A cable or stud detector shall be used when attempting to drill into walls, floors or ceilings.	P App C
Individual switches on panels and on multiple socket outlets should be well separated.	M 4.29
Switches and socket outlets for lighting and other equipment should be located so that they are easily reachable.	M 8.2
Switches and socket outlets for lighting and other equipment in habitable rooms should be located between 450 mm and 1200 mm from finished floor level (see Figure 2.11).	M 8.3

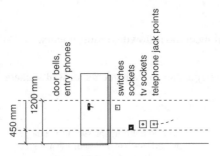

Figure 2.11 Heights of switches and sockets etc.

 The intention of this requirement is to help people with limited reach (e.g. seated in a wheelchair) access a dwelling's wall-mounted switches and socket outlets.

Outlets, controls and switches

The aim should be to ensure that all controls and switches are easy to operate, visible, free from obstruction and: • are located between 750 mm and 1200 mm above the floor • do not require the simultaneous use of both hands (unless necessary for safety reasons) to operate • switched socket outlets indicate whether they are 'on' • mains and circuit isolator switches clearly indicate whether they are 'on' or 'off' • individual switches on panels and on multiple socket outlets are well separated • controls that need close vision (e.g. thermostats) are located between 1200 mm and 1400 mm above the floor • front plates contrast visually with their backgrounds.	M 5.4i
The operation of all switches, outlets and controls should not require the simultaneous use of both hands (unless necessary for safety reasons).	M 4.30j
Where possible, light switches with large push pads should be used in preference to pull cords.	M 5.3
The colours red and green should not be used in combination as indicators of 'on' and 'off' for switches and controls.	M 5.3

Light switches

Light switches should:

- have large push pads
- align horizontally with door handles
- be within the 900 to 1100 mm from the entrance   M 4.30h and I
  door opening.

Switches and controls should be located between   M 4.30c and d
750 mm and 1200 mm above the floor.

Where possible, light switches with large push pads   M 5.3
should be used in preference to pull cords.

The colours red and green should not be used in   M 5.3
combination as indicators of 'on' and 'off' for
switches and controls.

Heat emitters

In toilets and bathrooms designed for disabled people, heat   M 5.10p
emitters (if located) should not restrict:

- the minimum clear wheelchair manoeuvring space
- the space beside a WC used to transfer from the
  wheelchair to the WC.

Heat emitters should either be screened or have their   M 5.4j
exposed surfaces kept at a temperature below 43°C.

Telephone points and TV sockets

All telephone points and TV sockets should be located   M 4.30a and b
between 400 mm and 1000 mm above the floor (or
400 mm and 1200 mm above the floor for
permanently wired appliances).

Thermostats

Controls that need close vision (e.g. thermostats) should   M 4.30f
be located between 1200 mm and 1400 mm above the floor.

Lighting circuits

All lighting circuits shall include a circuit protective conductor.	P App C

Light fittings

New or replacement non-metallic light fittings, switches or other components must not require earthing unless new circuit protective (earthing) conductors are provided.	App C

Fixed internal lighting (domestic buildings)

In order that dwelling occupiers may benefit from the installation of efficient electric lighting, whenever • a dwelling is extended, or • a new dwelling is created from a material change of use, or • an existing lighting system has been replaced as part of re-wiring works then the re-wiring works **must** comply with Part P.	L1B 43
Lighting fittings (including lamp, control gear, housing, reflector, shade, diffuser or other device for controlling the output light) should only take lamps with a luminous efficiency greater than 40 lumens per circuit watt.	L1A 42

 Note: Light fittings in less frequented areas such as cupboards and other storage areas do not count.

Fixed external lighting (i.e. lighting that is fixed to an external surface of the dwelling and which is powered from the dwelling's electrical system) should either:
 	
• have a lamp capacity not exceeding 150 W per light fitting that automatically switches off:
  – when there is enough daylight and
  – when it is not required at night or
• include sockets that can only be used with lamps, which have efficiency greater than 40 lumens per circuit watt.
	L1A 45

> Fixed energy-efficient light fittings (one per 25 m^2
> dwelling floor area (excluding garages) and one per four
> fixed light fittings) should be installed in the most
> frequented locations in the dwelling.
>
> L1B 44

 See GIL 20, Low energy domestic lighting, BRECSU, 1995 for further guidance.

Office, industrial and storage areas (non-domestic buildings)

> All areas that involve predominantly desk-based tasks
> (i.e. such as classrooms, seminar and conference
> rooms – **including** those in schools) shall have an average
> efficiency of not less than 45 luminare-lumens/circuit
> watt (averaged over the whole area).
>
> L2B 56

 Note: In other spaces (i.e. other than office or storage spaces) less efficient lamps may be used **provided** that the installed lighting has an average initial (100 hour) lamp plus ballast efficacy of not less than 50 lamp lumens per circuit watt.

Lighting controls (non-domestic buildings)

> The distance between the local switch and the luminaire it
> controls should generally be not more than six metres, or
> twice the height of the luminaire above the floor if this
> is greater.
>
> L2A 57
>
> Local switches should be:
>
> L2B 60
>
> - located in easily accessible positions within each
>   working area (or at boundaries between working areas
>   and general circulation routes)
> - operated by the deliberate action of the occupants
>   (referred to as occupant control), either manually
>   or remotely
> - located within six metres (or twice the height of the   L2B 61
>   light fitting above the floor if this is greater) of
>   any luminaire it controls.
>
> Occupant control of local switching may be supplemented   L2B 62
> by automatic systems which:
>
> - switch the lighting off when they sense the absence of
>   occupants or

- dim (or switch) the lighting off when there is sufficient daylight.

Automatically switched lighting systems **shall** be subject to a risk assessment.	L2A 54
Lighting controls should be provided to switch off the lighting during daylight hours and when the area is unoccupied.	L2A 54
If the space is daylit space served by side windows, the perimeter row of luminares should be separately switched.	L2A 57
Manually operated local switches should be in easily accessible positions within each working area, at boundaries between working areas, and at general circulation routes.	L2A 55
Local (manual) switching can be supplemented by automatic controls which:	L2A 58

- switch the lighting off when they sense the absence of occupants or
- dim (or switch off) the lighting when there is sufficient daylight.

Display lighting in all types of space (non-domestic buildings)

Display lighting should have an average initial (100 hour) efficiency of not less than 15 lamp-lumens per circuit watt.	L2A 60 L2B 64 and 65

 Note: In spaces where it would be reasonable to expect cleaning and restocking outside public access hours, general lighting should also be provided.

Where possible, display lighting should be connected in dedicated circuits that can be switched off at times when people will not be inspecting exhibits or merchandise or attending entertainment events.	L2A 60 L2B 66

 Note: In a retail store, for example, this could include timers that switch off the display lighting outside store opening hours.

Emergency escape lighting (non-domestic buildings)

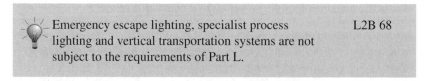 Emergency escape lighting, specialist process   L2B 68
lighting and vertical transportation systems are not
subject to the requirements of Part L.

General lighting efficiency in all other types of space (non-domestic buildings)

Lighting (over the whole of these areas) should have an   L2A 51
average initial efficacy of not less than 45 luminare-
lumens/circuit watt.

Lighting systems serving other types of space may use   L2A 53
lower powered and less efficient lamps.

Limiting the effects of solar gains in summer – buildings other than dwellings

In occupied spaces that are not served by a comfort   L2A 64a
cooling system, the combined solar and internal casual
gains (people, lighting and equipment) per unit floor
area averaged over the period of daily occupancy
should not be greater than 35 W/m^2 calculated over
a perimeter area not more than 6 m from the window
wall and averaged during the period 06.30–16.30 hrs GMT.

Consequential improvements (non-domestic buildings)

If a building has a total useful floor area greater than 1000 m^2 and the proposed
building work includes:

- an extension or
- the initial provision of any fixed building services or
- an increase to the installed capacity of any fixed building services

then:

any general lighting system serving an area greater than   L2B 18 4
100 m^2 which has an average lamp efficacy of less than
40 lamp-lumens per circuit watt, should be upgraded
with new luminares or improved controls.

Controlled services (non-domestic buildings)

Where the work involves the provision of a controlled service:

new lighting systems should be provided with controls that achieve reasonable standards of energy efficiency.	L2B 41c

Inspection and commissioning of the Building Services Systems (non-domestic buildings)

When Building Services Systems are commissioned: the systems and their controls shall be left in their intended working order and are capable of operating efficiently regarding the conservation of fuel and power.	L2A 77–78 L2B 70
Systems should be provided with meters to efficiently manage energy use and to ensure that at least 90% of the estimated annual energy consumption of each fuel is assigned to the various end-use categories (heating, lighting etc.).	L2A 43 L2B 67
Whenever a cooling plant (i.e. a chiller) is being replaced, cooling loads should (if practical and cost-effective) be improved through solar control and/or more efficient lighting.	L2B 45 and L2B 46
Air handling systems should be capable of achieving a specific fan power at 25% of design flow rate.	L2A 47

External lighting fixed to the building

External lighting (including lighting in porches, but not lighting in garages and carports) should:

Automatically extinguish when there is enough daylight, and when not required at night.	L1 45a
Have sockets that can only be used with lamps having an efficacy greater than 40 lumens per circuit watt (such as fluorescent or compact fluorescent lamp types, and **not** GLS tungsten lamps with bayonet cap or Edison screw bases).	L1 45b

Emergency alarms

Emergency assistance alarm systems should have:	M 5.4 h
• visual and audible indicators to confirm that an emergency call has been received, a reset control reachable from a wheelchair, WC, or from a shower/ changing seat, a signal that is distinguishable visually and audibly from the fire alarm.	
Emergency alarm pull cords should be:	M 4.30e
• coloured red • located as close to a wall as possible • have two red 50 mm diameter bangles.	
Front plates should contrast visually with their backgrounds.	M 4.30 m
The colours red and green should **not** be used in combination as indicators of 'on' and 'off' for switches and controls.	M 4.28

Smoke alarms – dwellings

Smoke alarms should normally be positioned in the circulation spaces between sleeping spaces and places where fires are most likely to start (e.g. kitchens and living rooms) to pick up smoke in the early stages, while also being close enough to bedroom doors for the alarm to be effective when occupants are asleep.

General

There should be at least one smoke alarm on every storey of a house (including bungalows).	B1 1.12
If more than one smoke alarm is installed in a dwelling then they should be linked so that if a unit detects smoke it will operate the alarm signal of all the smoke detectors.	B1 1.13
There should be a smoke alarm in the circulation space within 7.5 m of the door to every habitable room.	B1 1.14a
If a kitchen area is not separated from the stairway or circulation space by a door, there should also be an additional heat detector in the kitchen, that is interlinked to the other alarms.	B1 1.14b

Installation

> Smoke alarms should ideally be ceiling-mounted,   B1 1.14c–d
> 25 mm and 600 mm below the ceiling (25–150 mm in
> the case of heat detectors) and at least 300 mm from
> walls and light fittings.

 Units designed for wall mounting may also be used provided that the units are above the level of doorways opening into the space.

Smoke alarms should **not** be fitted:

> * over a stair shaft or any other opening between floors   B1 1.15
> * next to or directly above heaters or air conditioning   B1 1.16
>   outlets
> * in bathrooms, showers, cooking areas or garages, or   B1 1.16
>   any other place where steam, condensation or fumes
>   could give false alarms
> * in places that get very hot (such as a boiler room),   B1 1.16
>   or very cold (such as an unheated porch)
> * to surfaces which are normally much warmer or   B1 1.16
>   colder than the rest of the space.

Power supplies

The power supply for a smoke alarm system

> * should be derived from the dwelling's mains   B1 1.17
>   electricity supply via a single independent circuit at the
>   the dwelling's main distribution board (consumer unit)
> * should include a stand-by power supply that will   B1 1.17–18
>   operate during mains failure
> * should preferably **not** be protected by an RCD.   B1 1.20

 Smoke alarms may be interconnected using radio-links, provided that this does not reduce the lifetime or duration of any stand-by power supply.

Smoke alarms – other than dwellings

> The type of fire alarm/smoke detection system required   B1 1.23
> particular building depends on the type of occupancy and
> the means of escape strategy (e.g. simultaneous, phased
> or progressive horizontal evacuation).

Fire alarms

> Fire alarms should emit an audio and visual signal to warn occupants with hearing or visual impairments.   M 5.4g

All fire detection and fire warning systems shall be properly designed, installed and maintained.

- all buildings should have arrangements for detecting fire   B1 1.25
- all buildings should be fitted with suitable electrically operated fire warning system (in compliance with BS 5839) or have means of raising an alarm in case of fire (e.g. rotary gongs, handbells or shouting 'FIRE')   B1 1.26
- the fire warning signal should be distinct from other signals which may be in general use   B1 1.28
- in premises used by the general public, e.g. large shops and places of assembly, a staff alarm system (complying with BS 5839) may be used.   B1 1.29

Power-operated doors

Power-operated doors and gates should:

- be provided with a manual or automatic opening device in the event of a power failure where and when necessary for health or safety   K5 5.2d
- have safety features to prevent injury to people who are struck or trapped (such as a pressure-sensitive door edge which operates the power switch)
- have a readily identifiable and accessible stop switch.

Power-operated entrance doors

> Doors to accessible entrances shall be provided with a power-operated door opening and closing system if a force greater than 20 N is required to open or shut a door.   M 2.13a

Once open, all doors to accessible entrances should be wide enough to allow unrestricted passage for a variety of users, including wheelchair users, people carrying luggage, people with assistance dogs, and parents with pushchairs and small children.

The effective clear width through a single leaf door (or one leaf of a double leaf door) should be in accordance with Table 2.7.	M 2.13b

Table 2.7 Minimum effective clear widths of doors

Direction and width of approach	New buildings (mm)	Existing buildings (mm)
Straight-on (without a turn or oblique approach)	800	750
At right angles access route at least 1500 mm wide	800	750
At right angles to an access route at least 1200 mm wide	825	775
External doors to buildings used by the general public	1000	775

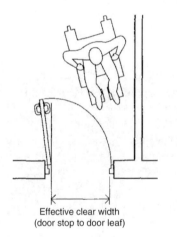
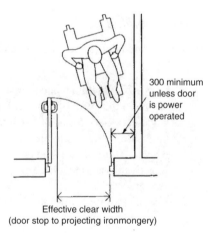

300 minimum unless door is power operated

Effective clear width
(door stop to door leaf)

Effective clear width
(door stop to projecting ironmongery)

Figure 2.12 Effective clear width and visibility requirements of doors

Power-operated entrance doors should have a sliding, swinging or folding action controlled manually (by a push pad, card swipe, coded entry, or remote control) or automatically controlled by a motion sensor or proximity sensor such as a contact mat.

Power-operated entrance doors should:

• open towards people approaching the doors	M 2.21a
• provide visual and audible warnings that they are operating (or about to operate)	M 2.21c
• incorporate automatic sensors to ensure that they open early enough (and stay open long enough) to permit safe entry and exit	M 2.21c
• incorporate a safety stop that is activated if the doors begin to close when a person is passing through	M 2.21b
• revert to manual control (or fail safe) in the open position in the event of a power failure	M 2.21d
• when open, should **not** project into any adjacent access route	M 2.21e
• ensure that its manual controls – are located between 750 mm and 1000 mm above floor level – are operable with a closed fist	M 2.21f
• be set back 1400 mm from the leading edge of the door when fully open	M 2.21g
• be clearly distinguishable against the background	M 2.21g
• contrast visually with the background.	M 2.19 and 2.21g

Note: Revolving doors are **not** considered 'accessible' as they create particular difficulties (and possible injury) for people who are visually impaired, people with assistance dogs or mobility problems and for parents with children and/or pushchairs.

Cellars or basements

LPG storage vessels and LPG fired appliances fitted with automatic ignition devices or pilot lights must **not** be installed in cellars or basements.	J 3.5i

Lecture/conference facilities

Artificial lighting should be designed to

• give good colour rendering of all surfaces	M 4.9 and 4.34
• be compatible with other electronic and radio frequency installations	M 4.12.1
• be compatible with other electronic and radio frequency installations.	M 4.36f

Swimming pools and saunas

Swimming pools and saunas are subject to special requirements specified in Part 6 of BS 7671: 2001.	P App A

Inspection and test

Electrical installations must be inspected and tested:

• during, at the end of installation and before they are taken into service to verify that they are reasonably safe **and** that they comply with BS 7671: 2001	P 1.6
• to verify that they meet the relevant equipment and installation standards.	P 0.1b

All electrical work should be inspected (during installation as well as on completion) to verify that the components have:

• been selected and installed in accordance with BS 7671	P 1.8a ii
• been made in compliance with appropriate British Standards or harmonised European Standards	P 1.8a i
• been evaluated against external influences (such as the presence of moisture)	P 1.8a ii
• not been visibly damaged (or are defective) so as to be unsafe	P 1.8a iii
• been tested to check satisfactory performance with respect to continuity of conductors, insulation resistance, separation of circuits, polarity, earthing and bonding arrangements, earth fault loop impedance and functionality of all protective devices including residual current devices	P 1.8b
• been inspected for conformance with section 712 of BS 7671: 2001	· P 1.9
• been tested as per section 713 of BS 7671: 2001	P 1.10
• been tested using appropriate and accurate instruments	P 1.10
• had their test results recorded using the model in Appendix 6 of BS 7671	P 1.10
• had their test results compared with the relevant performance criteria to confirm compliance.	P 1.10

Note: Inspections and testing of DIY work should **also** meet the above requirements.

Inspection and testing of non-notifiable work

Although it is not necessary for non-notifiable electrical installation work to be checked by a Building Control Body, it nevertheless **must** be carried out in accordance with the requirements of BS 7671: 2001.	P 1.30
Installers who are qualified to complete BS 7671 installation certificates and who carry out non-notifiable work should issue the appropriate electrical installation certificate for all but the simplest of like-for-like replacements.	P 1.32

Certification

BS 7671 Installation certificates

Compliance with Part P can be demonstrated by the issue of the appropriate BS 7671 electrical installation certificate.	P 1.8
An electrical installation certificate may **only** be issued by the installer responsible for the installation work.	P 1.28
Inspection and testing should be carried out in compliance with BS 7671: 2001.	P 1.9

 Section 712 of BS 7671: 2001 provides a list of all the inspections that may be necessary whilst Section 713 provides a list of tests.

Tests should be carried out using appropriate and accurate instruments under the conditions given in BS 7671, and the results compared with the relevant performance criteria to confirm compliance.	P 1.15
A copy of the installation certificate should be supplied to the person ordering the work.	P 1.9
Certificates may only be made out and signed by someone with the appropriate qualifications, knowledge and experience to carry out the inspection and test procedures.	P 1.9

Certificates should show that the electrical installation
work has been:

- inspected during erection as well as on completion and   P 1.11a
  that components have been
    - made in compliance with appropriate British Standards or
      harmonised European Standards
    - selected and installed in accordance with BS 7671: 2001
      (including consideration of external influences such
      as the presence of moisture)
    - not visibly damaged or defective so as to be unsafe
- tested for continuity of conductors, insulation   P 1.11b
  resistance, separation of circuits, polarity, earthing
  and bonding arrangements, earth fault loop impedance
  and functionality of all protective devices (including
  residual current devices).

A full Electrical Installation Certificate should be used   P 1.13
for the replacement of consumer units.

Minor works certificate

Appropriate tests (according to the nature of the work)   P 1.16
should be carried out.

A minor works certificate should be issued whenever   P 1.13
inspection and testing has been carried out, irrespective of
the extent of the work undertaken.

A minor works certificate shall **not** be used for the   P 1.13
replacement of consumer units or similar items.

Building Regulations compliance certificates

The following are additional certificates that are issued   P 1.17
on completion of notifiable works as evidence of compliance
with the Building Regulations:

- a Building Regulations Compliance Certificate (issued
  by Part P competent person scheme installers) –
  see Appendix 1 to this chapter
- completion certificates (issued by local authorities)
- final notices (issued by approved inspectors).

These documents are **different** documents to the BS 7671 installation certificate and are used to attest compliance with **all** relevant requirements of the Building Regulations – not just to Part P.

Certification of notifiable work

Only installers registered with a Part P competent person self-certification scheme are qualified to complete BS 7671 installation certificates and should do so in respect of every job they undertake.	P 1.18
A copy of the certificate should always be given to the person ordering the electrical installation work.	P 1.18
A Building Regulations Compliance Certificate must be issued to the occupant (and the Building Control Body) either by the installer or the installer's registration body within 30 days of the work being completed.	P 1.19
If notifiable electrical installation work is going to be carried out by a person **not** registered with a Part P competent person self-certification the work should be notified to a Building Control Body (the local authority or an approved inspector) before work starts.	P 1.21

 Note: The Building Control Body then becomes responsible for making sure the work is safe and complies with all relevant requirements of the Building Regulations.

On satisfactory completion of all work, the Building Control Body will issue a Building Regulation Completion Certificate (if they are the local authority) or a final certificate (if they are an approved inspector).	P 1.23
If notifiable electrical installation work is going to be carried out by installers who are not qualified to issue BS 7671 completion certificates (e.g. subcontractors or DIYers), then the Building Control Body must be notified before the work starts.	P 1.24
The Building Control Body then becomes responsible for making sure that the work is safe and complies with all relevant requirements of the Building Regulations – **but**, not at the householder's expense!.	P 1.26

Third party certification

Unregistered installers should not themselves arrange for a third party to carry out final inspection and testing.	P 1.28
A third party may only sign a BS 7671: 2001 Periodic Inspection Report (or similar) to indicate that electrical safety tests had been carried out on the installation which met BS 7671: 2001 criteria.	P 1.29
Third parties are not entitled to verify that the installation complies, fully, with BS 7671 requirements – for example with regard to routing of hidden cables.	

Provision of information

Sufficient information should be left with the occupant to ensure that persons wishing to operate, maintain or alter an electrical installation can do so with reasonable safety.	P 1.33
This information should include:	P 1.34

- all items called for by BS 7671: 2001
- an electrical installation certificate (describing the installation and giving details of work carried out)
- permanent labels (e.g. on earth connections and bonds and on items of electrical equipment such as consumer units and RCDs)
- operating instructions and log books
- detailed plans (but only for unusually large or complex installations).

2.6.2 Where can I get more information about the requirements for Electrical Safety?

Further guidance concerning the requirements for electrical safety is available from:

- The IET (Institution of Engineering and Technology) at www.theiet.org
- The NICEIC (National Inspection Council for Electrical Installation Contracting) at www.niceic.org.uk
- The ECA (Electrical Contractors' Association) at www.niceic.org.uk or www.eca.co.uk
- The Building Regulations in Brief (Fourth Edition). ISBN 0-7506-8058-X.

3

Mandatory requirements

In this chapter you will find a list all of the mandatory and the more essential requirements for the design, installation and maintenance of safe electrical installations that are called up in the Wiring Regulations and the Building Regulations. Critical requirements are sprinkled throughout the two Regulations and can be a bit of nightmare for you to find out what is required for a particular function. An attempt has been made, therefore, to pool the individual requirements of the various Chapters and Parts of these Regulations into separate logical sections so as to make it easier to find exactly what is required for a particular function, equipment, component or installation.

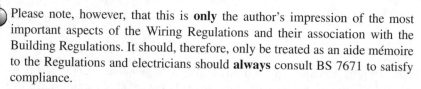

Please note, however, that this is **only** the author's impression of the most important aspects of the Wiring Regulations and their association with the Building Regulations. It should, therefore, only be treated as an aide mémoire to the Regulations and electricians should **always** consult BS 7671 to satisfy compliance.

As this chapter is primarily intended for reference purposes, explanations have been kept to an absolute minimum as more detail is available elsewhere in this book. For your assistance, this listing also includes a reference to the relevant chapters of the standard and has been split up into the following sections:

- design considerations
- electrical installations
- isolation and switching
- wiring systems
- earthing
- protective measures
- equipment
- components and accessories
- sockets
- cables and conductors
- power supplies
- circuits
- test and inspection
- special installations and locations.

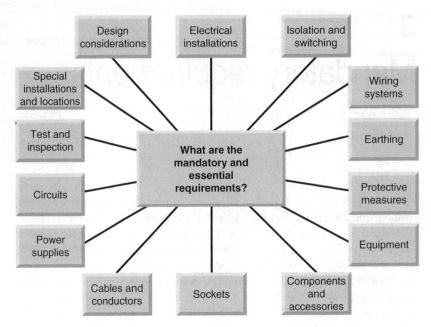

Figure 3.1 Mandatory and essential requirements of the Wiring and Building Regulations

 Note: Whilst the requirements from the Wiring Regulations are normally prefaced by the 'shall' (meaning that this section **is** a mandatory requirement), you will notice that the Building Regulations use the words 'should' (i.e. recommended), 'may' (i.e. permitted) or 'can' (i.e. possible).

The reason for this is that Approved Documents reproduce the actual *requirements* contained in the Building Regulations relevant to a particular subject area. This is then followed by *practical and technical guidance* (together with examples) showing how the requirements can be met in some of the more common building situations. There may, however, be alternative ways of complying with the requirements to those shown in the Approved Documents and you are, therefore, under no obligation to adopt any particular solution in an Approved Document – you may prefer to meet the requirement(s) in some other way, but you **must** meet the requirement!

 If any reader (particularly practising electrical engineers) has any thoughts about the contents of this chapter (such as areas where perhaps they feel I have not given sufficient coverage, omissions and/or mistakes etc.) then please let me know by emailing me at ray@herne.org.uk and I will make suitable amendments in the next edition of this book.

But first:

3.1 What are the mandatory requirements?

Protection against electric shock **shall** be provided.   410-01-01

Protective safety measures **shall** be applied in every   470-01-01
installation, part installation and/or equipment.

Electrical installations **shall** be designed so as to provide   131-01-01
for the:

- protection for persons, livestock and property
- proper functioning of the electrical installation.

There **shall** be no detrimental influence between   470-01-02
various protective measures used in the same
installation, part installation or equipment.

Equipment **shall** comply with the requirements of the   511-01-01
applicable British Standard or Harmonised Standard
for the intended use of the equipment.

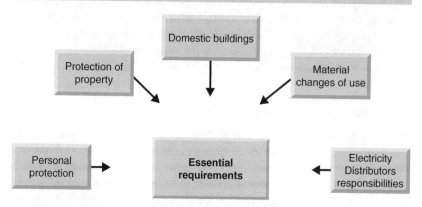

Figure 3.2 Mandatory requirements

3.1.1 Personal protection

Persons and livestock shall be protected:

- against dangers arising from contact with live parts   130-02-01
  of the installation

- from contact with exposed conductive parts during   130-02-02
  a fault
- from receiving burns from electrical equipment   130-03-01
- from being injured by excessive temperatures and/or   130-04-01
  electromagnetic stresses caused by overcurrents taking
  place in live conductors
- against overvoltages arising due to atmospherics   130-06-02
  and/or switching
- against heat and/or thermal radiation emitted by   130-03-02
  adjacent electrical equipment
- against injury due to a fault between live parts of   130-06-01
  circuits supplied at different voltages.

3.1.2 Protection of property

Property shall be protected:

- from being damaged by excessive temperatures and/or   130-04-01
  electromagnetic stresses caused by overcurrents
  happening in live conductors
- against harmful effects caused by a fault occurring   130-06-01
  between live parts of circuits supplied at different
  voltages.

3.1.3 Domestic buildings

Domestic buildings (as explained in Chapter 2) are covered by the Building
Act 1984 (as amended), which ensures that the health, welfare and convenience of persons living in or working in (or merely nearby) buildings is secured.
The intention of the Act is also to assist in the conservation of fuel and power,
prevent waste, undue consumption and the misuse and contamination of
water. To this end, all domestic buildings must meet the following, mandatory,
requirements.

Part P – Electrical safety

Reasonable provision shall be made in the design,   *Approved*
installation, inspection and testing of electrical   *Document*
installations in order to protect persons from fire or injury.   *P1*

Sufficient information shall be provided so that persons wishing to operate, maintain or alter an electrical installation can do so with reasonable safety.

Approved Document P2

Part M – Access and facilities for disabled people

 In addition to the requirements of the Disability Discrimination Act 1995 precautions need to be taken to ensure that:

Approved Document M

- *new non-domestic buildings and/or dwellings (e.g. houses and flats used for student living accommodation etc.)*
- *extensions to existing non-domestic buildings*
- *non-domestic buildings that have been subject to a material change of use (e.g. so that they become a hotel, boarding house, institution, public building or shop)*

are capable of allowing people, regardless of their disability, age or gender to:

- *gain access to buildings*
- *gain access within buildings*
- *be able to use the facilities of the buildings (both as visitors and as people who live or work in them)*
- *use sanitary conveniences in the principal storey of any new dwelling.*

Part L – Conservation of fuel and power

Energy efficiency measures shall be provided which:

- *provide lighting systems that utilise energy-efficient lamps with manual switching controls or, in the case of external lighting fixed to the building, automatic switching, or both manual and automatic switching controls as appropriate, such that the lighting systems can be operated effectively as regards the conservation of fuel and power*

Approved Document L1

- *provide information, in a suitably concise and understandable form (including results of performance tests carried out during the works), that shows building occupiers how the heating and hot water services can be operated and maintained.*

3.1.4 Extensions, material alterations and material changes of use

Where any electrical installation work is classified as an extension, a material alteration or a material change of use, the work must consider and include:

- confirmation that the mains supply equipment is suitable (and can) carry the additional loads envisaged   P 2.1b–P2.2
- the rating and the condition of existing equipment (belonging to both the consumer and the electricity distributor)   P 2.2a
- the amount of additions and alterations that will be required to the existing fixed electrical installation in the building   P 2.1a
- the necessary additions and alterations to the circuits which feed them   P 2.1a
- the protective measures required to meet the requirements   P 2.1a–P2.2b
- the earthing and bonding systems required are satisfactory and meet the requirements.   P 2.1a–P2.2c

3.1.5 Electricity distributors' responsibilities

The electricity distributor is responsible for:

- installing the cut-out and meter in a safe location   P 1.3
- ensuring that it is mechanically protected and can be safely maintained   P 1.3
- evaluating and agreeing proposals for new installations or significant alterations to existing ones.   P 1.4

Distributors are required to:

- provide an earthing facility for new connections   P 3.8
- maintain the supply within defined tolerance limits   P 3.8
- provide certain technical and safety information to   P 3.8
  consumers to enable them to design their installations.

Distributors and meter operators must ensure that their
equipment on consumers' premises

- is suitable for its purpose   P 3.9
- is safe in its particular environment   P 3.9
- clearly shows the polarity of the conductors.   P 3.9

Distributors are:

- prevented by the Regulations from connecting installations P 3.12
  to their networks which do not comply with BS 7671
- may disconnect consumers' installations which are a   P 3.12
  source of danger or cause interference with their
  networks or other installations.

3.2 Design considerations

Electrical installations shall be designed and installed
so that they:

- provide adequate protection against mechanical and   P 0.1a
  thermal damage
- are suitably enclosed (and appropriately separated)   P 1.2
  to provide mechanical and thermal protection
- do not present an electric shock or fire hazard to people P 0.1a
- provide adequate protection for persons against the   P 1.2
  risks of electric shock, burn or fire injuries
- meet the requirements of the Building Regulations.   P 3.1

The design of any electrical system will have to take into consideration not just
the requirements of the end customer but also the cost-effectiveness of supplying
and managing that system. The goal when designing electrical installations is
first to establish the user requirement and then design the circuit to transport the
energy (via distribution boards, wiring systems, circuit breakers, switches and
transformers etc.) as efficiently as possible to the required location – while at the
same time taking into account economic factors, network safety and redundancy.

The main areas that need to be considered by the designer include:

- type of demand
- type of equipment
- type of conductors
- provision of supplies
- emergency supplies
- emergency control
- protective devices and switches
- protective equipment
- insulation and switching
- accessibility of electrical equipments
- mutual interference
- environmental conditions

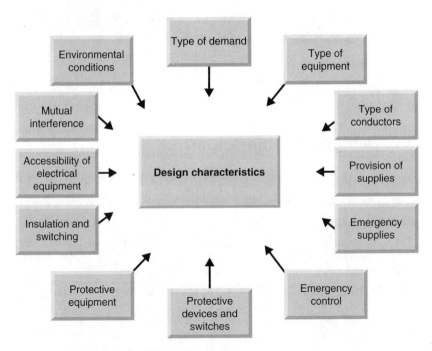

Figure 3.3 Design characteristics

and the design of any electrical installations needs to take into consideration the:

- type of electrical installation
- installation method
- type (and selection) of equipment
- type of wiring system
- choice of earthing system
- type of conductors
- safety protection and safety protective measures

- type of protective equipment required
- environmental conditions
- type of supply
- type of demand
- choice of supply
- characteristics of available supplies
- supplies for safety services and stand-by purposes
- emergency control and emergency switching
- safety service emergency supplies
- isolation and switching
- protective devices and switches
- auxiliary control units
- disconnecting devices
- prevention of mutual (detrimental) influence
- fire precautions
- installations (many)
- inspection and testing
- identification and notices
- future additions and alterations
- maintenance and repair.

The requirements that have to be followed in order to meet the Regulations are sprinkled throughout the standard. The main ones to remember, however, are shown in the following sections.

3.2.1 Equipment categories

For convenience the equipment has been split up into four categories.

Table 3.1 Examples of various impulse category equipment

Category	Example
I	Equipment intended to be connected to the fixed electrical installation where protection against transient overvoltage is external to the equipment, either in the fixed installation or between the fixed installation and the equipment. Examples of equipment are household appliances; portable tools and similar loads intended to be connected to circuits in which measures have been taken to limit transient overvoltages.
II	Equipment intended to be connected to the fixed electrical installation, e.g. household appliances, portable tools and similar loads, the protective means being either within or external to the equipment.
III	Equipment which is part of the fixed electrical installation and other equipment where a high degree of availability is expected, e.g. distribution boards, circuit breakers, wiring systems and equipment for industrial uses, stationary motors with permanent connection to the fixed installation.
IV	Equipment to be used at (or in the proximity of) the origin of the electrical installation upstream of the main distribution board, e.g. electricity meter, primary overcurrent device, ripple control unit.

Note: The Regulations (i.e. 528-01) stipulate that:

- Band I and Band II circuits shall not be contained within the same wiring system if that wiring system has a nominal voltage exceeding low voltage unless all cables have been insulated for the highest voltage likely to be present or:
  - each cable is enclosed within an earthed metallic screen or
  - the cables are insulated for their respective system voltages or
  - the cables are installed in a separate compartment of a cable ducting or cable trunking system or
  - the cables have an earthed metallic covering or
  - each conductor in a multicore cable is insulated for the highest voltage present in the cable
- and that in Band I and Band II voltage circuits, the cables and connections of circuits of different voltage Bands are segregated by a partition
- Band I circuits shall not be contained in the same wiring system as Band II voltage circuits unless:
  - every cable is insulated for the highest voltage present or
  - in a multicore cable (or cord) the cores of the Band I circuit have been:
    - insulated for the highest voltage present in the Band II circuit
    - separated from the cores of the Band II circuit by an earthed metal screen
- and the cables:
  - are insulated for their system voltage or
  - are installed in a separate compartment of a cable ducting or trunking system or
  - are installed on a tray or ladder separated by a partition or
  - use a separate conduit, trunking or ducting system.

3.2.2 External influences

General

All electrical equipment shall be capable of withstanding the stresses, environmental conditions and the characteristics of its location, or be provided with additional protection.	132-01-07
Conductors and cables shall be capable of withstanding all foreseen electromechanical forces (including fault current) during service.	521-03-01
The degree of protection of selected equipment shall be suitable for its intended use and mode of installation.	512-06-01
The mutual effect of a number of different external influences occurring simultaneously shall be taken into account.	512-06-02

Ambient temperature

Components

> Components (including cables and wiring enclosures) shall   522-01-02
> be installed in accordance with the temperature limits set by
> the relevant product specification or by the manufacturer.

Earthing

> Earthing arrangements shall be such that they are   542-01-07
> sufficiently robust (or have additional mechanical
> protection) to external influences.

 Note: This particularly applies with respect to thermal, thermomechanical and electromechanical stresses.

Electrical appliances

> Electric appliances producing hot water or steam shall be   424-02-01
> protected against overheating.

Fixed electrical equipment

> Heat generated from fixed electrical equipment shall not   422-01-01
> damage (or cause danger to) adjacent materials.

People

> Persons, fixed equipment and fixed materials that are
> adjacent to electrical equipment shall be protected against:
>
> • heat and/or thermal radiation that is emitted by that   130-03-02
>   electrical equipment
> • the harmful effects of: 421-01-01
>   – developed (and burns caused) by electrical equipment
>   – thermal radiation.

Electromagnetic compatibility

Electrical installations

Electrical installations shall be arranged so that they do not mutually interfere or cause electromagnetic interference with other electrical installations and non-electrical installations in a building.	131-11-01

Equipment

Equipment chosen:	
• shall be immune to the electromagnetic influences of other equipment and installations	515-02-01
• shall not cause electromagnetic influences to other equipment and/or installations.	515-02-02
Equipment shall comply with the appropriate EN or HD or implementing National Standard or (in certain circumstances) the appropriate IEC Standard or National Standard of another country.	132-01-01
Electrical equipment shall be selected and erected so as to avoid any harmful influence between the electrical installation and any non-electrical installations envisaged.	515-01-01

Property

Property shall be protected from being damaged by electromagnetic stresses caused by overcurrents occurring in live conductors.	130-04-01

Fire

All fixed electrical equipment with a surface temperature that is liable to cause a fire or damage adjacent materials shall either be:	422-01-02
• mounted on a support (or placed within an enclosure) that can withstand such temperatures as may be generated or	
• screened by material which can withstand the heat emitted by the electrical equipment or	
• mounted so that the heat can be safely dissipated.	

All fixed equipment that focuses or concentrates heat shall be isolated from other fixed objects or building elements that could be effected by high temperatures.	422-01-06
Cables shall comply with the requirements of BS EN 50265-1 for flame propagation.	527-01-03
Fire alarm and emergency lighting circuits shall be segregated from all other cables and from each other in accordance with BS 5839 and BS 5266.	528-01-04

Fixed equipment enclosures shall be constructed: 422-01-07

- using heat and fire-resistant materials, or
- be capable of withstanding the highest temperature likely to be produced by the electrical equipment in normal use.

Fixed equipment that is liable to emit an arc or high   422-01-03
temperature particles shall be:

- totally enclosed in arc-resistant material, or
- screened by arc-resistant material from materials that could be effected by emissions or
- mounted so as to allow safe extinction of the emissions at a sufficient distance from material that could be affected by emissions.

 Note: To provide mechanical stability, arc-resistant material used for this protective measure shall be:

- non-ignitable
- of low thermal conductivity
- of adequate thickness.

Live conductors (or joints between them) shall be terminated within an enclosure.	422-01-04
Sealing arrangement shall be visually inspected during erection, to verify that it conforms to the manufacturer's erection instructions. Details shall be recorded.	527-04-01
Sealing that has been disturbed during alterations work shall be reinstated as soon as practicable.	527-03-02
Temporary sealing arrangements shall be provided during the erection of a wiring system.	527-03-01

The heat generated from fixed electrical equipment shall not damage (or cause danger to) adjacent fixed material.	422-01-01
The risk of spread of fire shall be minimised by selection of an appropriate material and erection.	527-01-01
The wiring system shall not reduce the structural performance and fire safety of a building.	527-01-02
Where the risk of fire is high special precautions shall be taken.	527-01-03
Wiring systems (such as conduit, cable ducting, cable trunking, busbar or busbar trunking) that penetrate a building construction that has specified fire resistance, shall be internally sealed to maintain the degree of fire resistance.	527-02-02
Wiring systems that pass through floors, walls, roofs, ceilings, partitions or cavity barriers shall be sealed according to the degree of fire resistance required of the element concerned (if any).	527-02-01

 Note: See BS 476-23 for details concerning the sealing of wiring systems.

Cables

Cables shall not be run in a lift (or hoist) shaft unless it forms part of the lift installation.	528-02-06

Fire alarm and emergency lighting

Fire alarm and emergency lighting circuits shall be segregated from all other cables and from each other in accordance with BS 5839 and BS 5266.	528-01-04

Fireman's switch

All exterior installations that are located on single premises shall (wherever practicable) be controlled by a single fireman's switch.	476-03-06
All firemen's switches shall meet the requirements of the Regulations plus any requirements of the local fire authority.	476-03-07

Lighting points

Ceiling roses shall not be installed in circuits whose operating voltage is greater than 250 volts.	553-04-02

Live supply conductors

Circuits shall be capable of being isolated from all live supply conductors by a linked switch or linked circuit breaker.	482-02-08

Safety services

Alarm and control devices shall be clearly indicated and identified.	563-01-06
Protection against electric shock and fault current shall be met for each source.	565-01-02
Safety service circuits shall have adequate fire resistance for the locations through which it passes.	563-01-02

Wiring systems

Wiring systems installed near non-electrical services shall be arranged so as not to cause damage to each other.	528-02-04
Wiring systems routed near a service liable to cause condensation (e.g. water, steam or gas services) shall be protected.	528-02-03

Humidity

Wiring systems shall be capable of withstanding condensation and/or ingress of water during installation, use and maintenance.	522-03-01

Mechanical

Buried cables, conduits and ducts shall be at a sufficient depth to avoid being damaged.	522-06-03

Care shall be taken when selecting and using fixings and   522-07-01
connections for cables which are part of a wrong system
and which are fixed (or supported by) a structure
equipment is subject to vibration.

Conductors

Conductors shall not be subjected to excessive mechanical   522-12-01
stress.

Earthing arrangements

Earthing arrangements shall be sufficiently robust   542-01-07
(or have additional mechanical protection) to external
influences.

Enclosures

The insulating enclosure shall be capable of resisting   413-03-05
mechanical, electrical and thermal stresses likely
to be encountered.

Equipment

Equipment shall be capable of withstanding all   412-02-01
mechanical stresses normally encountered during
service.

Maintenance

Equipment shall not be capable of becoming   462-01-03
unintentionally or inadvertently reactivated during
mechanical maintenance.

Protective devices

An emergency stopping facility shall be provided where mechanical movement of electrically actuated equipment is considered dangerous.	463-01-05

Pollutants and contaminants

Smoke or fumes

Wiring systems shall not be installed near a service which produces smoke or fumes unless protected by shielding.	528-02-02

Solar radiation

Wiring systems shall withstand solar radiation and/or ultraviolet radiation.	522-11-01

 Note: Special precautions may need to be taken for equipment subject to ionising radiation.

Weather and precipitation

Equipment

Electric appliances producing hot water or steam shall be protected against overheating.	424-02-01
Equipment likely to be exposed to weather, corrosive atmospheres or other adverse conditions shall be constructed (or protected) to prevent danger arising from such exposure.	131-05-01

Wiring systems

Wiring systems:

- installed near water and subjected to waves shall be protected — 522-03-03
- routed near a service liable to cause condensation (e.g. water, steam or gas services) shall be protected — 528-02-03

- shall be capable of withstanding condensation   522-03-01
  and/or ingress of water during installation, use and
  maintenance
- shall withstand flora and fauna   522-10-01
- shall withstand mould growth.   522-09-01

3.3 Electrical installations

Installations shall comply with the requirements for: 400-01-01

- protection against electric shock
- protection against thermal effects
- protection against overcurrent
- protection against fault current
- protection against undervoltage
- isolation and switching.

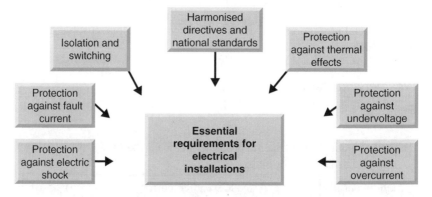

Figure 3.4 Essential requirements for electrical installations

All electrical installations **shall**:

- be designed and installed (suitably enclosed and   P 1.2
  appropriately separated) to provide mechanical and
  thermal protection
- provide adequate protection for persons against the   P 1.2
  risks of electric shock, burn or fire injuries
- must meet the requirements of the Building Regulations P 3.1

- be inspected and tested during installation, at the end
of installation and before they are taken into service to
verify that they
  - are safe **and** that they comply with BS 7671   P 1.6
  - meet the relevant equipment and installation standards. P 0.1b

3.3.1 Electrical installation work

All electrical installation work:

- is to be carried out professionally   P 1.1
- is to comply with the Electricity at Work Regulations   P 1.1
1989 as amended
- may only be carried out by persons that are competent   P 3.4a
to prevent danger and injury while doing it, or who
are appropriately supervised.

3.3.2 Number of circuits

A separate circuit shall be provided for each part of the   314-01-02
installation which needs to be separately controlled.

3.3.3 Final circuits

Final circuits (of installations with more than one final   314-01-04
circuit) shall be separately connected to a distribution
board.

The number of final circuits required shall take into   314-01-03
consideration the requirements for:

- overcurrent protection
- isolation and switching
- the current-carrying capacities of conductors.

The wiring of each final circuit shall be electrically   314-01-04
separate from other final circuits.

3.3.4 Domestic cable installation

The following requirements are emphasised in the Building Regulations.

Cables installed under a floor (or above a ceiling)

These shall not be liable to damage from that floor or the ceiling (or their fixings).

Cables passing through floor timber joist used for floors and ceilings

These shall:

- be at least 50 mm from the top, or bottom, of the joist, or
- use an earthed armour or metal sheath as a protective conductor, or
- be of insulated concentric construction, or
- be protected by an earthed, steel conduit enclosure, or
- be protected by some form of mechanical protection.

Cables that are concealed in a wall or partition less than 50 mm from the surface

These shall either be:

- protected by an earthed metallic conductor complying with BS 5467, BS 6346, BS 6724, BS 7846, BS EN 60702-1 or BS 8436, or
- of an insulated concentric construction complying with BS 4553-1, BS 4553-2 or BS 4553-3, or
- enclosed in earthed conduit, trunking or ducting, or
- mechanically protected against penetration of the cable by nails, screws etc., or
- installed within 150 mm from the top of the wall or partition, or
- within 150 mm of an angle formed by two adjoining walls or partitions.

3.3.5 Alarm systems

Fire alarms should emit an audio and visual signal to warn occupants with hearing or visual impairments.   M (5.4g)

Emergency assistance alarm systems should have: M (5.4h)

- visual and audible indicators to confirm that an emergency call has been received
- a reset control reachable from a wheelchair, WC, or from a shower/changing seat
- a signal that is distinguishable visually and audibly from the fire alarm.

Emergency alarms

Emergency alarm pull cords should be:	M (4.30e)

- coloured red
- located as close to a wall as possible
- have two red 50 mm diameter bangles.

Front plates should contrast visually with their backgrounds.	M (4.30m)
The colours red and green should **not** be used in combination as indicators of 'ON' and 'OFF' for switches and controls.	M (4.28)

Fire alarms

All fire detection and fire warning systems shall be properly designed, installed and maintained and the following facilities shall be provided:

- all buildings should have arrangements for detecting fire   B1 (1.25)
- all buildings should be fitted with suitable electrically   B1 (1.26)
  operated fire warning system (in compliance with BS
  5839) or have means of raising an alarm in case of fire
  (e.g. rotary gongs, handbells or shouting 'FIRE')
- the fire warning signal should be distinct from other   B1 (1.28)
  signals which may be in general use
- in premises used by the general public (e.g. large   B1 (1.29)
  shops and places of assembly) a staff alarm system
  (complying with BS 5839) may be used.

Smoke alarms – dwellings

Smoke alarms should normally be positioned in the circulation spaces between sleeping spaces and places where fires are most likely to start (e.g. kitchens and living rooms) to pick up smoke in the early stages, while also being close enough to bedroom doors for the alarm to be effective when occupants are asleep.

General

There should be at least one smoke alarm on every storey of a house (including bungalows).	B1 (1.12)

If more than one smoke alarm is installed in a dwelling then they should be linked so that if a unit detects smoke it will operate the alarm signal of all the smoke detectors.	B1 (1.13)
There should be a smoke alarm in the circulation space within 7.5 m of the door to every habitable room.	B1 (1.14a)
If a kitchen area is not separated from the stairway or circulation space by a door, there should also be an additional heat detector in the kitchen that is interlinked to the other alarms.	B1 (1.14b)

Installation

Smoke alarms should ideally be ceiling-mounted, 25 mm and 600 mm below the ceiling (25–150 mm in the case of heat detectors) and at least 300 mm from walls and light fittings.	B1 (1.14c–d)
Smoke alarms should not be fitted or fixed:	B1 (1.15)
• over a stair shaft or any other opening between floors	
• next to or directly above heaters or air conditioning outlets	B1 (1.16)
• in bathrooms, showers, cooking areas or garages, or any other place where steam, condensation or fumes could give false alarms	B1 (1.16)
• in places that get very hot (such as a boiler room), or very cold (such as an unheated porch)	B1 (1.16)
• to surfaces which are normally much warmer or colder than the rest of the space.	B1 (1.16)

Power supplies

The power supply for a smoke alarm system:

• should be derived from the dwelling's mains electricity supply from a single independent circuit at the dwelling's main distribution board (consumer unit)	B1 (1.17)
• should include a stand-by power supply that will operate during mains failure	B1 (1.17–18)
• should preferably not be protected by any Residual Current Device (RCD).	B1 (1.20)

Smoke alarms may be interconnected using radio-links, provided that this does not reduce the lifetime or duration of any stand-by power supply.

Smoke alarms – buildings other than dwellings

The type of fire alarm/smoke detection system required for a particular building depends on the type of occupancy and the means of escape strategy (e.g. simultaneous, phased or progressive horizontal evacuation).	B1 (1.23)

3.3.6 Lighting

External lighting fixed to the building

External lighting (including lighting in porches, but not lighting in garages and carports) should:

• automatically extinguish when there is enough daylight, and when not required at night	L1 (1.57a)
• have sockets that can only be used with lamps having an efficacy greater than 40 lumens per circuit watt (such as fluorescent or compact fluorescent lamp types, and not GLS tungsten lamps with bayonet cap or Edison screw bases).	L1 (1.57b)

Fixed lighting

In locations where lighting can be expected to have most use, fixed lighting (e.g. fluorescent tubes and compact fluorescent lamps – but not GLS tungsten lamps with bayonet cap or Edison screw bases) with a luminous efficacy greater than 40 lumens per circuit watt should be available.	L1 (1.54)

High-voltage discharge lighting installations

High-voltage electric signs and high-voltage luminous discharge tube installations shall meet the requirements of BS 559.	554-02-01

3.3.7 Electrical connections

Connections and joints shall be accessible for inspection, testing and maintenance.	526-04-01
Connections between a conductor and equipment shall provide durable electrical continuity and ample mechanical strength.	526-01-01
Enclosure shall provide mechanical protection and protection against external influences.	526-03-01
Non-sheathed cables (and cores of sheathed cables from which the sheath has been removed) shall be enclosed in an equipment enclosure or an accessory complying with the appropriate British Standard.	526-03-03
Terminations and joints in a live conductor (or a PEN conductor) shall only be made inside an enclosure (e.g. equipment enclosure or an accessory complying with the appropriate British Standard).	526-03-02
The method used for electrical connections will depend on the type and number of conductors being connected together and the ambient conditions.	526-02-01

Sealing arrangements

Details of sealing arrangements shall be recorded.	527-04-01
Sealing arrangement shall be visually inspected during erection, to verify that it conforms to the manufacturer's erection instructions. Details shall be recorded.	527-04-01
Sealing that has been disturbed during alterations work shall be reinstated as soon as practicable.	527-03-02
Temporary sealing arrangements shall be provided during the erection of a wiring system.	527-03-01

3.3.8 Power-operated doors

Doors to accessible entrances shall be provided with a power-operated door opening and closing system if a force greater than 20 N is required to open or shut a door.	M (2.13a)

Once open, all doors to accessible entrances should be wide enough to allow unrestricted passage for a variety of users, including wheelchair users, people carrying luggage, people with assistance dogs, and parents with pushchairs and small children.   M2

Power-operated doors and gates should:

- have safety features to prevent injury to people who are struck or trapped (such as a pressure sensitive door edge which operates the power switch)   K5 (5.2d)
- be provided with a manual or automatic opening device in the event of a power failure where and when necessary for health or safety   K5 (5.2d)
- have a readily identifiable and accessible stop switch.   K5 (5.2d)

Power-operated entrance doors should:

- open towards people approaching the doors   M (2.21a)
- provide visual and audible warnings that they are operating (or about to operate)   M (2.21c)
- incorporate automatic sensors to ensure that they open early enough (and stay open long enough) to permit safe entry and exit   M (2.21c)
- incorporate a safety stop that is activated if the doors begin to close when a person is passing through   M (2.21b)
- revert to manual control (or fail safe) in the open position in the event of a power failure   M (2.21d)
- when open, should not project into any adjacent access route   M (2.21e)
- ensure that its manual controls: M (2.21f)
  - are located between 750 mm and 1000 mm above floor level
  - are operable with a closed fist
- be set back 1400 mm from the leading edge of the door when fully open   M (2.21g)
- be clearly distinguishable against the background   M (2.21g)
- contrast visually with the background.   M (2.19 and 2.21g)

3.3.9 Proximity to other electrical services

A Band I circuit shall **not** be contained in the same wiring system as Band II voltage circuits.   528-01-02

Band I and Band II circuits shall **not** be contained within the same wiring system if that wiring system has a nominal voltage exceeding that of low voltage.	528-01-01
Cables shall **not** be run in a lift (or hoist) shaft unless it forms part of the lift installation.	528-02-06
Fire alarm and emergency lighting circuits shall be segregated from all other cables and from each other.	528-01-04
Telecommunication circuits shall be segregated.	528-01-04
The cables and connections of circuits of different voltage Bands shall be segregated by a partition.	528-01-07

 If metal, this partition shall be earthed.

Wiring systems:

• installed near a non-electrical services shall not damage each other	528-02-04
• located in close proximity to a non-electrical service shall be protected from indirect contact and hazards likely to arise from that other service	528-02-01
• routed near a service liable to cause condensation (e.g. water, steam or gas services) shall be protected	528-02-03
• shall not be installed near a service which produces heat, smoke or fume unless protected by shielding.	528-02-02

3.3.10 Telephone points and TV sockets

All telephone points and TV sockets should be located between 400 mm and 1000 mm above the floor (or 400 mm and 1200 mm above the floor for permanently wired appliances).	M (4.30a and b)

3.3.11 Verification

Sealing arrangements shall be visually inspected during erection, to verify that they conform to the manufacturer's erection instructions. Details shall be recorded.	527-04-01

3.4 Isolation and switching

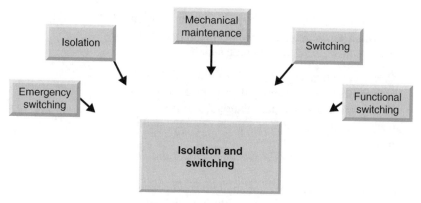

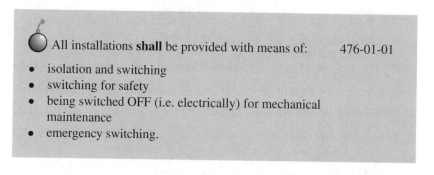

Figure 3.5 Isolation and switching devices

🔴 All installations **shall** be provided with means of: 476-01-01

- isolation and switching
- switching for safety
- being switched OFF (i.e. electrically) for mechanical maintenance
- emergency switching.

For compliance with the Building Regulations, the aim should be to ensure that all controls and switches are easy to operate, visible and free from obstruction and:

- they are located between 750 mm and 1200 mm above the floor
- they do not require the simultaneous use of both hands (unless necessary for safety reasons) to operate
- switched socket outlets indicate whether they are 'ON'
- mains and circuit isolator switches clearly indicate whether they are 'ON' or 'OFF'
- individual switches on panels and on multiple socket outlets are well separated
- controls that need close vision (e.g. thermostats) are located between 1200 mm and 1400 mm above the floor
- front plates contrast visually with their backgrounds.

A main linked switch (or linked circuit breaker) shall be   460-01-02
provided as near as practicable to the origin of every
installation.

Note: For d.c. systems, all poles shall be provided with a means of isolation.

A main switch that is intended to be operated by an unskilled person (e.g. householder) shall interrupt both live conductors of a single-phase supply.	476-01-03
Circuits and final circuits shall be provided with a means of switching for interrupting the supply on load (particularly circuits and parts of an installation that (for safety reasons) need to be switched independently of other circuits and/or installations).	476-01-02

 Note: This regulation does not apply to short connections between the origin of the installation and the consumer's main switchgear.

If an installation is supplied from more than one source: • a main switch shall be provided for each supply source • a notice shall be prominently placed warning operators that more than one switch needs to be operated.	460-01-02
Isolation devices or switching shall **not** be inserted in the outer conductor of a concentric PEN cable.	546-02-06
Switches (unless supplied from more than one energy source) shall not break a protective conductor or a PEN conductor.	460-01-03
The supply to electrical installations and electrically powered equipment shall have non-automatic isolation and switching facilities.	460-01-01

3.4.1 Isolation

Circuits **shall** be capable of being isolated from live supply conductors.	461-01-01
Equipment shall **not** be capable of being inadvertently or unintentionally energised.	461-01-02
Semiconductor devices shall **not** be used as isolating devices.	537-02-03
Off-load isolators shall **not** be used for functional switching.	537-05-03
Isolators shall **not** break a protective conductor or a PEN conductor.	460-01-03

All motor circuits shall be provided with a disconnector   476-02-03
to disconnect the motor and equipment.

If an isolating device for a particular circuit is some distance   476-02-02
away from the equipment to be isolated, then that isolating
device shall be capable of being secured in the open position.

 Note: When this is achieved by a lock or removable handle, the key or handle
shall be non-interchangeable with any others used for a similar purpose within
the premises.

Isolation devices shall:

- indicate the installation or circuit it isolates   537-02-09
- isolate all live supply conductors (less PEN conductors)   537-02-01
  from the circuit concerned
- prevent unintentional closure caused by mechanical   537-02-06
  shock or vibration.

Isolators, when used with a circuit breaker for isolating   476-02-01
main switchgear for maintenance, shall be interlocked
with that circuit breaker or placed and/or guarded so that
it can only be operated by skilled persons.

Off-load isolation devices shall be secured against   537-02-07
inadvertent and unauthorised operation.

Plug and socket outlets may be used as a means of isolation.   537-02-10

Self-contained luminaries (or circuits supplying luminaries)   476-02-04
with an open circuit voltage exceeding low voltage shall have:

- an interlock (in addition to the switch normally used
  for controlling the circuit) that automatically disconnects
  the live parts of the supply before access to these live
  parts is permitted
- some form of isolatator (in addition to the switch
  normally used for controlling the circuit) to isolate the
  circuit from the supply
- a switch (with a lock or removable handle) or a
  distribution board which can be locked.

The position of isolation contacts shall be either externally   537-02-04
visible or clearly indicated.

Where possible, groups of circuits shall be capable of being   461-01-01
isolated from live supply conductors, by a common means.

3.4.2 Switching

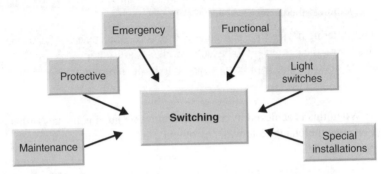

Figure 3.6 Switching for electrical installations

Switchgear shall **not** be connected to conductors intended to operate at a temperature exceeding 70°C.   512-02-01

Emergency switching

Installations **shall** be provided with an emergency switching device where there is a risk of an electric shock occurring.	463-01-01 and 463-01-05
Plug and socket outlet shall **not** be used as a device for emergency switching.	537-04-02
Resetting the emergency switching device shall **not** re-energise the equipment concerned.	537-04-05
Plug and socket outlet shall **not** be used as a device for emergency switching.	537-04-02

A readily accessible, easily operated, emergency stopping facility shall be provided in every place where an electrically driven machine may give rise to danger.	476-03-02
All electrically driven machines that could be dangerous shall be supplied with an emergency stopping facility that is readily accessible and easily operated.	476-03-02

Note: Where more than one means of starting the machine is provided and danger could be caused by unexpected restarting, then means shall be provided to prevent such restarting.

Emergency switching devices shall use a latching type operation and be capable of being restrained in the OFF or STOP position.   537-04-05

Emergency switching devices used to interrupt the supply shall be capable of cutting off the full load current of the relevant part of the installation.   537-04-01

Emergency switching shall consist of: 537-04-02

- a single switching device that directly cuts off the incoming supply, or
- a combination of several items of equipment that are operated by a single action that cuts off the appropriate supply.

Emergency switching shall:

- act as directly as possible on the appropriate supply conductors   463-01-02
- require only one initiative action
- not introduce a further hazard when activated   463-01-03
- be readily accessible and durably marked.   463-01-04

Where inappropriate operation of an emergency switching device could be dangerous, then that switching device shall be organised so that it can only be operated by skilled personnel.   476-03-03

Where practicable emergency switching devices that directly interrupt the main circuit shall be manually operated.   537-04-03

Functional switching

Off-load isolators, fuses and links shall **not** be used for functional switching.   537-05-03

Plug and socket outlets with a rating of more than 16 A may **not** be used as a switching device for d.c.   537-05-05

A functional switching device shall be provided for all circuits that need to be independently controlled.	464-01-01 464-01-03
All main switches that are likely to be operated by an unskilled person (e.g. householder) shall interrupt both live conductors of a single-phase supply.	476-01-03
Control circuits shall limit the possibility of a fault causing unintentional operation of the controlled equipment.	464-02-01
Functional switching devices shall be suitable for the most difficult duty intended.	537-05-01
Semiconductor functional switching devices may control the current without necessarily opening the corresponding poles.	537-05-02
The colours red and green should not be used in combination as indicators of 'on' and 'off' for switches and controls.	M (5.3)

Protective switching

Switches (unless linked) shall **not** be inserted in an earthed neutral conductor.	131-13-02

Single pole switches shall only be inserted in the phase conductor.	131-13-01

Light switches

Front plates should contrast visually with their backgrounds.	M (4.30m)
Individual switches on panels and on multiple socket outlets should be well separated.	M (4.29)
Light switches should: • have large push pads • align horizontally with door handles • be within 900 to 1100 mm from the entrance door opening.	M (4.30h and I)

Mains and circuit isolator switches should clearly indicate whether they are 'ON' or 'OFF'.   M (4.30l)

Switched socket outlets should indicate whether they are 'ON'.   M (4.30k)

Switches and controls should be located between 750 mm and 1200 mm above the floor.   M (4.30c and d)

Switches for lighting should be:

- located so that they are easily reachable   M2 (8.2)
- installed between 450 mm and 1200 mm from the finished floor level (see Figure 2.10).   M2 (8.3)

The colours red and green should **not** be used in combination as indicators of 'ON' and 'OFF' for switches and controls.   M (4.28)

Where possible, light switches with large push pads should be used in preference to pull cords.   M (5.3)

Switching off for mechanical maintenance

Equipment shall **not** be capable of becoming unintentionally or inadvertently reactivated during mechanical maintenance.   462-01-03

A device for switching OFF for mechanical maintenance (or a control switch for such a device) shall:

- be manually operated   537-03-02
- either be inserted in the main supply circuit or in the control circuit   537-03-01
- have an externally visible contact gap or clearly indicate the OFF or OPEN position.   537-03-02

If there is a risk of burns or injury from mechanical movement during maintenance, a switching off device shall be provided.   462-01-01

This device shall be suitably located and identified.   462-01-02

Special Installations and locations

LPG storage vessels and LPG fired appliances fitted   J (3.5i)
with automatic ignition devices or pilot lights must **not**
be installed in cellars or basements.

Swimming pools and saunas are subject to special   P AppA 1
requirements specified in Part 6 of BS 7671: 2001.

3.5 Wiring systems

Cables and/or conductors used as fixed wiring shall:

- be supported   522-08-05
- not be exposed to undue mechanical strain –   522-08-05
  particularly at terminations.

Cables:

- concealed in floors and walls (in certain circumstances) P AppA 2d
  are required to have an earthed metal covering, be
  enclosed in steel conduit, or have additional mechanical
  protection (see BS 7671 for more information)
- to an outside building (e.g. garage or shed) if run   P AppA 2d
  underground, should be routed and positioned so as to
  give protection against electric shock and fire as a
  result of mechanical damage to a cable.

Conductors shall not subjected to excessive mechanical   522-12-01
stress.

Flexible wiring systems shall be:

- installed so that excessive tensile and torsional stresses 522-08-06
  to the conductors and connections are avoided
- used for flexible or unstable structures.   522-12-02

Heat-resisting flexible cables are required for the final   P AppA 2d
connections to certain equipment (see makers'
instructions).

The radius of bends in a wiring system shall be such   522-08-03
that conductors and cables shall not suffer damage.

Where a conductor or a cable is not continuously   522-08-04
supported it shall be supported at appropriate intervals
to ensure that the conductor or cable is not damaged
because of its own weight.

Wiring systems:

- (if buried in the structure) shall have a conduit or   522-08-02
  cable ducting system for each circuit which shall be
  completely erected before the cable is drawn in
- shall be selected and erected so as to minimise (during 522-08-01
  installation, use and maintenance) their sheath, cable
  insulation and terminations being damaged
- shall not penetrate a load bearing element of a building. 522-12-03

Withdrawable wiring systems shall have be adequate   522-08-02
means of access for drawing cable in or out.

3.6 Earthing

In a TN installation, all exposed conductive parts   413-02-06
shall be connected (by a protective conductor) to the main
earthing terminal of the installation and that terminal shall
be connected to the earthed point of the supply source.

In a TT installation, all exposed conductive parts – which 413-02-18
are protected by a single protective device – **shall** be
connected, via the main earthing terminal, to a common
earth electrode.

In an IT installation, live conductors shall **not** be directly 413-02-21
connected to earth.

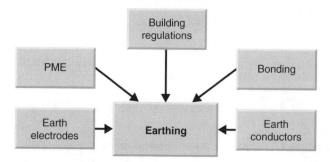

Figure 3.7   Earthing requirements

Earthing arrangements may be used jointly or   542-01-06
separately for protective and functional purposes,
according to the requirements of the installation.

3.6.1 Requirements from the Building Regulations

All electrical installations shall be properly earthed.	P App C1
All lighting circuits shall include a circuit protective conductor.	P App C
All socket outlets which have a rating of 32 A or less, and which may be used to supply portable equipment for use outdoors, shall be protected by an RCD.	P App C
Distributors are required to provide an earthing facility for all new connections.	P 3.8
It is **not** permitted to use a gas, water or other metal service pipe as a means of earthing for an electrical installation (this does not rule out, however, equipotential bonding conductors being connected to these pipes).	P App C
New or replacement, non-metallic light fittings, switches or other components must not require earthing (e.g. non-metallic varieties) unless new circuit protective (earthing) conductors are provided.	P App C
Socket outlets that will accept unearthed (2-pin) plugs must not use supply equipment that needs to be earthed.	P App C
Where electrical installation work is classified as an extension, a material alteration or a material change of use, the work must consider and include the earthing and bonding systems are satisfactory and meet the requirements.	P 2.1a– P 2.2c

3.6.2 Protective multiple earthing

Socket outlets of caravan pitch supply equipment   608-13-05
must **not** be bonded to the PME terminal.

3.6.3 Earth electrodes

Metalwork belonging to gas, water or other services shall **not** be used as a protective earth electrode.   542-02-04

Earth electrodes shall be designed and constructed so that they can: 542-02-03

- withstand damage
- take into account a possible increase in resistance due to corrosion.

3.6.4 Earthing conductors

All earthing conductors **shall** meet the requirements for protective conductors.   542-03-01

3.6.5 Equipotential and supplementary bonding conductors

Main equipotential bonding conductors are required for all water service pipes, gas installation pipes, oil supply pipes plus any other 'earthy' metalwork that may be present on the premises.   P App C

Supplementary bonding conductors connecting   547-03-01 and 547-03-02

- two exposed conductive parts or
- an exposed conductive part to an extraneous conductive part

shall have a conductance (if sheathed or mechanically protected) not less than the protective conductor connected to an exposed conductive part.

Supplementary bonding to a fixed appliance shall be provided by a supplementary conductor or a conductive part of a permanent and reliable nature.   547-03-04

Supplementary equipotential bonding conductors are required for installations and locations where there is an increased risk of electric shock (e.g. bathrooms and shower rooms).   P App C

The minimum size of supplementary equipotential bonding conductors (without mechanical protection) is 4 mm².   P App C

3.7 Protective measures

Protection against electric shock **shall** be provided.   410-01-01

Bare live parts (other than an overhead line) shall **not** be within 2.5 m of: 412-05-02

- an exposed conductive part
- an extraneous conductive part
- a bare live part of any other circuit.

Functional Extra-Low Voltage (FELV) alone shall **not** be used as a protective measure.   411-01-02

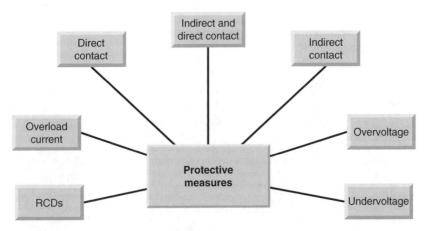

Figure 3.8   Protective measures

Bare (or insulated) overhead lines being used for distribution between buildings and structures shall be installed in accordance with the Electricity Safety, Quality and Continuity Regulations 2002.   412-05-01

Bare live parts (other than overhead lines) shall **not** be within arm's reach.	412-05-02
Exposed parts of electrical equipment shall be located (or guarded) so as to prevent accidental contact and/or injury to persons or livestock.	133-01-06
Faults occurring in one circuit of an equipment that is supplied by two different circuits shall not alter the degree of protection against electric shock nor the correct operation of the other circuit.	564-01-01

Live parts shall be completely covered with insulation which: 412-02-01

- is capable of durably withstanding electrical, mechanical, thermal and chemical stresses normally encountered during service
- can only be removed by destruction.

Live parts shall be inside enclosures (or behind barriers) protected to at least IP2X or IPXXB.	412-03-01

Persons and/or livestock shall be protected against dangers that may arise from:

- contact with live parts of the installation   130-02-01
- contact with exposed conductive parts during a fault.   130-02-02

Simultaneously accessible exposed conductive parts shall be connected to the same earthing system individually, in groups, or collectively.	413-02-03

 Note: If an overcurrent protective device is used for protection against electric shock, the protective conductor shall be incorporated in the same wiring system as the live conductors or be located in their immediate proximity.

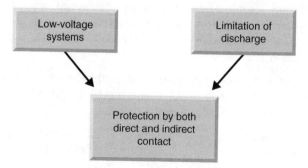

Figure 3.9 Direct and indirect contact

Low-voltage systems

There are three low-voltage systems to be considered. These are:

- SELV (Safety Extra-Low Voltage)
- FELV (Functional Extra-Low Voltage)
- PELV (Protective Extra-Low Voltage).

Protection by SELV

A system which does not use a device such as an   411-02-03
autotransformer, potentiometer, semiconductor device etc.,
to provide electrical separation, shall **not** be deemed
to be a SELV system.

Functional Extra-Low Voltage alone shall **not** be used as a   411-01-02
protective measure.

For protection by SELV: 411-02-01

- the nominal circuit voltage shall not exceed extra-low
  voltage
- the supply shall be from a safety source.

All live parts of a SELV system shall: 411-02-05

- be electrically separated from that of any other
  higher-voltage system
- not be connected to earth
- not be connected to a live part or a protective conductor
  forming part of another system.

Circuit conductors for each SELV system shall either be: 411-02-06

- physically separated from those of any other system or
- insulated for the highest voltage present or
- enclosed in an insulating sheath additional to their basic
  insulation.

Conductors of systems with a higher voltage than SELV   411-02-06
shall be separated from the SELV conductors by an earthed
metallic screen or an earthed metallic sheath.

Electrical separation between live parts of a SELV system   411-02-06
(including relays, contactors and auxiliary switches) and
any other system shall be maintained.

SELV circuit conductors that are contained in a multicore   411-02-06
cable with other circuits having different voltages shall be
insulated, individually or collectively, for the highest
voltage present in the cable or grouping.

The socket outlet of a SELV system shall: 411-02-10

- be compatible with the plugs used for other systems in
  use in the same premises
- not have a protective conductor contact.

PELV

PELV systems which are used for parts of a low-voltage system that are not
connected to:

- earth
- an exposed conductive part of another system
- a protective conductor of any system
- an extraneous conductive part

shall provide protection against direct contact by either:

- barriers or enclosures (with a degree of protection of at   471-14-02
  least 1P2X or IPXXB), or
- insulation capable of withstanding 500 V a.c. rms for
  60 seconds.

FELV

If an extra-low-voltage system does not comply with the requirements for
SELV in some respect (other than that specified for PELV above) then:

protection against direct contact shall be provided by either: 471-14-03

- barriers or enclosures or
- insulation corresponding to the minimum voltage
  required for the primary circuit.

 Note: All socket outlets and luminaire supporting couplers in a FELV system
shall use a plug that is dimensionally different with those used for any other
system in use in the same premises.

Limitation of discharge

When an equipment incorporates a method for limiting the amount of current which can pass through the body of a person (or livestock) to a non-dangerous level, then protection against both direct and indirect contact shall be deemed to be provided. In which case:

all live parts of a SELV system shall: 411-02-05

- be physically separated from those of any other system
- be electrically separated from that of any other higher-voltage system
- not be connected to earth
- not be connected to a live part or a protective conductor forming part of another system.

3.7.1 Protection against direct contact

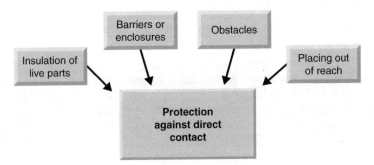

Figure 3.10 Protection against indirect contact

🔔 RCDs may **not** be used as the sole means of   412-06-01 and
protection against direct contact.   412-06-02

One of the following protective methods are required for protection against direct contact:

- barriers and/or enclosures
- insulation of live parts
- obstacles
- placing out of reach.

Barriers and/or enclosures

Live parts shall be inside enclosures (or behind barriers) protected to at least IP2X or IPXXB.	412-03-01 chg2
The horizontal top surface of easily assessable barriers or enclosures shall be protected to at least 1P4X.	412-03-02

Insulation of live parts

Live parts shall be completely covered with insulation which can only be removed by destruction.	412-02-01
Equipment shall be capable of withstanding all mechanical, chemical, electrical and thermal influences stresses normally encountered during service.	412-02-01
Obstacles shall be secured to prevent unintentional removal.	412-04-02

Obstacles

Obstacles may **not** be removed without using a key or tool.	412-04-02

Placing out of reach

Protection against direct contact by placing out of reach may **only** be used in an area accessible to skilled persons.	471-06-01
Bare live parts (other than overhead lines) shall **not** be within arm's reach.	412-05-02

Bare (or insulated) overhead lines used for distribution between buildings and structures shall be installed in accordance with the Electricity Safety, Quality and Continuity Regulations 2002.	412-05-01

Bare live parts (other than overhead lines) shall not be:

- within arm's reach   412-05-02
- within 2.5 m of: 412-05-02
  - an exposed conductive part
  - an extraneous conductive part
  - a bare live part of any other circuit.

3.7.2 Protection against indirect contact

Protection against indirect contact **shall** be provided   551-04-01
for each supply source which can operate independently
of other sources.

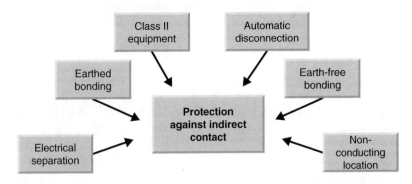

Figure 3.11 Protection against indirect contact

It is permissible to dispense with protective measures against indirect contact only if:

- overhead line insulator brackets (and metal parts connected to them) are not within arm's reach
- unearthed street furniture that is supplied from an overhead line is inaccessible whilst in normal use
- the steel reinforcement of steel reinforced concrete poles is not accessible
- exposed conductive parts (including small isolated metal parts such as bolts, rivets, nameplates not exceeding 50 mm × 50 mm and cable clips) cannot be gripped or cannot be contacted by a major surface of the human body
- there is no risk of fixing screws used for non-metallic accessories coming into contact with live parts
- inaccessible lengths of metal conduit not exceeding 150 mm
- metal enclosures mechanically protect equipment.

The following methods for protection against indirect contact may be used:

- automatic disconnection
- class II equipment
- earthed equipotential bonding
- earth-free local equipotential bonding
- electrical separation
- non-conducting location.

Automatic disconnection

Installations that provide protection against indirect contact by automatically disconnecting the supply shall have a circuit protective conductor run to (and terminated at) each point in the wiring and at each accessory.	471.08-08

 Except suspended lampholders, which have no exposed conductive parts.

Where automatic disconnection of TN, TT and/or IT systems cannot be achieved by using overcurrent protective devices, then either local supplementary equipotential bonding or a residual current device shall be used.	413-02-04

Class II equipment

Class II equipment shall **not** be the only protection between live parts of the installation and exposed metalwork of that equipment.	413-03-02
Cables with a non-metallic sheath or a non-metallic enclosure are **not** considered to be of Class II construction.	471-09-04

Protection shall be provided by one or more of the following: 413-03-01

- electrical equipment having double or reinforced insulation
- low-voltage switchgear
- low-voltage controlgear assemblies
- supplementary insulation
- reinforced insulation applied to uninsulated live parts.

Class II equipment shall not be the only protection between live parts of the installation and exposed metalwork of that equipment.   413-03-02

Circuits supplying Class II equipment shall have a circuit protective conductor run to, and terminated at, each point in the wiring and at each accessory.   471-09-02

 Except suspended lampholders, which have no exposed conductive parts.

The insulation of operational electrical equipment shall be at least 1P2X or IPXXB.   413-03-04

The insulating enclosure shall be capable of resisting mechanical, electrical and thermal stresses likely to be encountered.   413-03-05

The metalwork of exposed Class II equipment shall be mounted so that it is not in electrical contact with any other part of the installation that is connected to a protective conductor.   471-09-02

Note: Cables with a non-metallic sheath or a non-metallic enclosure are **not** considered to be of Class II construction.

Earthed equipotential bonding

Protection by earthed equipotential bonding and automatic disconnection of supply **shall** be applied according to the requirements for the type of system earthing in use (i.e. TN, TT or IT).   413-02-01

Main equipotential bonding conductors **shall** (for each installation) be connected to the main earthing terminal of that installation.   413-02-02

Main equipotential bonding conductors such as: 413-02-02

• water service pipes
• gas installation pipes

- other service pipes and ducting
- central heating and air conditioning systems
- exposed metallic structural parts of the building
- the lightning protective system

shall be connected to the main earthing terminal of that installation.

Equipotential bonding shall also be applied to any metallic   413-02-02
sheath of a telecommunication cable.

Where supplementary equipotential bonding is necessary,   413-02-27
it shall connect together the exposed conductive parts of
equipment in the circuits concerned.

Earth-free local equipotential bonding

This method shall **only** be used in special circumstances.   413-05-01

Earth-free local equipotential bonding **shall**:

- **only** be used in special situations which are earth-free   471-11-01
- **only** be used in special circumstances   413-05-01
- **not** be used in caravans, motor caravans and caravan   471-11-01
  parks where there is an increased shock risk.

Protection by earth-free local equipotential bonding is **not**   601-05-03
permitted in locations containing a bath or a shower.

Earth-free local equipotential bonding is intended to prevent the appearance of a dangerous voltage between simultaneously accessible parts in the event of failure of the basic insulation.

For some installations and locations (such as agricultural and horticultural buildings, saunas, caravans etc.) that are subject to an increased risk of shock, earth-free local equipotential bonding shall not be used.

Persons entering or leaving an equipotential location shall   413-05-04
not be exposed to a dangerous potential difference.

Electrical separation

Protection by electrical separation may be used to supply several items of equipment from a single separated source, but **only** for special situations and under effective supervision.   471-12-01

No live part of a separated circuit **shall** be connected (at any point) to another circuit or to earth.   413-06-03

The supply source to the circuit shall be either: 413-06-02

- an isolating transformer complying with BS 3535 or
- a motor generator.

Separated circuits shall, preferably, use a separate wiring system.   413-06-03

The voltage of an electrically separated circuit shall not exceed 500 V.   413-06-02

Protection by electrical separation may be applied to the supply of any individual item of equipment by means of a transformer complying with BS 3535 the secondary of which is not earthed, or a source affording equivalent safety.   471-12-01

Non-conducting location

This form of protection is intended to prevent simultaneous contact with parts which may be at different potentials through failure of the basic insulation of live parts.

Protection by non-conducting location is **not** permitted in locations containing a bath or a shower.   601-05-03

This form of protection shall **not** be used in installations and locations subject to increased risk of shock as covered in Part 6 of the Regulations.   471-10-01

This method of protection is **not** recognised for general application and may not be used in installations such as agricultural and horticultural buildings (saunas, caravans etc.) that are subject to an increased risk of shock.   413-04-01

3.7.3 Protection against overcurrent

Overcurrents caused by faults in live conductors shall   130-04-01
not damage or cause injury to persons, livestock or property.

The main considerations are fault current and overload current.

Fault current

Fault current protective devices shall be provided: 473-02-05

- at the supply end of each parallel conductor where two conductors are in parallel
- at the supply **and** load ends of each parallel conductor where more than two conductors are in parallel.

Fault current protective devices shall be placed (on the load   473-02-01
side) at the point where the current-carrying capacity of   473-02-02
the installation's conductors is likely to be lessened due to:

- the method of installation
- the cross-sectional area
- the type of cable or conductor used
- inherent environmental conditions.

A single protective device may be used to protect conductors   473-02-05
in parallel against the effects of fault current occurring.

Circuit protective devices shall break any fault current   434-01-01
flowing before the current can cause any danger due to
thermal or mechanical effects produced in circuit
conductors or associated connections.

Conductors shall be capable of carrying fault current   130-05-01
without overheating.

Construction of fault current protective devices shall be: 473-02-02

- less than 3 m in length between the point where the value of current-carrying capacity is reduced and the position of the protective device
- installed so as to minimise the risk of fault current
- installed so as to minimise the risk of fire or danger to persons.

Persons and livestock shall be protected from being injured by excessive temperatures and/or electromagnetic stresses that are caused by overcurrents occurring in live conductors.   130-04-01

Potential short-circuit and earth fault currents shall be assessed for each supply source.   551-02-02

Property shall be protected from being damaged by excessive temperatures and/or electromagnetic stresses that are caused by overcurrents occurring in live conductors.   130-04-01

The supply to all live conductors shall be automatically interrupted in the event of overload current and fault current.   431-01-01

Overload current

An overcurrent detection device **shall** be provided for each phase conductor which is capable of disconnecting only that particular conductor.   473-03-01

Overload protection devices **shall** be used at any point where the current-carrying capacity of the installation's conductors is reduced.   473-01-01

All circuits **shall** be designed so that a small overload of long duration is unlikely to occur.   433-01-01

Non-skilled personnel shall be prevented from changing the setting or calibration of an overcurrent release of a circuit breaker unless they use a key or a tool and this results in a visible indication of its setting (or calibration).   533-01-05

Overload protection devices need not be provided: 473-01-04

- for conductors situated on the load side, if the conductor is protected by a protective device placed on the supply side of that point
- for a conductor that is not likely to carry overload current
- at the origin of an installation, if the distributor has provided an overload device.

Overload protection devices need not be used for circuits   473-01-03
supplying current-using equipment such as:

- exciter circuits of rotating machines
- supply circuits of lifting magnets
- secondary circuits of current transformers
- circuits which supply fire extinguishing devices and
- where unexpected disconnection of the circuit could
  cause danger.

Protection against overcurrent shall take into account   533-03-01
minimum and maximum fault current conditions.

Protective devices shall be provided to break any overload   433-01-01
current flowing in the circuit conductors before the current
can damage insulation, joints, terminations and/or the
surroundings of the conductors.

3.7.4 Protection against overvoltage

 Persons, livestock and property **shall** be protected   130-06-02
against overvoltages arising due to atmospheric
phenomena and/or switching.

Note: If there is also a lightning protection system, reference shall be made to
BS 6651.

Installations that are supplied by (or include) low-voltage   443-02-03
overhead lines shall incorporate protection against
overvoltages of atmospheric origin if the location is
subject to more than 25 thunderstorm days per year.

Overvoltage protective devices shall be located as close   443-02-06
as possible to the origin of the installation.

Persons and livestock shall be protected against:

- injury due to a fault between live parts of circuits   130-06-01
  supplied at different voltages
- overvoltages arising due to atmospheric phenomena   130-06-02
  and/or switching.

Property shall be protected against any harmful effects caused by a fault occurring between live parts of circuits supplied at different voltages.	130-06-01

Protection against overvoltages of atmospheric origin shall be provided in the installation of the building by: 443-02-05

- a surge-protective device with a protection level not exceeding Category II or
- other means providing at least an equivalent attenuation of overvoltages.

3.7.5 Protection against undervoltage

Current-using equipment or installations that can be damaged by a drop in voltage (but without that damage being dangerous) shall: 451-01-02

- take precautions against the foreseen damage
- verify that the foreseen damage is an acceptable risk.

Precautions shall be taken where a reduction in voltage, or loss and subsequent restoration of voltage, could cause danger.   451-01-01

The automatic reclosure of a protective device is not permitted where that reclosure is likely to cause danger.   451-01-06

3.7.6 Residual current devices

Residual current devices may **not** be used as the sole means of protection against direct contact.   412-06-01

Although RCDs reduce the risk of electric shock, they may **not** be used as the sole means of protection against direct contact.   412-06-02

Automatic disconnection using an RCD shall **not** be applied to a circuit incorporating a PEN conductor.   471-08-07

If an RCD is used in a TN-C-S system, a PEN conductor shall **not** be used on the load side.   413-02-07

RCDs shall be located so that their operation will not be harmed by magnetic fields from other equipment.   531-02-07

RCDs that are associated with a circuit that would normally be expected to have a protective conductor, shall **not** be considered sufficient for protection against indirect contact if there is no such conductor.   531-02-05

RCDs that are used to provide protection against indirect contact, shall be capable of withstanding the likely thermal and mechanical stresses that it will probably be subjected to if fault occurred.   531-02-08

RCDs that can be operated by a non-skilled person shall be designed and installed so that their operating current and/or time delay mechanism cannot be altered without using a key or a tool.   531-02-10

RCDs shall:

- be capable of disconnecting all of the phase conductors of a circuit at the same time   531-02-01
- be located so that their operation will not be impaired by magnetic fields caused by other equipment   531-02-07
- ensure that any protective conductor current which may occur will be unlikely to cause unnecessary tripping of the device.   531-02-04

All alterations made to the setting and/or calibration of an RCD shall be visibly indicated.   531-02-10

If an IT system is protected against indirect contact by an RCD, each final circuit shall be separately protected.   413-02-25

If the installation is part of a TF or a TT system, all socket outlet circuits shall be protected by an RCD.   471-08-06

RCDs shall be capable of disconnecting all of the phase conductors of a circuit at the same time.   531-02-01

Socket outlets rated at 32 A or less that supply portable equipment for use outdoors shall be provided with an RCD.   471-16-01

The magnetic circuit of the transformer of an RCD shall enclose all the live conductors of the protected circuit.   531-02-02

The protective conductor:

- of an RCD shall be outside the magnetic circuit   531-02-02
- to the PEN conductor shall be on the source side of the RCD.   413-02-07

The residual operating current of the protective device shall comply with the requirements for protection against indirect contact. 531-02-03

Two or more RCDs connected in series shall be capable of operating separately. 531-02-09

3.8 Equipment

Electrical equipment shall **not** become a fire hazard to adjacent materials. 130-03-02

Electrical equipment shall **not**: 132-01-08

- harm other equipment
- damage (or weaken) the supply during normal service and/or switching operations.

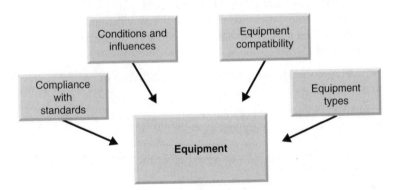

Figure 3.12 Equipment types and conditions

Electrical equipment shall be:

- fully accessible for operation, inspection, testing, maintenance, repair and later replacement 131-12-01
- selected and erected in compliance with the requirements of these Regulations. 510-01-01

Exposed parts of electrical equipment whose temperature is likely to injure persons or livestock shall be located (or guarded) so as to prevent accidental contact. 133-01-06

Fixed equipment shall be protected against heat and/or   130-03-02
thermal radiation emitted by adjacent electrical equipment.

Installed electrical equipment shall:

- minimise the risk of ignition of flammable materials   133-01-06
- be accessible for operational, inspection and maintenance 513-01-01
  purposes.

Switchgear, protective devices, accessories and other types   512-02-01
of equipment shall not be connected to conductors intended
to operate at a temperature exceeding 70°C.

The characteristics of electrical equipment shall not be   133-01-02
impaired by the process of erection.

3.8.1 Compliance with standards

All equipment shall comply with the appropriate EN or   132-01-01
HD or implementing British Standard or (in certain
circumstances) the appropriate IEC Standard or National
Standard of another country.

Equipment complying with a foreign national   511-01-02
standard may be used **only** if it provides the same degree
of safety afforded by a British or harmonised Standard.

3.8.2 Operational conditions and external influences

Equipment shall be selected and erected so that it will not: 512-05-01

- harm other equipment
- damage the supply during normal service (including
  switching operations).

Equipment shall be suitable for: 512-02-01

- the installation's nominal voltage
- the design current and
- the current likely to flow in abnormal conditions.

The rated frequency of the equipment shall be the same as   512-03-01
the nominal frequency of the supply to the circuit concerned.

The power characteristics of equipment shall be suitable   512-04-01
for the duty demanded of the equipment.   512-04-02

3.8.3 Equipment compatibility

 Electrical equipment shall **not** cause harmful effects   132-01-08
on other equipment.

Equipment shall be selected and erected so that it will not: 512-05-01

• harm other equipment
• damage the supply during normal service (including switching operations).

The characteristics of electrical equipment (e.g. harmonic   331-01-01
currents, fluctuating loads) that might have a harmful effect
on other electrical equipment and/or services, shall be assessed.

For telecommunication and data transfer circuits etc., consideration shall be given to electrical interference, both electromagnetic and electrostatic (see BS EN 50081 and BS EN 50082).

Also see 89/336/EC for details concerning electromagnetic compatibility.

Fixed materials shall be protected against heat and/or   130-03-02
thermal radiation emitted by adjacent electrical equipment.

3.8.4 Heating equipment

Heaters

Heaters for liquid or other substances shall have an   554-04-01
automatic device to prevent a dangerous rise in temperature.

Heat emitters

Heat emitters should either be screened or have their   M (5.4j)
exposed surfaces kept at a temperature below 43°C.

In toilets and bathrooms, heat emitters (if located) should   M (5.10p)
not restrict:

- the minimum clear wheelchair manoeuvring space
- the space beside a WC used to transfer from the
  wheelchair to the WC.

3.8.5 Portable equipment

Permanent protection shall be provided in all areas where   413-04-05
the use of mobile and/or portable equipment is envisaged.

For installations and locations where the risk of electric   471-16-02
shock is increased by a reduction in body resistance and/or
by contact with earth potential, circuits supplying portable
equipment for use outdoors and not connected to a socket
by means of flexible cable or cord (having a current-
carrying capacity of 32 A or less) shall be provided with
supplementary protection (i.e. an RCD).

RCD protection is required for all socket outlets which   P App C
have a rating of 32 A or less and which may be used to
supply portable equipment for use outdoors.

3.8.6 Rotating machines and motors

Emergency switching devices shall be capable of cutting off   537-04-01
the full load current of the relevant part of the installation.

Fixed electric motors shall be provided with a readily   131-14-02
accessible, efficient means of switching off, that is easily
operated and safely positioned.

Unsupervised motors which are automatically (or remotely)   482-02-11
controlled shall be protected against excessive temperature
by a protective device with manual reset.

Water heaters and electrode boilers

All metal parts that are in contact with the water shall be   554-05-02
solidly and metallically connected to the metal water pipe
that carries the water supply to that heater/ boiler.

Electrode boilers and water heaters shall:

- be earthed   554-03-03
- only be connected to an a.c. system   554-03-01
- include an RCD if they are directly connected to   554-03-04
  a supply that exceeds low voltage

No single-pole switch, non-linked circuit breaker or fuse   554-05-04
shall be fitted in the neutral conductor between the heater/
boiler and the origin of the installation.

The current-carrying capacity of the neutral conductor   554-03-05
shall be not less than that of the largest phase conductor
connected to the equipment.

The heater or boiler shall be permanently connected to the   554-05-03
electricity supply via a double-pole linked switch.

The metal water pipe that carries the water supply to that   554-05-02
heater/boiler should be connected to the main earthing
terminal by the circuit protective conductor.

The protective conductor shall be connected to the shell of   554-03-03
the heater or boiler.

The shell of electrode boilers and water heaters shall be   554-03-03
bonded to the metallic sheath and armour (if any) of the
incoming supply cable.

The shell of electrode water heaters/boilers that are   554-03-05
connected to a three-phase low-voltage supply shall be
connected to the neutral of the supply as well as to the
earthing conductor.

The supply to electrode boilers and water heaters shall be   554-03-02
controlled by a linked circuit breaker.

When an electrode water heater or boiler is connected to a   554-03-05
three-phase low-voltage supply, the shell of the heater/
boiler shall be connected to the neutral of the supply as
well as to the earthing conductor.

Where an electrode water heater/boiler is connected   554-03-06
to single-phase supply and one electrode is connected   554-03-07
to a neutral conductor earthed by the distributor,
the shell of the heater/boiler shall be connected to the
neutral of the supply as well as to the earthing
conductor.

3.9 Components and accessories

Components, cables and wiring enclosures shall be installed in accordance with the temperatures limits set by the relevant product specification or by the manufacturer. 522-01-02

Accessories shall not be connected to conductors intended to operate at a temperature exceeding 70°C. 512-02-01

There are a number of requirements within the regulations concerning components and accessories to electrical installations and the following is a selection of the most important ones that should be considered by every electrician.

3.9.1 Cable couplers

Except for a SELV or a Class II circuit, cable couplers shall: 553-02-01

- be non-reversible
- be capable of including a protective conductor
- comply with BS 196, BS EN 60309-2, BS 4491 or BS 6991 – as appropriate
- be fitted at the end of the cable remote from the supply. 553-02-02

Luminaire supporting couplers shall only be used to hold up one luminaire electrical connection. 553-04-04

3.9.2 Ceiling roses

Ceiling roses shall not be installed in circuits with an operating voltage greater than 250 volts. 553-04-02

Fixed lighting points shall use a ceiling rose to BS 67. 553-04-01

Unless specifically designed for multiple pendants, ceiling roses shall only have one outgoing flexible cord. 553-04-03

3.9.3 Circuit breakers

Non-skilled personnel shall be prevented from changing the setting or calibration of the overcurrent release of a circuit breaker unless they use a key or a tool and this results in a visible indication of its setting (or calibration). 533-01-05

3.9.4 Control gear

Control gear shall be clearly identified and grouped in locations accessible only to skilled or instructed persons.	563-01-05

3.9.5 Fuses

- Off-load fuses and links shall **not** be used for functional switching.   537-05-03
- Fuses shall **not** be inserted in the neutral conductor of TN or TT systems.   530-01-02
- Fuses shall **not** be inserted in an earthed neutral conductor.   131-13-02

3.9.6 Lampholders

 Filament lampholders shall **not** be installed in circuits operating in excess of 250 V.   553-03-02

Bayonet lampholders B15 and B22 shall:

- comply with BS EN 61184   553-03-03
- have a temperature rating T2 as described in BS EN 61184.   553-03-03

Except for E14 and E27 lampholders complying with BS EN 60238, the outer contact of all Edison screw or single-centre bayonet cap type lampholders in TN or a TT system circuit shall be connected to the neutral conductor.   553-03-04

This regulation also applies to track-mounted systems

Unless the wiring is enclosed in earthed metal or insulating material, lampholders with an ignitability characteristic 'P' (as specified in BS 476 Part 5) or where separate overcurrent protection is provided, shall not be connected to any circuit where the rated current of the overcurrent protective device exceeds the appropriate value stated in Table 55B (page 126) of the IEE Wiring Regulations.   553-03-01

3.9.7 Lighting points

Fixed lighting points shall use one of the following: 553-04-01

- a batten lampholder to BS 7895, BS EN 60238 or BS EN 61184
- a ceiling rose to BS 67
- a connection unit to BS 5733 or BS 1363-4
- a luminaire designed to be connected directly to the circuit wiring
- a luminaire supporting coupler to BS 6972 or BS 7001
- a suitable socket outlet.

Lighting installations shall be controlled by: 553-04-01

- a switch (or combination of switches) to BS 3676 and/or BS 5518, or
- a suitable automatic control system.

3.9.8 Luminaries

Luminaire supporting couplers which have a protective conductor contact shall **not** be installed in a:

- non-conducting location   413-04-03
- SELV system.   411-02-11

Extra-low-voltage luminaries that do not allow the   554-01-02
connection of a protective conductor shall only be
installed as part of a SELV system.

If a pendant luminaire is installed, the associated   554-01-01
accessory shall be suitable for the mass suspended.   522-08-06

3.9.9 Thermostats

Controls that need close vision (e.g. thermostats) should   M (4.30f)
be located between 1200 mm and 1400 mm above the floor.

3.10 Sockets

3.10.1 Socket outlets

🔔 All household socket outlets **shall** be of the shuttered type.	553-01-04
All potentially hazardous fixed or stationary appliances that are **not** connected to the supply by means of a plug and socket outlet (rated at less than 16 A) shall be provided with a means of interrupting the supply on load.	476-03-04
In a non-conducting location, socket outlets shall **not** incorporate an earthing contact.	413-04-03
Plug and socket outlets shall **not** be used as a device for emergency switching.	537-04-02
Plug and socket outlets with a rating of more than 16 A may **not** be used as a switching device for d.c.	537-05-05
The colours red and green should **not** be used in combination as indicators of 'on' and 'off' for switches and controls.	M (4.28)

All socket outlets should be wall-mounted.	M (4.30a and b)
Except for SELV circuits:	
• all plug and socket outlets shall be non-reversible and capable of including a protective conductor	553-01-02

• the pin of a plug shall not be capable of making contact with any live contact:
  – of its associated socket outlet when any other pin of the plug is completely exposed
  – of any socket outlet within the same installation other than the type of socket outlet for which the plug is designed.
	553-01-01

Front plates should contrast visually with their backgrounds.	M (4.30m)
Individual switches on panels and on multiple socket outlets should be well separated.	M (4.29)
Mains and circuit isolator switches should clearly indicate whether they are 'ON' or 'OFF'.	M (4.30l)

Older types of socket outlet designed non-fused plugs must not be connected to a ring circuit.	P App C
Plug and socket outlets may be used in single-phase a.c. or two-wire d.c. circuits that operate at a nominal voltage not exceeding 250 volts for electric clocks and electric shavers, provided that they comply with the relevant British Standard.	553-01-05
Sensitive RCD protection is required for all socket outlets which have a rating of 32 A or less and which may be used to supply portable equipment for use outdoors.	P App C

Socket outlets should:

• be located no nearer than 350 mm from room corners	M (4.30g)
• not be used to supply equipment that needs to be earthed if those socket outlets will accept unearthed (2-pin) plugs	P App C
• indicate whether they are 'ON'.	M (4.30k)

Switches and socket outlets for lighting should be:

• installed between 450 mm and 1200 mm from the finished floor level	M2 (8.3)
• located so that they are easily reachable.	M2 (8.2)
The voltage drop between the supply terminals of an installation and a socket outlet shall not exceed 4% of the nominal voltage of the supply.	525-01-02
Wall sockets and outlets shall be mounted above the floor and above working surfaces.	553-01-06
Where portable equipment is likely to be used, a conveniently accessible socket outlet shall be provided.	553-01-07

3.10.2 Wall-mounted socket outlets

A cable or stud detector shall be used when attempting to drill into walls, floors or ceilings.	P App C
Wall-mounted socket outlets should comply with the requirements of Part M (see 2.13.24).	P 1.5

Wall sockets shall meet the following requirements (see Table 3.2).

Table 3.2 Building Regulations requirements for wall sockets

Type of wall	Requirement	Section
Timber framed	Power points may be set in the linings provided there is a similar thickness of cladding behind the socket box	E (p14)
	Power points should not be placed back to back across the wall	E (p14)
Solid masonry	Deep sockets and chases should **not** be used in separating walls	E 2.32
	Stagger the position of sockets on opposite sides of the separating wall	E 2.32f
Cavity masonry	Stagger the position of sockets on opposite sides of the separating wall	E 2.65e
	Deep sockets and chases should **not** be used in a separating wall	E 2.65d2
	Deep sockets and chases in a separating wall should **not** be placed back to back	E 2.65d2
Framed walls with absorbent material	Sockets should: • be positioned on opposite sides of a separating wall • not be connected back to back • be staggered a minimum of 150 mm edge to edge	E 2.146b E 2.146b2 E 2.146b2

3.11 Cables

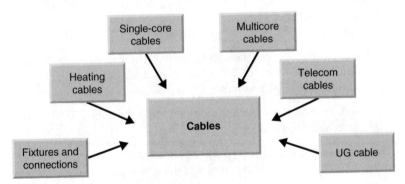

Figure 3.13 Types of cables

Bends and curves in a wiring system shall not damage cables.	522-08-03
Cables and cords shall comply with the requirements of BS EN 50265-2-1 or 2-2.	482-03-03
Cables that are used as fixed wiring shall be supported and not be exposed to undue mechanical strain, particularly at terminations.	522-08-05

Cables with a non-metallic sheath or non-metallic
enclosure are not considered to be a Class II construction.   471-09-04

Where a cable is not continuously supported it shall be
supported at appropriate intervals so that the cable is not
damaged by its own weight.   522-08-04

3.11.1 Fixtures and connections

Cable locations which contain long vertical runs or
bunched cables (and where the risk of flame propagation
is high) shall meet the requirements specified in BS 4066-3.   482-02-04

New or replacement non-metallic light fittings, switches
or other components must not require earthing (e.g. non-
metallic varieties) unless new circuit protective (earthing)
conductors are provided.   P App C

Particular care shall be taken in the selecting and using of
fixings and connections for cables when the wiring system
(fixed or supported by a structure equipment) is subject
to vibration.   522-07-01

3.11.2 Heating cables

Heating cables that are going to be laid (directly) in the
soil, road, or structure of a building shall be installed so
that they: 554-06-03

- are completely embedded in the substance it is intended
  to heat
- are not damaged by movement by it or the substance
  in which it is embedded
- comply in all respects with the maker's instructions
  and recommendations.

Heating cables that pass through (or are in close proximity
to) a fire hazard shall be: 554-06-01

- enclosed in material with an ignitability characteristic
  'P' as specified in BS 476 Part 5
- protected from any mechanical damage.

3.11.3 Single-core cables

Owing to possible electromagnetic effects, single-   521-02-01
core cables that are armoured with steel wire or tape shall
not be used for a.c. circuits.

The conductance of the outer conductor of a concentric   546-02-04
single-core cable shall not be less than the internal conductor.

The metallic sheaths (and/or non-magnetic armour) of single-   523-05-01
core cables in the same circuit shall either be bonded together

- at both ends of their run (solid bonding) or
- at one point in their run (single point bonding).

3.11.4 Multicore cables

A Band I circuit shall **not** be contained in the same wiring system
as a Band II voltage circuit.

Note: Consideration shall also be given to electrical interference, both electromagnetic and electrostatic (see BS EN 50081 and BS EN 50082).

3.11.5 Telecommunication cables

Equipotential bonding shall be applied to the metallic   413-02-02
sheath of a telecommunication cable.

3.11.6 UG cable

Unless installed in a conduit or duct, cables that are buried   522-06-03
in the ground shall be:

- marked by cable covers or marking tape
- either of an insulated concentric construction or
  protected by earthed armour or metal sheath
- at a sufficient depth to avoid being damaged by any   522-06-03
  reasonably foreseeable disturbance.

3.12 Conductors

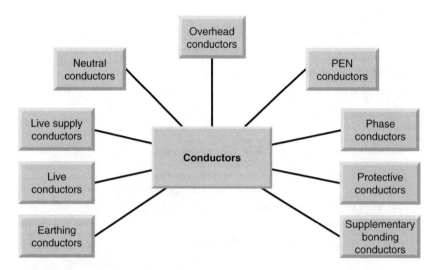

Figure 3.14 Conductors

3.12.1 Connection methods

The method of connection will depend on:

- the effectiveness of the insulation of the conductors
- the number and shape of the conductor's wires
- the conductor's material and insulation
- the cross-sectional area of the conductor
- the number of conductors being connected together
- the temperature of the terminals during normal service
- the type of locking arrangement for situations subject to vibration or thermal cycling

and for soldered connections:

- the mechanical stress and temperature rise under fault current conditions.

3.12.2 General

Conductors intended to operate at temperatures above   512-02-01
70°C shall **not** be connected to switchgear, protective
devices, accessories or other types of equipment.

Conductors in parallel **shall** be protected against overcurrent.   432-05-01

Conductors shall **not** be subjected to excessive mechanical   522-12-01
stress.

The metalwork of exposed Class II equipment shall be mounted so that it is not in electrical contact with any part of the installation connected to a protective conductor.	471-09-02
Earth-free local equipotential bonding shall **only** be used in special circumstances.	413-05-01
Conductors shall be capable of withstanding all foreseen electromechanical forces (including fault current) during service.	521-03-01
Conductors shall not be damaged by undue bends in the wiring system.	522-08-03
Conductors used as fixed wiring shall be supported and not be exposed to undue mechanical strain, particularly at terminations.	522-08-05
Conductors used as fixed wiring shall be: • supported • not exposed to undue mechanical strain (particularly at terminations).	522-08-05
Connections between conductors (and between a conductor and equipment) shall ensure electrical continuity and mechanical strength.	526-01-01
Where a conductor is not continuously supported it shall be supported at appropriate intervals so that the conductor or cable does not suffer damage by its own weight.	522-08-04

3.12.3 Identification of conductors

Green shall **not** be used on its own.	514-04-05

Conductors may be identified by numbers, with the number 0 indicating the neutral or mid-point conductor.	514-05-04
Bare conductors (or busbars) used as protective conductors shall be identified by equal green and yellow stripes which are 15 mm to 100 mm wide and where no more than 70% of the surface area is covered by one colour.	514-04-02

A permanent label/warning notice – with the words   514-13-01
'Safety Electrical Connection – Do Not Remove' shall
be permanently fixed at or near to the:

- bonding conductor's connection point to an extraneous part
- connection of earthing conductors to the earth electrode.

Single-core cables used as protective conductors shall be   514-04-02
coloured green-and-yellow throughout their length.

Protective conductors shall be a bi-colour combination of   514-04-02
green and yellow. Neither colour shall cover more than
70% of the surface being coloured.

 Note: This combination of green and yellow shall **not** be used for any other purpose.

3.12.4 Earthing conductors

Earthing conductors shall be capable of being   542.04.02
disconnected to enable the resistance of the earthing
arrangements to be measured.

 Note: If electrical earth monitoring is used, the operating coil shall be connected in the pilot conductor and **not** in the protective earthing conductor.

3.12.5 Live conductors

Bare live conductors **shall** be installed on insulators.   521-07-02

Exposed bare live conductors shall **not** be used in locations   482-02-09
where there is a risk of fire owing to the nature of processed
or stored materials.

In an IT system, all exposed conductive parts **shall** be   413-02-23
earthed.

Live supply conductors **shall** be capable of being isolated from circuits.	461-01-01

Live conductors (or joints between them) shall be terminated within an enclosure.	422-01-04
The supply to live conductors shall be protected by an automatic interruption device.	431-01-01

3.12.6 Live supply conductors

Live supply conductors shall be capable of being isolated from circuits.	461-01-01
Isolation devices shall isolate all live supply conductors from the circuit concerned.	537-02-01

3.12.7 Neutral conductors

Functional switching devices shall **not** be placed solely in the neutral conductor.	464-01-02

3.12.8 Overhead conductors

Hard-drawn copper conductors may be used to suspend non-flexible cables and flexible cords (not forming part of a portable appliance or luminaire and which are sheathed with lead, PVC or an elastomeric material) and for aerial use.	521-01-01

3.12.9 PEN conductors

PEN conductors shall **not** be isolated or switched.	460-01-04
Automatic disconnection using an RCD shall **not** be applied to a circuit incorporating a PEN conductor.	471-08-07
Isolators and switches shall **not** break a PEN conductor.	460-01-03
A separate metal cable enclosure shall **not** be used as a PEN conductor.	543-02-10

Where an RCD is used in a TN-C-S system:

• a PEN conductor shall **not** be used on the load side	413-02-07
• a separate metal cable enclosure shall **not** be used as a PEN conductor.	543-02-10

3.12.10 Phase conductors

An overcurrent detection device shall be provided for each phase conductor which shall be capable of only disconnecting that particular conductor – unless the disconnection of one phase could cause danger or damage.	473-03-01
The neutral conductor of switchgear that is used to disconnect live conductors shall not be capable of being:	530-01-01

- disconnected before the phase conductors
- reconnected before (or at the same time as) the phase conductors.

3.12.11 Protective conductors

Flexible or pliable conduit shall **not** be selected as a protective conductor.	543-02-01
Gas pipes, oil pipes and flexible (or pliable) conduit may **not** be used as a protective conductor.	543-02-01
Exposed conductive parts of equipment shall **not** be used as a protective conductor for other equipment.	543-02-08
If electrical earth monitoring is used, the operating coil shall be connected in the pilot conductor and **not** in the protective earthing conductor.	543-03-05
If the protective conductor forms part of a cable, then this shall only be earthed in the installation containing the associated protective device.	542-01-09
In installations and locations where the risk of an electric shock is increased by a reduction in body resistance and/or by contact with earth potential, all plugs, socket outlets and	471-15-07

cable couplers of a reduced low-voltage system shall have
a protective conductor contact.

Joints that can be disconnected for test purposes are permitted in a protective conductor circuit.	543-03-04
Metal water pipes of water heaters which have immersed and uninsulated heating elements should not be connected to the main earthing terminal by the circuit protective conductor.	554-05-01 554-05-02
Portable generators (and other installations which are not permanently fixed) shall be provided with protective conductors that are part of the cord or cable, between separate items of equipment.	551-04-06
Protective conductors shall be suitably covered against mechanical and chemical deterioration and electrodynamic effects.	543-03-01
Switching devices shall not be inserted in a protective conductor except for multipole linked switching or plug-in devices where the protective conductor circuit:	543-03-04

- cannot be interrupted before the live conductors and
- has to be re-established before the live conductors are re-connected.

3.12.12 Supplementary bonding conductors

Supplementary bonding conductors connecting:	547-03-01 and 547-03-02

- two exposed conductive parts or
- an exposed conductive part to an extraneous conductive part

shall have a conductance (if sheathed or mechanically protected) not less than the protective conductor connected to an exposed conductive part.

 Note: If mechanical protection is not provided, then the cross-sectional area shall be greater than 4 mm^2.

3.13 Power supplies

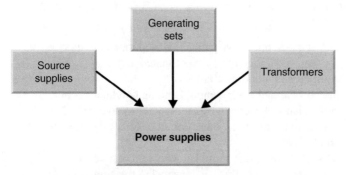

Figure 3.15 Power supplies for electrical installations

3.13.1 Source supplies

The installation shall conform to the design specifications for the following supply characteristics:

- nature of current: a.c. and/or d.c.
- Purpose and number of conductors (see Table 3.3)

Table 3.3 Conductors for source supplies

For a.c.	For d.c.
• phase conductor(s) • neutral conductor • protective conductor • PEN conductor	• outer conductor • middle conductor • earthed conductor • live conductor • protective conductor • PEN conductor

- values and tolerances:
    - nominal voltage and voltage tolerances
    - nominal frequency and frequency tolerances
    - maximum current allowable
    - prospective short-circuit current
    - earth fault loop impedance
    - protective measures inherent in the supply (e.g. earth, neutral or mid wire)
- particular requirements of the distributor.

Source supplies to the circuit (that are fed from a fixed   413-06-05
installation) may supply more than one item of equipment
provided that:

- all exposed conductive parts of the separated circuits
  are connected together by an insulated and non-earthed
  equipotential bonding conductor
- the non-earthed equipotential bonding conductor is not
  connected to a protective conductor, or to an exposed
  conductive part of any other circuit or to any extraneous
  conductive part
- all socket outlets are provided with a protective
  conductor contact (that is connected to the equipotential
  bonding conductor)
- all flexible equipment cables (other than Class II
  equipment) have a protective conductor for use as an
  equipotential bonding conductor
- exposed conductive parts which are fed by conductors
  of different polarity (and which are liable to a double fault
  occurring) are fitted with an associated protective device
- any exposed conductive part of a separated circuit
  cannot come in contact with an exposed conductive part
  of the source.

Consumer units

Accessible consumer units should be fitted with a child-   P 1.5
proof cover or installed in a lockable cupboard.

All circuits and final circuits shall be provided with a   476-01-02
means of switching for interrupting the supply on load.

The voltage drop between the supply terminals of an   525-01-02
installation and a socket outlet (or the terminals of a fixed
current-using equipment) shall not exceed 4% of the
nominal voltage of the supply.

TN systems

Where an RCD is used in a TN-C-S system, a PEN   413-02-07
conductor shall **not** be used on the load side.

TT systems

Exposed conductive parts that are protected by a single protective device shall be connected (via the main earthing terminal) to a common earth electrode.	413-02-18

IT system

Live conductors shall **not** be directly connected with earth. 413-02-21

3.13.2 Generating sets

The safety and proper functioning of all other sources of supply shall **not** be weakened by the generating set. 551-02-01

Protection against indirect contact shall be provided for each supply source which is capable of operating independently of other sources. 551-04-01

When a generator set is provided with overcurrent detection, this shall be located as near as practicable to the generator terminals. 551-05-01

Where a generating set is intended to operate in parallel with a distributor's:

- circulating harmonic currents shall be limited so that the thermal rating of conductors is not exceeded 551-05-02
- the generating set shall be capable of being isolated from the distributor's network. 551-07-04

3.13.3 Transformers

The neutral (i.e. star) point of the secondary windings of three-phase transformers (or the mid-point of the secondary windings of single-phase transformers) shall be connected to earth. 471-15-04

3.13.4 Autotransformers and step-up transformers

Step-up autotransformers shall not be connected to an IT system.	555-01-02

3.14 Circuits

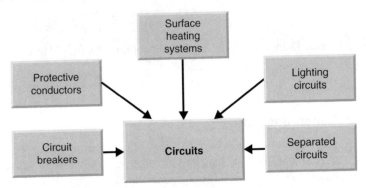

Figure 3.16   Electrical installations – circuits

3.14.1 Circuit breakers

Single-pole fuses, switches or circuit breakers shall only be inserted in the phase conductor.	131-13-01
Switches, circuit breakers (except where linked) and/or fuses shall be inserted in an earthed neutral conductor.	131-13-02
Any linked switch or linked circuit breaker that has been inserted in an earthed neutral conductor shall be capable of breaking all of the related phase conductors.	131-13-02

3.14.2 Circuit protective conductors

Circuit protective devices shall break any fault current flowing before such current causes any danger owing to the thermal or mechanical effects that are produced in circuit conductors or associated connections.	434-01-01

> Installations that provide protection against indirect contact   471-08-08
> by automatically disconnecting the supply shall have a
> circuit protective conductor run to (and terminated at) each
> point in the wiring and at each accessory.

 Except suspended lampholders which have no exposed conductive parts.

3.14.3 Electric surface heating systems

> The equipment, system design, installation and testing of   554-07-01
> an electric surface heating system shall meet the
> requirements of BS 6351.

3.14.4 Lighting circuits

> All lighting circuits shall include a circuit protective   P App C
> conductor.

3.14.5 Separated circuits

> Separated circuits shall, preferably, use a separate   413-06-03
> wiring system.

 Note: If this is not feasible, multicore cables (without a metallic sheath) or insulated conductors (in an insulating conduit) may be used.

3.15 Test and inspection

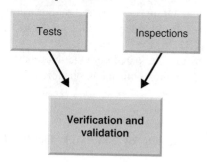

Figure 3.17 Tests and inspections

3.15.1 Inspections

Electrical installations **must** be inspected and tested   P 1.6
during, at the end of installation and before they are taken
into service to verify that they:

- are safe
- comply with BS 7671: 2001
- meet the relevant equipment and installation standards.   P 0.1b

All electrical work should be inspected (during installation as well as on completion) to verify that the components have:

- been selected and installed in accordance with   P 1.8a ii
  BS 7671
- been made in compliance with appropriate British   P 1.8a i
  Standards or harmonised European Standards
- been evaluated against external influences (such as the P 1.8a ii
  presence of moisture)
- not been visibly damaged (or are defective) so as to be P 1.8a iii
  unsafe
- been tested to check satisfactory performance with   P 1.8b
  respect to continuity of conductors, insulation
  resistance, separation of circuits, polarity, earthing
  and bonding arrangements, earth fault loop impedance
  and functionality of all protective devices including
  residual current devices
- been inspected for conformance with section 712 of   P 1.9
  BS 7671: 2001
- been tested as per section 713 of BS 7671: 2001   P 1.10
- been tested using appropriate and accurate instruments   P 1.10
- had their test results recorded using the model in   P 1.10
  Appendix 6 of BS 7671
- had their test results compared with the relevant   P 1.10
  performance criteria to confirm compliance.

 Note: Inspections and testing of DIY work should **also** meet the above requirements.

3.15.2 Tests

> The insulation resistance between live conductors and   713-04-01
> between each live conductor and earth shall be measured
> before the installation is connected to the supply.
>
> A polarity test shall be made to verify that:
>
> - fuses and single-pole control and protective devices are   713-09-01
>   only connected in the phase conductor
> - circuits (other than BS EN 60238 E14 and E27
>   lampholders) which have an earthed neutral conductor
>   centre contact bayonet (and Edison screw lampholders)
>   have their outer or screwed contacts connected to the
>   neutral conductor
> - wiring has been correctly connected to socket outlets
>   and similar accessories.

3.16 Special installations and locations

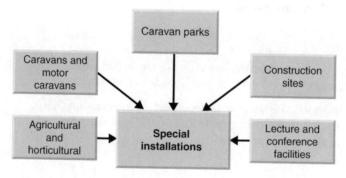

Figure 3.18 Special installations and locations

3.16.1 Agricultural and horticultural premises

> Metallic grids laid in the floor for supplementary bonding   605-08-03
> shall be connected to the protective conductors of the
> installation.
>
> Circuits (less SELV) supplying a socket outlet shall be   605-03-01
> protected by an RCD.

3.16.2 Caravans and motor caravans

 All extraneous conductive parts of a caravan or motor   608-03-04
caravan that are likely to become live in the event of a fault
shall be bonded to the circuit protective conductor by a
conductor with a minimum cross-sectional area of 4 mm².

All protective conductors, regardless of cross-sectional   608-06-03
area, **shall** be insulated.

Low-voltage socket outlets shall include a protective   608-08-02
contact (unless supplied by an individual winding of an
isolating transformer).

Plugs designed for extra-low-voltage socket outlets shall   608-08-03
be incompatible with low-voltage socket outlets.

3.16.3 Caravan parks

 All overhead conductors used in caravan parks **shall**   608-12-03
be completely covered with insulation which is:

- at least 2 m away from the boundary of any caravan pitch
- not less than 6 m in vehicle movement areas and 3.5 m
  in all other areas.

Note: Poles and other overhead wiring supports shall be protected against any
reasonably foreseeable vehicle movement.

A polarity test shall be made to verify that wiring has been   713-09-01
correctly connected to socket outlets and similar accessories.

At least one socket outlet shall be provided for each pitch.   608-13-02

Socket outlets must not be bonded to the PME terminal.   608-13-05

Socket outlets shall be protected (individually or in groups   608-13-05
of no more than three) by an RCD complying with
BS 4293, BS EN 61008-1 or BS EN 61009-1.

3.16.4 Construction site installations

The following are additional requirements for installations and locations where the risk of an electric shock is increased by a reduction in body resistance and/or by contact with earth potential.

Circuits supplying portable equipment for use outdoors shall be provided with an RCD as supplementary protection.	471-16-02
Plug and socket outlets shall comply with BS EN 60309-2.	604-12-02

3.16.5 Lecture/conference facilities

Artificial lighting should be designed to:

• give good colour rendering of all surfaces	M (4.9) and in (4.34)
• be compatible with other electronic and radio frequency installations	M (4.12.1) and M (4.36f)

4

Earthing

From an electrical point of view, the world is effectively a huge conductor at zero potential and is used as a reference point which is called 'earth' (in the UK) or 'ground' in the USA. People and animals are normally in contact with the earth and so if another part, which is open to touch, becomes charged at a different voltage from earth, then a shock hazard will exist.

This chapter reminds the reader about the different types of earthing systems and earthing arrangements. It then lists the main requirements for safety protection (direct and indirect contact), protective conductors and protective equipment before briefly touching on the test requirements for earthing.

 Similar to other chapters, it should be noted that these lists of requirements are **only** the author's impression of the most important aspects of the Wiring Regulations, and electricians should **always** consult BS 7671 to satisfy compliance.

4.1 What is earth?

In electrical terms, 'earth' is defined as:

> *The conductive mass of the earth, whose electric potential at any point is conventionally taken as zero.*

From an astrological and geophysical point of view, on the other hand:

> *Earth (also known as the Earth, Terra, and – mostly in the 19th century – Tellus) is the third planet outward from the Sun. It is the largest of the solar system's terrestrial planets and the only planetary body that modern science confirms as harbouring life. The planet formed around 4.57 billion (4.57 $\times$ 10^9) years ago and 'shortly' thereafter (i.e. about 4.533 billion years ago!) acquired its single natural satellite, the moon. Its astronomical symbol consists of a circled cross, representing a meridian and the equator. (Source Wikipedia)*

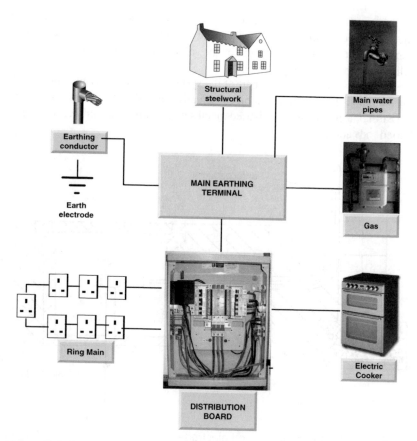

Figure 4.1 Bonding and earthing

4.2 What is meant by 'earthing' and how is it used?

> **Definition**
>
> **Earthing:** *Connection of the exposed conductive parts of an installation to the main earthing terminal of that installation.*

'*Earthing*' is a process that is used to connect all of the parts that could become charged to the general mass of the earth and in so doing, provide a path for fault currents which will hold these parts as close as possible to earth (i.e. zero) potential. In doing so, this will prevent a potential difference happening between the earth and earthed parts, as well as letting the flow of fault current to operate the protective systems.

An 'earthing system', on the other hand, defines the electrical potential of the conductors relative to that of the earth's conductive surface. The choice of earthing system has implications for the safety and electromagnetic compatibility of the power supply.

An 'earth electrode' is the part of the system that is directly in contact with the earth and this can be just a metal (usually copper) rod or stake driven into the earth or a connection to a buried metal service, pipe or a complex system of buried rods and wires.

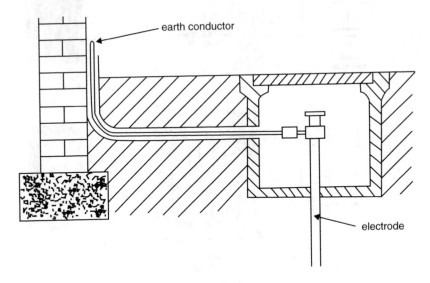

earth conductor

electrode

Figure 4.2 Earth conductor and electrode

The resistance of the electrode-to-earth connection will determine its quality and this can be improved by:

- increasing the surface area of the electrode that is in contact with the earth
- increasing the depth to which the electrode is driven
- using several connected ground rods
- increasing the moisture of the soil
- improving the conductive mineral content of the soil, and
- increasing the land area covered by the ground system.

A Protective Earth (PE) connection ensures that all exposed conductive surfaces are at the same electrical potential as the surface of the earth and thus avoiding the risk of an electrical shock if a person, or an animal, touches an equipment (or device) in which an insulation fault has occurred. PE also ensures that if an insulation fault occurs, then a high fault current will flow which will trigger an overcurrent protection device (e.g. a fuse) that will disconnect the power supply.

A Functional Earth (FE) connection, as well as providing protection against electrical shock, can also carry a current during the normal operation of a device – a facility that is often required by devices such as surge suppression and electromagnetic-compatibility filters, some types of antennas as well as a number of measuring instruments.

In a mains (i.e. a.c. power) wiring installation, the '*ground*' wire is (directly or indirectly) connected to one or more earth electrodes and carries currents away under fault conditions. These earth electrodes may be located locally or in the supplier's network some distance away, and the ground wire is also usually bonded to pipework so as to keep it at the same potential as the electrical ground during a fault.

The standard method of attaching the electrical supply system to earth is to make a direct connection between the two at the supply transformer so that the neutral conductor (often the star point of a three-phase supply – see Figure 4.3) is connected to earth using an earth electrode or the metal sheath and/or armouring of a buried cable.

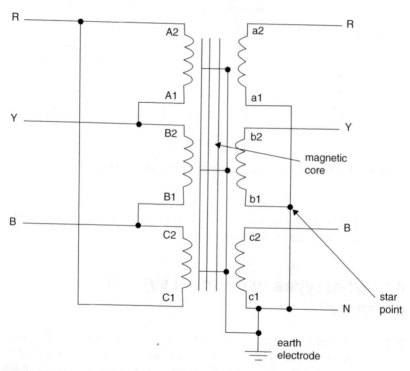

Figure 4.3 Three-phase delta/star transformer showing earthing arrangements

 Note: Lightning conductor systems must be bonded to the installation earth with a conductor that is **not** larger (i.e. in cross-sectional area) than that of the earthing conductor (541-01-03 and BS 6651).

4.3 Advantages of earthing

The main advantage to earthing is that the whole electrical system is tied to the potential of the general mass of the earth and cannot 'float' at another potential. By connecting earth to metalwork (that is not intended to carry current), a path is provided for fault current which can be detected by a protective conductor and, if necessary, broken. The path for this fault current is shown in Figure 4.4.

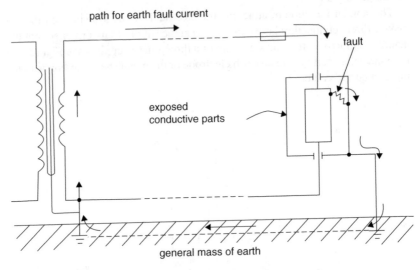

Figure 4.4 Path for earth fault current (shown by arrows)

The disadvantage of earthing is mainly the cost: of having to provide protective conductors and earth electrodes, etc.

4.4 What types of earthing system are there?

In the Regulations, an electrical system is defined as consisting 'of a single source of electrical energy and an installation' and the type of system depends on the link between the source and the exposed conductive parts of the installation, to earth.

 Note: In this context, an 'exposed conductive part' means a conductive part of an equipment which can be touched and which is not (i.e. currently) a live part, but which 'may' become live under fault conditions.

The four basic systems available are classified as TN (which is then further subdivided into TN-C, TN-C-S and TN-S), TC, TT and IT (as shown in Figure 4.5).

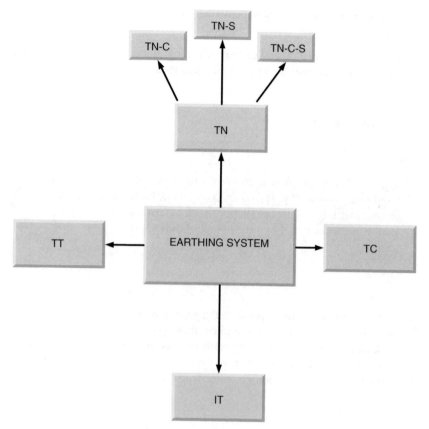

Figure 4.5 Earthing systems

4.4.1 System classification

In order to identify the different systems, a unique four-lettered code is used.
The **first letter** indicates the type of supply earthing, so that:

- **T** indicates that one or more points of the supply are directly earthed (for example, the earthed neutral at the transformer)
- **I** indicates either that the supply system is not earthed (at all) or that the earthing includes a deliberately inserted impedance, the purpose of which is to limit fault current.

The **second letter** indicates the earthing arrangement in the installation, so that:

- **T** indicates that all exposed conductive metalwork is connected directly to earth
- **N** indicates that all exposed conductive metalwork is connected directly to an earthed supply conductor provided by the electricity supply company.

The **third and fourth** letters indicate the arrangement of the earthed supply conductor system and:

- S – ensure that neutral and earth conductor systems are quite separate; and
- C – ensure that neutral and earth are combined into a single conductor.

4.4.2 TN system

In a TN installation, all exposed conductive parts   413-02-06
shall be connected (by a protective conductor) to the main earthing terminal of the installation and that terminal shall be connected to the earthed point of the supply source.

In a TN system, one or more points in the generator or transformer is connected to earth (usually the star point in a three-phase system). The body of the electrical device is then connected with earth via this earth connection at the transformer and exposed conductive parts of the installation are then connected to that point by protective conductors.

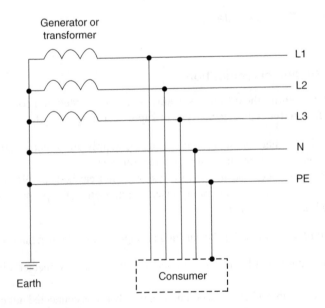

Figure 4.6 TN system

The conductor that connects together the exposed metallic parts of the consumer installation is called the *Protective Earth* (*PE*) whilst the conductor that connects to the star point in a three-phase system (or which carries the return current in a single-phase system) is called the *Neutral* (*N*).

There are three variants of TN systems – TN-C, TN-S and TN-C-S.

4.4.3 TN-C system

A TN-C system in one where the Neutral (N) and Protective Earth (PE) functions are combined in a single conductor (i.e. a PEN or protective multiple earth conductor) throughout the system and this combined neutral and earth wiring is then used both by the supply and from within the installation itself.

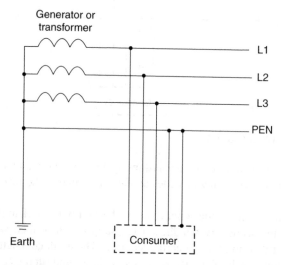

Figure 4.7 TN-C system

There are a number of disadvantages with using a TN-C network. For example:

- RCDs (residual current devices) are far less likely to detect an insulation fault
- they are vulnerable to unwanted triggering caused by contact between earths of circuits, on different RCDs or with real ground
- any connection between the combined neutral and earth core and the body of the earth could end up carrying significant current under normal conditions
- if there is a contact problem in the PEN conductor, then all parts of the earthing system beyond the break will raise to the potential of the live conductor(s).

 TN-C systems normally use an earthed concentric system, which can only be installed under special conditions.

4.4.4 TN-S system

TN-S systems have separate Protective Earth (PE) and Neutral (N) conductors from the transformer to the consuming device and these conductors remain separated throughout the system (see Figure 4.8).

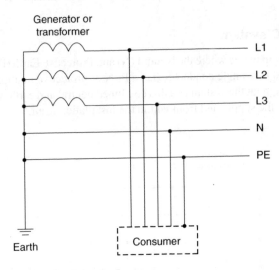

Figure 4.8 TN-S system

The TN-S is the most common earthing system in the UK and one where the electricity supply company provides an earth terminal at the incoming mains position.

This earth terminal is then connected by the supply PE back to the star point (neutral) of the secondary winding of the supply transformer, which is also connected, at that point, to an earth electrode. The earth conductor is usually the armour and sheath (if applicable) of the underground supply cable. In TN-S systems an RCD can be used as an additional protection.

Electromagnetic compatibility

One of the advantages of a TN-S system concerns electromagnetic compatibility whereby the consumer has a low-noise connection to earth and does not suffer from the voltage that appears on the neutral conductor as a result of the return currents and the impedance of that conductor. This is of particular importance with some types of telecommunication and measurement equipment.

 TN-S networks also save costs by having a fairly low-impedance earth connection near each consumer.

4.4.5 TN-C-S system

A TN-C-S system is one that uses a combined PEN conductor between the transformer and the building distribution point substation and the entry point into the building and then splits up into separate PE and N lines within the building (see Figure 4.9) to fixed indoor wiring and flexible power cords.

Note: In the UK, this system is also known as Protective Multiple Earthing (see below) as it connects the combined neutral and earth to real earth at many locations to thereby reduce the risk of broken neutrals.

Although TN-C networks save the cost of an additional conductor to separate N and PE connections, special cable types and lots of connections to earth are required.

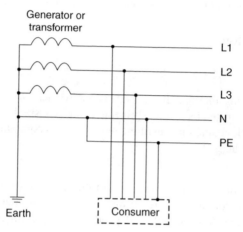

Figure 4.9 TN-C-S system

As any connection between the combined neutral and earth core and the body of the earth could end up carrying significant current under normal conditions, the use of TN-C-S is not recommended for locations such as petrol stations etc., where there is a combination of lots of buried metalwork and explosive gases.

In addition, owing to the possibility of a lost neutral, the use of TN-C-S supplies is banned for caravans and boats in the UK and it is often recommended to make outdoor wiring TT with a separate rod.

Protective multiple earthing

Socket outlets of caravan pitch supply equipment must **not** be bonded to the PME terminal   608-13-05

Protective Multiple Earthing (PME) is an earthing arrangement, found in TN-C-S systems, where the supply neutral conductor is used to connect the earthing conductor of an installation with earth (see Figure 4.10).

Requirements

> Where PME conditions apply, the main equipotential   547-02-01
> bonding conductor shall be selected in accordance with
> the neutral conductor of the supply and in accordance
> with the values shown in Table 54H (p. 120) of the IEE
> Wiring Regulations.

 Note: Local distributor's network conditions may require a larger conductor.

4.4.6 TT system

> In a TT installation, all exposed conductive parts –   413-02-18
> which are protected by a single protective device – **shall**
> be connected, via the main earthing terminal, to a common
> earth electrode.

A TT system is one which has one point of the energy source directly earthed
and the exposed conductive parts of the consumer's installation are provided
with a local connection to earth, independent of any earth connection at the
generator (see Figure 4.10).

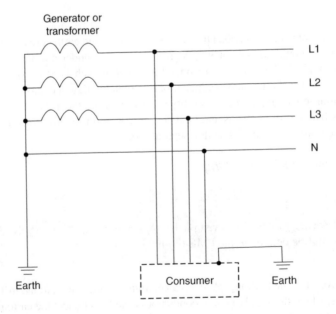

Figure 4.10 TT system

This type of installation is usually found in rural locations where the system is not provided with an earth terminal by the electricity supply company and the installation is fed from an overhead supply. Neutral and earth (protective) conductors must be kept quite separate throughout the installation and the final earth terminal must be connected to an earth electrode – via an earthing conductor.

TT systems (similar to TN-S systems) have a low-noise connection to earth, which is particularly important with some types of telecommunication and measurement equipment.

4.4.7 IT system

> In an IT installation, live conductors **shall** not be directly connected to earth.   413-02-21

An IT system is one which has no direct connection between live parts and earth and where the exposed conductive parts of the electrical installation are earthed (see Figure 4.11).

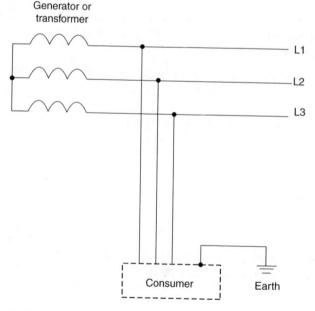

Figure 4.11 IT system

An IT system is similar to a TT system except that the supply earthing in an IT system can either be from an unearthed supply or one which (although not totally earthed) is connected to earth through a current-limiting impedance.

This lack of earth will usually mean that normal protective methods cannot be used and for this reason, IT systems are not normally allowed in the UK public supply system – except for hospitals and other medical locations where it is recommended for use with circuits supplying medical equipment intended for life-support of patients.

As previously mentioned, in the UK the type of system is either TT, TN-S or TN-C-S and their prime difference concerns their particular earthing arrangements shown in Table 4.1.

4.4.8 Requirements

General

The type of earthing system chosen (i.e. TN-C, TN-S, TN-C-S, TT or IT) depends on the characteristics of the energy, the source of energy and the availability of existing earthing facilities.	312-03-01

In locations with risks of fire due to the nature of processed or stored materials:

wiring systems (less those using mineral insulated cables and busbar trunking systems) shall be protected against earth insulation faults as follows: • in TN and TT systems, by RCDs having a rated residual operating current (IAn) not exceeding 300 mA • in IT systems, by insulation monitoring devices with audible and visible signals.	482-02-06

TN system

Where an RCD is used in a TN-C-S system, a PEN conductor shall **not** be used on the load side.	413-02-07

Exposed conductive parts of an installation shall be connected (by a protective conductor) to the main earthing terminal of the installation and that terminal	413-02-06

Table 4.1 Earth provision

System	Provision of earth	Remarks
TT Earthing conductor	Earthing is provided by the consumer's own installation earth electrode.	No earthing facility is made available to the consumer by the distributor or (if such a facility is made available) it is not used.
TN-S Earthing conductor	Earthing is provided by the distributor.	This is usually provided either by a direct connection to the supply cable sheath or via a separate protective conductor in the form of a split-concentric cable or overhead conductor.
TN-C-S Earthing conductor	Earthing is provided by the distributor.	This is connected to the incoming supply neutral to give a protective multiple earth (PME) supply, where the supply neutral and protective conductors are in the form of a combined neutral and earth (CNE) conductor.

shall be connected to the earthed point of the supply source depending on which type of system it is (i.e. TN-S or TN-C-S).

In a TN system, one or more of the following types of protective device shall be used: 413-02-07

- an overcurrent protective device
- an RCD.

The protective conductor to the PEN conductor shall be on the source side of the RCD.   413-02-07

The maximum disconnection times to a circuit supplying socket outlets and to other final circuits which supply portable equipment that is intended for manual movement during use (or hand-held Class I equipment) shall not exceed those shown in Table 41A (p. 44) of the IEE Wiring Regulations.   413-02-09

 Note: This requirement does not apply to a final circuit which supplies an item of stationary equipment that is connected via a plug and socket outlet and where precautions have already been taken to prevent the use of the socket outlet for supplying hand-held equipment, nor to a reduced low-voltage circuit.

Where a fuse is used, the maximum values of earth fault loop impedance shall be in accordance with Table 41B1 (page 45) of the IEE Wiring Regulations.   413-02-10

 Note: See appropriate British Standard for types and rated currents of fuses other than those mentioned in the Regulations, in Table 4.2 below.

A disconnection time of 5 s is allowed for a final circuit that is only supplying stationary equipment.   413-02-13

When a part of a TN system does not meet the requirements for automatic disconnection, then: 531-03-01

- that part may be protected by an RCD
- exposed conductive parts of that part shall be connected to the TN

Table 4.2 British Standards for fuse links

	Standard	Current rating	Voltage rating	Breaking capacity	Notes
1	BS 2950	Range 0.05–25 A	Range 1000 V (0.05 A) to 32 V (25 A) a.c. and d.c.	Two or three times current rating	Cartridge fuse links for telecommunication and light electrical apparatus. Very low breaking capacity
2	BS 646	1, 2, 3 and 5 A	Up to 250 V a.c. and d.c.	1000 A	Cartridge fuse intended for fused plugs and adapters to BS 546: 'round-pin' plugs
3	BS 1362 cartridge	1, 2, 3, 5, 7, 10 and 13 A	Up to 250 V a.c.	6000 A	Cartridge fuse primarily intended for BS 1363: 'flat pin' plugs
4	BS 1361 HRC cut-out fuses	5, 15, 20, 30, 45 and 60 A	Up to 250 V a.c.	16,500 A 33,000 A	Cartridge fuse intended for use in domestic consumer units. The dimensions prevent interchangeability of fuse links which are not of the same current rating
5	BS 88 motors	Four ranges, 2–1200 A	Up to 660 V, but normally 250 or 415 V a.c. and 250 or 500 V d.c.	Ranges from 10,000 to 80,000 A in four a.c. and three d.c. categories	Part 1 of Standard gives performance and dimensions of cartridge fuse links, whilst Part 2 gives performance and requirements of fuse carriers and fuse bases designed to accommodate fuse links complying with Part 1
6	BS 2692	Main range from 5 to 200 A; 0.5 to 3 A for voltage transformer protective fuses	Range from 2.2 to 132 kV	Ranges from 25 to 750 MVA (main range) 50 to 2500 MVA (VT fuses)	Fuses for a.c. power circuits above 660 V
7	BS 3036 rewirable	5, 15, 20, 30, 45, 60, 100, 150 and 200 A	Up to 250 V to earth	Ranges from 1000 to 12,000 A	Semi-enclosed fuses (the element is a replacement wire) for a.c. and d.c. circuits
8	BS 4265	500 mA to 6.3 A 32 mA to 2 A	Up to 250 V a.c.	1500 A (high breaking capacity) 35 A (low breaking capacity)	Miniature fuse links for protection of appliances of up to 250 V (metric standard)

- earthing system's protective conductor or to a
separate earth electrode.

If protection is provided by an RCD then: 413-02-16

$Z_s \times I_{\lambda n} \leqslant 50$ V, where:

- Z_s is the earth fault loop impedance in ohms
- $I_{\lambda n}$ is the rated residual operating current of the
protective device in amperes.

The following are additional requirements for installations where a generating set acts as a switched alternative (i.e. as a stand-by system) to the distributor's network.

When a generator is operating as a switched alternative   551-04-03
to a TN system, then protection by automatic disconnection
of supply shall not rely on the connection to the earthed
point of the distributor's network.

For TN-S systems, RCDs shall be positioned to avoid   551-06-02
incorrect operation owing to the possibility of a parallel
neutral earth path.

TT system

Exposed conductive parts that are protected by a single   413-02-18
protective device shall be connected (via the main
earthing terminal) to a common earth electrode.

IN a TT system, one or more of the following types of   413-02-19
protective device shall be used:

- an RCD (the preferred method) or
- an overcurrent protective device.

Each circuit shall meet the following requirement: 413-02-20

$RA \times I_a < 50$ V

where:

- RA is the sum of the resistances of the earth electrode
and the protective conductor(s) connecting it to the
exposed conductive part
- I_a is the current causing the automatic operation of
the protective device within 5 s.

IT system

Live conductors shall not be directly connected with earth.   413-02-21

Precautions shall be taken to guard against the risk of   413-02-22
electric shock in the event of two faults existing
simultaneously by using one (or both) of the following
devices:

- an overcurrent protective device
- an RCD

All exposed conductive parts shall be earthed so that   413-02-23
(for each circuit):

$$RA \times I_d \leqslant 50 \, V$$

where:

- RA is the sum of the resistances of the earth
  electrode and the protective conductor connecting
  it to the exposed conductive parts
- I_d is the fault current of the first fault of negligible
  impedance between a phase conductor and an
  exposed conductive part.

An insulation monitoring device shall be provided to   413-02-24
indicate the first fault from a live part to an exposed
conductive part or to earth.

Where exposed conductive parts are earthed collectively, 413-02-25
then protection shall be as for a TN system

Where exposed conductive parts are earthed   413-02-25
individually or in groups, then the conditions
for protection shall be as for a TT systems.

Where protection is provided by an RCD (and   531-05-01
disconnection following a first fault is not envisaged) the
non-operating residual current of the device shall be at
least equal to the current which circulates on the first
fault to earth.

> The maximum disconnection time for a second fault   413-02-26
> in IT systems shall be in accordance with Table 4.3.

Table 4.3 Maximum disconnection time in IT systems (2nd fault)

Installation	Maximum disconnection time	
	Neutral not disturbed	Neutral disturbed
120/240	0.8 s	5.0 s
230/240	0.4 s	0.8 s
400/690	0.2 s	0.4 s
580/1000	0.1 s	0.2 s

4.5 Earthing arrangements

The main earthing terminal shall be connected to earth as shown below.

System	Main earthing terminal
TN-S	Connected to the earthed point of the energy source.
TN-C-S	Connected (by the distributor) to the neutral of the energy source.
TT or IT	Connected via an earthing conductor to an earth electrode.

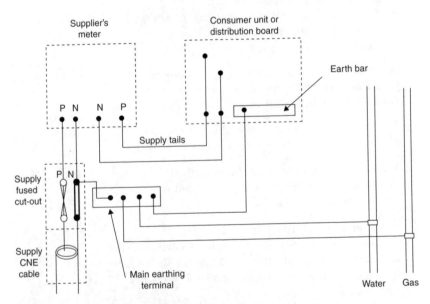

Figure 4.12 Domestic earthing arrangement

4.5.1 General

Earthing arrangements may be used jointly or separately for protective and functional purposes, according to the requirements of the installation.   542-01-06

Earthing arrangements shall ensure that: 542-01-07

- they are sufficiently robust (or have additional mechanical protection) to external influences
- the impedance from the consumer's main earthing terminal to the earthed point of the supply meets the protective and functional requirements of the installation
- earth fault currents and protective conductor currents that may occur are carried without danger.

 Note: This particularly applies with respect to thermal, thermomechanical and electromechanical stresses.

Precautions shall be taken against possible damage to other metallic parts through electrolysis.   542-01-08

If a number of installations have separate earthing arrangements then any protective conductor that is common to one of these installations: 542-01-09

- shall either be capable of carrying the maximum fault current likely to flow through them or
- earth one installation and be insulated from the earthing arrangements of other installation(s).

 If the protective conductor forms part of a cable, then this shall only be earthed in the installation containing the associated protective device.

4.5.2 Earth electrodes

Metalwork belonging to gas, water or other services shall **not** be used as a protective earth electrode.   542-02-04

Earth electrodes shall be designed and constructed   542-02-03
so that they can:

- withstand damage
- take into account a possible increase in resistance
  due to corrosion.

An earth electrode is a conductor, or group of conductors, that are in close contact with (and provide an electrical connection to) earth. The following types of earth electrode may be used for electrical installations:

- earth rods or pipes
- earth tapes or wires
- earth plates
- lead sheaths and other metal cable covers
- other suitable underground metalwork
- structural metalwork embedded in foundations
- welded metal reinforced concrete (except prestressed concrete) embedded in the earth.

Requirements

The type and embedded depth of an earth electrode   542-02-02
shall be such that soil drying and freezing will not
increase its resistance above 200 Ω.

Lead sheaths or other metal cable covers may be used   542-02-05
provided that:

- arrangements have been made to warn the owner
  of the electrical of any proposed change to the
  cable that might affect its suitability as an earth
  electrode
- precautions have been taken to prevent excessive
  deterioration by corrosion
- the sheath or covering is in effective contact with earth
- the consent of the cable owner has been obtained.

 Note: An earthing conductor is a protective conductor that connects the main earthing terminal of an installation to an earth electrode or to other means of earthing.

4.5.3 Earth fault loop impendence

As shown in Figure 4.13, the earth fault loop starts at the point of the fault and comprises:

- the Circuit Protective Conductor (CPC)
- the consumer's earthing terminal and earthing conductor
- (for TN systems) the metallic return path or (for TT and IT systems) the earth return path
- the path through the earthed neutral point of the transformer
- the transformer winding
- the phase conductor from the transformer to the point of fault.

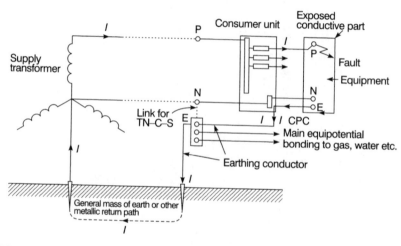

Figure 4.13 Earth fault loop impendence path

4.5.4 Earthing conductors

All earthing conductors **shall** meet the requirements   542-03-01
for protective conductors.

The earthing conductor is an important part of the earth fault loop impedance as it is a protective conductor that connects the main earthing terminal of an installation to an earth electrode, or to some other means of earthing.

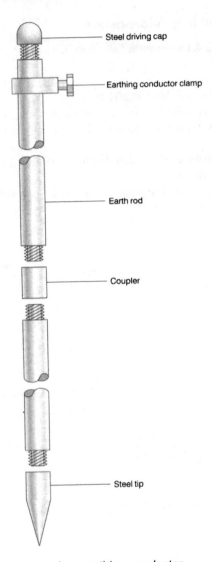

Figure 4.14 An example of an earthing conductor

Requirements

| Buried earthing conductors shall have a cross-sectional area not less than the values shown in Table 54A (p. 113) of the IEE Wiring Regulations. | 542-03-01 |

Where PME (protective multiple earthing) exists,   542-03-01
the requirements for main equipotential bonding
conductors (i.e. for the cross-sectional area of a
main equipotential bonding conductor) shall
be met.

The thickness of tape or strip conductors shall be capable   542-03-01
of withstanding mechanical damage and corrosion (see
BS 7430).

The connection of an earthing conductor to the earth   542-03-03
electrode shall be:

- soundly made
- electrically and mechanically satisfactory
- suitably labelled
- suitably protected against corrosion.

4.5.5 Main earthing terminals

All installations shall have a main earthing terminal (or   542-04-01
bar) to connect the earthing conductor to:

- the circuit protective conductors
- the main bonding conductors
- functional earthing conductors (if required)
- the lightning protection system bonding conductor
  (if any).

The earthing conductor shall be capable of being   542-04-02
disconnected so as to enable the resistance of the
earthing arrangements to be measured.

Joints in an earthing conductor shall: 542-04-02

- be capable of being disconnected only by means of
  a tool
- be mechanically strong
- ensure the maintenance of electrical continuity.

4.5.6 Earthing points

It has been proved that the resistance area around an earth electrode depends on the size of the electrode and the type of soil and that this electrode resistance is particularly important with regard to the voltage at the surface of the ground. For example, as shown in Figure 4.15, for a 2 m rod with its top at ground level, approximately 80–90% of the voltage appearing at the electrode under fault conditions, will be dropped in the first 2.5–3 m from the electrode.

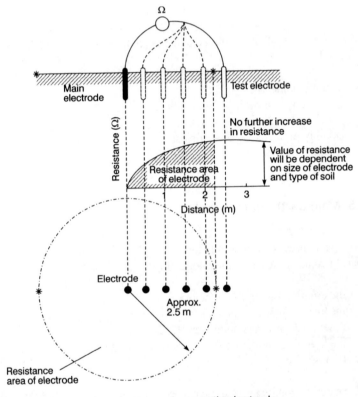

Figure 4.15 The resistance area of an earth electrode

This can be particularly dangerous where livestock is concerned. For example, in some circumstance a grazing cow might have its forelegs inside the resistive area, whilst its hindlegs are outside of the area. Bearing in mind that a potential difference of 25 V can be lethal, measures have to be taken to reduce this risk. One method is to house the earth electrode in a pit that is below ground level (as shown in Figure 4.16).

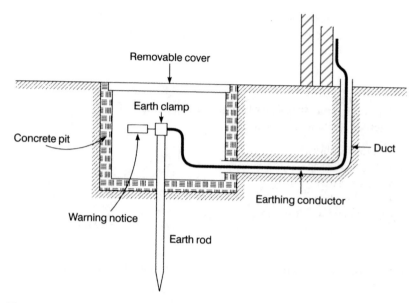

Figure 4.16 An earth electrode protected by a pit below ground level

4.6 Safety protection

Around 1000 electrical accidents at work are reported to HSE every year and about 30 people die of their injuries. Many of these deaths and injuries arise from:

- use of poorly maintained electrical equipment
- work near overhead power lines
- contact with underground power cables during excavation work
- work on or near 230 V domestic electricity supplies
- use of unsuitable electrical equipment in explosive areas such as car paint spraying booths.

In addition, fires started by poor electrical installations and faulty electrical appliances cause many additional deaths and injuries.

For this reason, protection against electric shock and safety protection methods are an essential part of the Regulations and the following mandatory requirement, therefore, needs to be observed:

All installations shall comply with the requirements for safety protection in respect of:

471

- electric shock
- thermal effects
- overcurrent
- undervoltage
- isolation and switching.

Note: 'Installation' in this context is taken to mean either as a whole or in its several parts.

4.6.1 Earthing arrangements for protective purposes

If an overcurrent protective device is used for protection against electric shock, the protective conductor shall be incorporated in the same wiring system as the live conductors or be located nearby.

544-01-01

4.6.2 Earthing arrangements for protective and functional purposes

Protection by non-conducting location and earth-free local equipotential bonding is **not** permitted in locations containing a bath or a shower.

601-05-03

Where earthing is required for protective as well as functional purposes, then the requirements for protective measures shall take precedence.

546-01-01

4.7 Protection against direct contact

RCDs may **not** be used as the sole means of protection against direct contact.   412-06-01 and 412-06-02

A person may perform work involving direct contact with electrical parts only if the electrical part:

- is isolated from all sources of electricity
- is tested to ensure its isolation from all sources of electricity and
- is earthed if it is of high voltage.

Work may be performed by a person, operating plant or vehicle coming within the exclusion zone for an electrical part, only if the electrical part:

- is isolated from all sources of electricity
- is tested to ensure its isolation from all sources of electricity and
- is earthed if it is of high voltage.

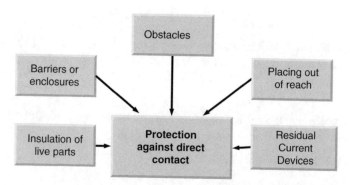

Figure 4.17 Protection against direct contact

To meet these requirements, the Regulations state that one of the following, basic, measures shall be used for protection against indirect contact:

- insulating live parts
- using a barrier or an enclosure
- using obstacles
- placing equipment out of reach
- by using an RCD.

4.7.1 Protection by insulation of live parts

As the title suggests, this is a basic form of insulation protection and is intended to prevent contact with a live part of an electrical installation from direct contact. Paint, lacquers and varnishes do **not** provide adequate protection.

Requirements

Live parts shall be completely covered with insulation   412-02-01
which can only be removed by destruction.

Equipment shall be capable of withstanding all   412-02-01
mechanical, chemical, electrical and thermal influences
stresses normally encountered during service.

4.7.2 Protection by barriers or enclosures

This intention of this form of protection is to prevent or deter any contact with a live part. Whilst, generally speaking, this method is for protection against direct contact, it also provides a degree of protection against indirect contact.

Requirements

Live parts **shall** be inside enclosures (or behind   412-03-01
barriers) protected to at least IP2X or IPXXB.

The horizontal top surface of easily assessable barriers   412-03-02
or enclosures shall be protected to at least 1P4X.

All barriers and/or enclosures shall be: 412-03-03

- firmly secured in place
- have sufficient stability and durability to maintain
  the required degree of protection and separation
  from any known live part.

If it is necessary to remove a barrier or open an enclosure   412-03-04
or remove a part of an enclosure, the removal or opening
shall be possible only:

- by using a key or tool
- after disconnecting the supply
- if an1P2X or IPXXB intermediate barrier is used.

This regulation does not apply to:

- a ceiling rose complying with BS 67
- a pull cord switch complying with BS 3676
- a bayonet lampholder complying with BS EN 61184
- an Edison screw lampholder complying with BS EN 60238.

4.7.3 Protection by obstacles

The intention of this form of protection is to prevent unintentional contact with a live part, but **not** an intentional contact caused by deliberately circumnavigating the obstacle.

Requirements

Obstacles shall be secured to prevent unintentional removal.	412-04-02
Obstacles may be removable without using a key or tool.	412-04-02

4.7.4 Protection by placing out of reach

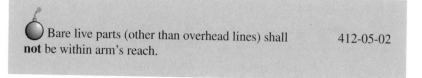

Bare live parts (other than overhead lines) shall **not** be within arm's reach.	412-05-02

This intention of this form of protection is to prevent unintentional direct contact with a live part.

Requirements

Bare (or insulated) overhead lines used for distribution between buildings and structures shall be installed in accordance with the Electricity Safety, Quality and Continuity Regulations 2002.	412-05-01
Bare live parts (other than an overhead line) shall not be within 2.5 m of:	412-05-02

- an exposed conductive part
- an extraneous conductive part
- a bare live part of any other circuit.

Protection against direct contact by placing out of reach   600
may only be used in an area accessible to skilled persons.
It may **not** be used for some special installations and
locations (e.g. swimming pools etc.) where an increased
risk of shock exists.

4.7.5 RCDs

Definition

RCD: *A mechanical switching device or association of devices*
intended to cause the opening of the contacts when the residual
current attains a given value under specified conditions.

The residual operating current of the protective   531-02-03
device **shall** comply with the requirements for
protection against indirect contact.

Automatic disconnection using an RCD shall **not** be   471-08-07
applied to a circuit incorporating a PEN conductor.

RCDs that are associated with a circuit that would   531-02-05
normally be expected to have a protective conductor,
shall **not** be considered sufficient for protection against
indirect contact if there is no such conductor.

RCDs **shall:**

- be capable of disconnecting all of the phase   531-02-01
  conductors of a circuit at the same time.
- be located so that their operation will not be impaired   531-02-07
  by magnetic fields caused by other equipment.
- ensure that any protective conductor current   531-02-04
  which may occur will be unlikely to cause
  unnecessary tripping of the device.

RCDs that provide protection against indirect contact   531-02-08
and which are used separately from an overcurrent
protective device, **shall** be capable of withstanding the
likely thermal and mechanical stresses that it is likely
to be subjected to if fault occurred.

RCDs that can be operated by a non-skilled person **shall** be designed and installed so that their operating current and/or time delay mechanism cannot be altered without using a key or a tool.   531-02-10

All alterations made to the setting and/or calibration of an RCD **shall** be visibly indicated.   531-02-10

RCDs which are powered from an auxiliary source and which do not operate automatically in the case of failure of the auxiliary source **shall** only be used if: 531-02-06

- protection against indirect contact is ensured
- the device is part of a supervised (tested and inspected) installation.

The protective conductor to the PEN conductor **shall** be on the source side of the RCD.   413-02-07

Two or more RCDs connected in series **shall** be capable of operating separately.   531-02-09

If an IT system is protected against indirect contact by an RCD, each final circuit **shall** be separately protected.   413-02-25

If an RCD is used in a TN-C-S system, a PEN conductor shall **not** be used on the load side.   413-02-07

If the installation is part of a TT system, all socket-outlet circuits **shall** be protected by an RCD.   471-08-06

If an installation includes an RCD, then it **shall** have a notice (fixed in a prominent position) that reads as follows: 514-08-01

This installation, or part of it, is protected by a device which automatically switches off the supply if an earth fault develops. Test quarterly by pressing the button marked 'T' or 'Test'. The device should switch off the supply and should then be switched on to restore the supply. If the device does not switch off the supply when the button is pressed, seek expert advice.

Figure 4.18 RCD safety notice

In electrical installations, an RCD or an RCB (Residual Current circuit Breaker) is a circuit breaker that operates to disconnect a (or rather their particular) circuit whenever they detect that current leaking out of that circuit (such as current leaking to earth through a ground fault) exceeds safety limits.

Figure 4.19 illustrates the construction of an RCD and works on the principle that in a normal (i.e. healthy) circuit, the magnetic effects of the phase and neutral currents will cancel out because the same current will pass through the phase coil, the load, and then back through the neutral coil. In a faulty circuit where the phase or the neutral are to earth, the currents will no longer be equal and the out-of-balance current will produce some residual magnetism in the core. As the magnetism will be alternating, it will link with the turns of the search coil and induce an EMF in it which will drive a current through the trip coil and cause the tripping mechanism to operate.

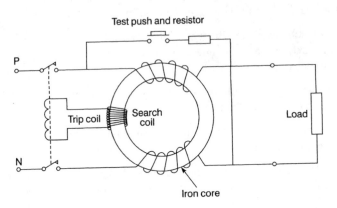

Figure 4.19 RCD

Although RCDs reduce the risk of electric shock, they may **not** be used as the sole means of protection against direct contact (412-06-01 and 412-06-02).

Requirements

Circuits supplying portable equipment for use outdoors and not connected to a socket outlet via a flexible cable or cord (having a current-carrying capacity of 32 A or less) shall be protected by an RCD.	471-16-02
Installations which are part of a TT system and which are protected by a single RCD shall (unless the part of the installation between the origin used Class II equipment or an equivalent insulation) be placed at the origin of the installation.	531-04-01

 Where there is more than one origin this requirement applies to each origin.

Socket outlets that are rated at 32 A or less and that are supplying portable equipment that is being used outdoors, shall be provided with supplementary protection in the form of an RCD.   471-16-01

 This regulation does not apply to a socket outlet supplied by a circuit that is protected by:

- SELV
- electrical separation
- automatic disconnection and reduced low-voltage systems.

The magnetic circuit of the transformer of an RCD shall enclose all the live conductors of the protected circuit.   531-02-02

The protective conductor of an RCD shall be outside the magnetic circuit.   531-02-02

When an IT system is protected by an RCD (and disconnection following a first fault is not expected), the non-operating residual current of the device shall be at least equal to the current which circulates on the first fault to earth.   531-05-01

When an RCD is used in a TN-S system, it shall be positioned so as to avoid incorrect operation owing to the possible existence of a parallel neutral earth path.   551-06-02

Where an RCD is used for protection against electric shock, the product of the rated residual operating current and the earth fault loop impedance shall not exceed 50 Ω.   471-15-06

Where an RCD is used in a TN system for automatic disconnection of a circuit which extends beyond the conductive parts, then exposed conductive parts need not be connected to the TN system's protective earthed equipotential zone, **provided** that they are connected to an earth electrode which has a resistance appropriate to the operating current of the RCD.   413-02-17

Where automatic disconnection of TN, TT and/or IT systems cannot be achieved by using overcurrent protective devices, then either local supplementary equipotential bonding or an RCD shall be used.   413-02-04

Wiring (other than mineral insulated cables and busbar trunking systems) in TN and TT systems shall be protected against earth insulation faults by RCDs having a rated residual operating current (IAn) not exceeding 300 mA.	482-02-06
For installations and locations where there is an increased risk of shock (such as agricultural and horticultural premises, building sites, bathrooms, swimming pools etc.), additional measures may be required, such as automatic disconnection of supply by means of an RCD with a rated residual operating current not exceeding 30 mA.	471-08-01

4.8 Protection against indirect contact

Indirect contact (i.e. when part of the body touches or is in dangerous proximity to any object that is in contact with energised electrical equipment or exposed conductive parts which might become live under fault conditions) has always been a potential problem to the unwary when installing, maintaining or inspecting electrical installations. It should also be remembered that as voltages increase, the potential for arcing increases and through arcing, injuries and/or fatalities will often occur **even** if actual bodily contact with high-voltage lines and/or equipment is not made!

Requirements

Protection against indirect contact **shall** be provided for each supply source which can operate independently of other sources.	551-04-01

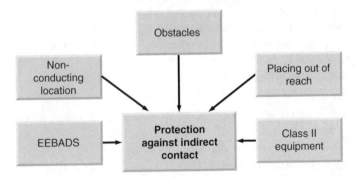

Figure 4.20 Protection against indirect contact

To meet these requirements, the Regulations (ie 413-01-01) state that one of the following, basic, measures shall be used for protection against indirect contact:

- Earthed Equipotential Bonding and Automatic Disconnection of Supply (EEBADS)
- Class II equipment or equivalent insulation
- non-conducting location
- earth-free local equipotential bonding
- electrical separation.

4.8.1 Protection by Earthed Equipotential Bonding and Automatic Disconnection of Supplies (EEBADS)

An earthed equipotential zone is a zone within which exposed conductive parts and extraneous conductive parts are maintained at substantially the same potential by bonding, such that, under fault conditions, the differences in potential between simultaneously accessible exposed and extraneous conductive parts will not cause electric shock.

Earthed equipotential bonding provides a very good form of protection against indirect contact by joining together (i.e. bonding) all of the metallic parts together and then connecting them to earth. This ensures that all metalwork is at (or near) zero volts and so under fault conditions, all metalwork will rise to a similar potential and simultaneous contact with two metal parts will not result in electric shock as there is no significant PD between them.

For installations and locations with an increased risk of shock (such as bathrooms and saunas etc.) additional measures may be required, such as:

- automatic disconnection of supply by means of an RCD with a rated residual operating current (Isn) not exceeding 30 mA
- supplementary equipotential bonding
- reduction of maximum fault clearance time.

Note: The application of protection by earthed equipotential bonding (and automatic disconnection of supply) will depend on the requirements of the type of system earthing in use (e.g. TN, TT or IT).

Requirements

The earth loop impedance of all circuits which supply fixed equipment that is outside of the earthed equipotential zone (and which have exposed conductive parts that could be touched by a person who has direct contact with earth) shall ensure that disconnection occurs within the time stated in Table 41A (p. 44) of the IEE Wiring Regulations.	471-08-03

Installations that provide protection against indirect   471-08-08
contact by automatically disconnecting the supply
shall have a circuit protective conductor run to (and
terminated at) each point in the wiring and at each
accessory.

 Except suspended lampholders, which have no exposed conductive parts.

4.8.2 Protection by Class II equipment or equivalent insulation

Cables with a non-metallic sheath or a non-metallic   471-09-04
enclosure are **not** considered to be of Class II
construction.

Class II equipment shall **not** be the only protection   413-03-02
between live parts of the installation and exposed
metalwork of that equipment.

Class II equipment is unique in that as well as providing the basic insulation for
live parts, it also has a second layer of insulation, which can be used to either prevent contact with exposed conductive parts or to make sure that there can never
be any contact between such exposed conductive parts and live parts.

Class II protection is provided by one or more of the following:

- electrical equipment having double or reinforced insulation
- low-voltage switchgear
- low-voltage control gear assemblies
- supplementary insulation
- reinforced insulation applied to uninsulated live parts.

Requirements

Circuits supplying Class II equipment shall have a circuit   471-09-02
protective conductor that shall be run to (and terminated
at) each point in the wiring and at each accessory.

 Except suspended lampholders, which have no exposed conductive parts.

The insulation of operational electrical equipment shall be at least 1P2X or IPXXB.	413-03-04
The insulating enclosure shall be capable of resisting mechanical, electrical and thermal stresses likely to be encountered.	413-03-05
The metalwork of exposed Class II equipment shall be mounted so that it is not in electrical contact with any other part of the installation that is connected to a protective conductor.	471-09-02

4.8.3 Protection by non-conducting location

| This method of protection is **not** recognised for general application and may not be used in installations such as agricultural and horticultural buildings (saunas, caravans etc.) that are subject to an increased risk of shock. | 413-04-01 |

This method of protection is intended to prevent simultaneous contact with parts which may be at different potentials (i.e. through the failure of the basic insulation of live parts) and a 'non-conducting location' is a location where there is no earthing or protective system because:

- there is nothing which needs to be earthed
- exposed conductive parts are arranged so that it is impossible to touch two of them (or an exposed conducting part and an extraneous conductive part) at the same time.

Requirements

Exposed conductive parts which might attain different potentials through failure of the basic insulation of live parts shall be arranged so that a person will not come into simultaneous contact with:	413-04-02
• two exposed conductive parts or	
• an exposed conductive part and any extraneous conductive part.	

This may be achieved if the location has an insulating floor and insulating walls and one or more of the following arrangements applies: 413-04-07

- if protective obstacles (that are not connected to earth or to exposed conductive parts and which are made out of insulating material) are used between exposed conductive parts and extraneous conductive parts
- the distance between any separated exposed conductive parts (and between exposed conductive parts and extraneous conductive parts) is not less than 2.5 m (1.25 m for parts out of arm's reach)
- the insulation is of acceptable electrical and mechanical strength.

In a non-conducting location: 413-04-03

- there shall be no protective conductors
- socket outlets shall not include an earthing contact
- luminaire supporting couplers which have a protective conductor contact shall not be installed.

The resistance of an insulating floor or wall shall not be less than: 413-04-04

- 50 kΩ where the voltage to earth does not exceed 500 V
- 100 kΩ where the voltage to earth exceeds 500 V, but does not exceed low voltage.

Permanent protection shall be provided in all areas where the use of mobile or portable equipment is anticipated.   413-04-05

It shall not be possible for a potential on an extraneous conductive part (inside a location) to be transmitted outside that location.   413-04-06

4.8.4 Protection by earth-free local equipotential bonding

Earth-free local equipotential bonding shall:

- only be used in special situations which are earth-free   471-11-01
- only be used in special circumstances   413-05-01
- not be used in caravans, motor caravans and caravan parks where there is an increased shock risk.   471-11-01

'Earth-free bonding' is, as the name implies, an area in which all exposed metal parts are connected together, but not to earth, thus ensuring that inside the area there can be no danger (even if the voltage to earth is very high) as all of the metalwork that can be touched will be at the same potential. This form of protection is intended to prevent a dangerous voltage appearing between simultaneously accessible parts in the event of failure of the basic insulation.

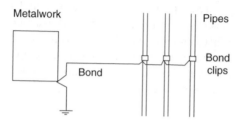

Figure 4.21 Earth-free bonding

Requirements

An equipotential bonding conductor shall be used to connect together simultaneously accessible exposed conductive parts and extraneous conductive parts.	413-05-02
Local equipotential bonding conductors shall not be in electrical contact with earth.	413-05-03
Persons entering or leaving an equipotential location shall not be exposed to a dangerous potential difference.	413-05-04
Earth-free local equipotential bonding shall only be used in special situations, which are earth-free.	471-11-01
A warning notice, warning that earth-free local equipotential bonding is being used, shall be fixed in a prominent position adjacent to every point of access to the location concerned.	514-13-02

4.8.5 Protection by electrical separation

Protection by electrical separation may be used to supply several items of equipment from a single separated source, but **only** for special situations and under effective supervision.	471-12-01

> **No** live part of a separated circuit shall be connected   413-06-03
> (at any point) to another circuit or to earth.

This form of protection is intended for an individual circuit and is aimed at preventing a shock current occurring by contact with an exposed conductive part that might be energised by a fault in the basic insulation of that circuit. The electrical system is completely separated from others (so that there isn't a whole circuit through which shock current could flow). Protection by electrical separation may be applied to the supply of any individual item of equipment by means of a transformer complying with BS 3535 (the secondary of which is not earthed) or a source affording equivalent safety.

Requirements

> Source supplies may supply more than one item of   413-06-05
> equipment provided that:
>
> - the non-earthed equipotential bonding conductor is not connected to a protective conductor, or to an exposed conductive part of any other circuit or to any extraneous conductive part and
>   - all exposed conductive parts of the separated circuit are connected together by an insulated and non-earthed equipotential bonding conductor
>   - all socket outlets are provided with a protective conductor contact (that is connected to the equipotential bonding conductor)
>   - all flexible equipment cables (other than Class II equipment) have a protective conductor for use as an equipotential bonding conductor
>   - exposed conductive parts which are fed by conductors of different polarity (and which are liable to a double fault occurring) are fitted with an associated protective device meeting the requirements of Regulation 413-02-04
> - any exposed conductive part of a separated circuit cannot come in contact with an exposed conductive part of the source.
>
> If the primary circuit of the functional extra-low-voltage   471-14-05
> source is protected by electrical separation, then the
> exposed conductive parts of equipment in that functional
> extra-low-voltage system shall be connected to the
> non-earthed protective conductor of the primary circuit.

A warning notice (warning that protection by electrical separation is being used) shall be fixed in a prominent position adjacent to every point of access to the location concerned.	471-12-01 and 514-13-02

4.9 Protection against both direct and indirect contact

Figure 4.22 Protection against direct and indirect contact

The Regulations (411-01-01) state that one of the following, basic, measures shall be used for protection against both direct contact and indirect contact:

- SELV or
- limitation of discharge of energy.

4.9.1 Protection by SELV

A system which does not use a device such as an autotransformer, potentiometer, semiconductor device etc., to provide electrical separation, shall **not** be deemed to be a SELV system.	411-02-03
Functional extra-low voltage alone shall **not** be used as a protective measure.	411-01-02

SELV (Separated Extra-Low Voltage) is an extra-low-voltage system that is electrically separated from earth and from other systems so that a single fault cannot give rise to the risk of electric shock. PELV (Protective Extra-Low Voltage) is an extra-low-voltage system which is not electrically separated from earth, but which otherwise satisfies all the requirements for SELV.

SELV is a term used to describe the highest voltage level that can be contacted by a person without causing injury. It is usually defined as 60 V d.c.

Requirements

To meet the requirements for protection by SELV, the nominal circuit voltage shall not exceed extra-low voltage.	411-02-01

The SELV source shall be from one of the following: 411-02-02

  (i) a safety isolating transformer complying with BS 3535
 (ii) a motor-generator with windings providing electrical separation equivalent to that of the safety isolating transformer specified in (i) above
(iii) a battery or other form of electrochemical source
 (iv) a source independent of a higher voltage circuit (e.g. an engine-driven generator)
  (v) electronic devices which (even in the case of an internal fault) restrict the voltage at the output terminals so that they do not exceed extra-low voltage.

All live parts of a SELV system shall: 411-02-05

- be electrically separated from that of any other higher voltage system
- not be connected to earth
- not be connected to a live part or a protective conductor from another system.

Circuit conductors for each SELV system shall be physically separated from those of any other system.   411-02-06

Where this proves impracticable, SELV circuit conductors shall be:

- insulated for the highest voltage present
- enclosed in an insulating sheath (that is additional to their basic insulation).

The conductors of systems with a higher voltage than SELV shall be physical separated from the SELV conductors by an earthed metallic screen or an earthed metallic sheath   411-02-06

The system shall not be deemed to be SELV if any exposed conductive part of an extra-low-voltage system is capable of coming into contact with an exposed conductive part of any other system.   411-02-08

SELV circuit conductors that are contained in a multicore cable with other circuits that have different voltages shall   411-02-06 be insulated, individually or collectively, for the highest voltage present in the cable or grouping.

Electrical separation between live parts of a SELV system (including relays, contactors and auxiliary switches) and any other system shall be maintained.   411-02-06

Exposed-conductive parts of a SELV system shall **not** be connected to: 411-02-07

- earth
- an exposed conductive part of another system
- a protective conductor of any system
- an extraneous conductive part.

Note: Except where that electrical equipment is mainly required to be connected to an extraneous conductive part (in which case, measures shall be incorporated so that the parts cannot attain a voltage exceeding extra-low voltage).

If the nominal voltage of a SELV system exceeds 25 V a.c. RMS or 60 V ripple-free d.c., protection against direct contact shall be provided by one or more of the following: 411-02-09

- a barrier (or an enclosure) capable of providing protection to at least 1P2X or IPXXB
- insulation capable of withstanding a type-test voltage of 500 V a.c. RMS for 60 seconds.

SELV socket outlets shall: 411-02-10

- be compatible with the plugs used for other systems in use in the same premises
- not have a protective conductor contact.

Luminaire supporting couplers which have a protective conductor contact shall **not** be installed in a SELV system.   411-02-11

If an extra-low-voltage system complies with the requirements for SELV but needs to be connected to earth (or a live part or a protective conductor of another system) then protection against direct contact shall be provided by either: 471-14-02

- barriers or enclosures with a degree of protection of at least 1P2X or IPXXB, or

- insulation capable of withstanding 500 V a.c. RMS for 60 seconds.

If the primary circuit of the functional extra-low-voltage source is protected by electrical separation, then the exposed conductive parts of equipment in that functional extra-low-voltage system shall be connected to the non-earthed protective conductor of the primary circuit.   471-14-05

4.9.2 Limitation of discharge of energy

 This type of protection may also be used for electric fences supplied from electric fence controllers complying with BS EN 61011 or BS EN 6101 1-1.

Requirement

Protection against both direct and indirect contact shall be deemed to be provided when the equipment includes some means of limiting the amount of current which can pass through the body of a person or livestock to a value lower than that likely to cause danger.   411-04-01

This form of protection shall only be applied to: 471-03-01

- an individual item of current-using equipment (complying with an appropriate British Standard)
- an equipment which can safely limit the current that can flow from the equipment through the body of a person or livestock.

4.10 Protection against fault current

Circuit breakers used as fault current protection devices shall be capable of making any fault current up to and including the prospective fault current.   432-04-01

Circuit protective devices shall break any fault current flowing before such current causes danger due to thermal or mechanical effects that are produced in circuit conductors or associated connections.   434-01-01

Conductors shall be able to carry fault current without overheating.   130-05-01

Fault current protection devices shall be capable of breaking.	432-04-01
Fault currents (under both short-circuit and earth fault conditions) shall be calculated either by enquiry or by measurement.	434-02-01
The breaking capacity rating of each device shall be not less than the prospective short-circuit current or earth fault current at the point at which the device is installed.	434-03-01

 Note: A lower breaking capacity is allowed if another protective device is installed on the supply side.

4.11 Protection against overvoltage

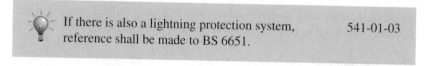

If there is also a lightning protection system, reference shall be made to BS 6651.	541-01-03

Overvoltage is the hazardous condition that occurs when the voltage in a circuit (or part of a circuit) is suddenly raised over its upper limit. An overvoltage incident can be permanent or transient and it is often referred to as a 'voltage spike'. A typical example of a naturally occurring transient overvoltage is by

Figure 4.23 Lightning

lightning – whereas man-made sources are usually electromagnetic induction when switching on (or off) inductive loads (e.g. electric motors or electromagnets). Transient overvoltage might last microseconds and reach hundreds of volts, sometimes thousands of volts, in amplitude.

In accordance with the Regulations, additional protection against overvoltages of atmospheric origin is not necessary for:

- installations that are supplied by low-voltage systems which do not contain overhead lines
- installations that are supplied by low-voltage networks which contain overhead lines and their location is subject to less than 25 thunderstorm days per year
- installations that contain overhead lines and their location is subject to less than 25 thunderstorm days per year

provided that they meet the required minimum equipment impulse to withstand voltages shown in Table 44A (p. 61) of the IEE Wiring Regulations.

Suspended cables with insulated conductors that have earthed metallic coverings are considered to be an 'underground cable'. (443-02-01)

Requirements

Persons, livestock and property shall be protected against overvoltages arising due to atmospheric phenomena and/or switching.	130-06-02
Persons and livestock shall be protected against injury due to a fault between live parts of circuits supplied at different voltages.	130-06-01
Property shall be protected against any harmful effects caused by a fault occurring between live parts of circuits supplied at different voltages.	130-06-01
Installations that are supplied by (or include) low-voltage overhead lines shall incorporate protection against overvoltages of atmospheric origin if the location is subject to more than 25 thunderstorm days per year.	443-02-03
Protection against overvoltages caused by lightening shall be provided in the installation of the building by: • a surge protective device with a protection level not exceeding Category II or • other means providing at least an equivalent attenuation of overvoltages.	443-02-05
Overvoltage protective devices shall be located as close as possible to the origin of the installation.	443-02-06

4.12 Additional requirements

The following are additional requirements for installations and locations where the risk of electric shock is increased by a reduction in body resistance and/or by contact with earth potential.

The supply source to a reduced low-voltage circuit shall be one of the following:	471-15-03
• a double wound isolating transformer complying with BS 3535-2, or	
• a motor-generator whose windings provide isolation equivalent to an isolating transformer, or	
• a source independent of other supplies such as an engine-driven generator.	
The neutral (i.e. star) point of the secondary windings of three-phase transformers and generators (or the mid-point of secondary windings from single-phase transformers and generators) shall be connected to earth.	471-15-04
All exposed-conductive parts of a reduced low-voltage system shall be connected to earth.	471-15-06
The earth fault loop impedance at all points (including socket outlets) shall ensure that the disconnection time does not exceed 5 seconds.	471-15-06
Where a circuit breaker is used, the maximum value of earth fault loop impedance shall conform to Table 471A (page 71) of the IEE Wiring Regulations.	471-15-06
Where a fuse is used, the maximum value of earth fault loop impedance shall allow a disconnection time of less than 5 seconds.	471-15-06
Where an RCD is used, the product of the rated residual operating current (in amperes) and the earth fault loop impedance (in ohms) shall not exceed 50.	471-15-06

4.13 Protective bonding conductors

Equipotential bonding ensures that protective devices will operate and remove dangerous potential differences, before a hazardous shock can be delivered. This is achieved by making sure that all of the installation's earthed metalwork (i.e. exposed conductive parts) is connected to other metalwork (i.e. extraneous

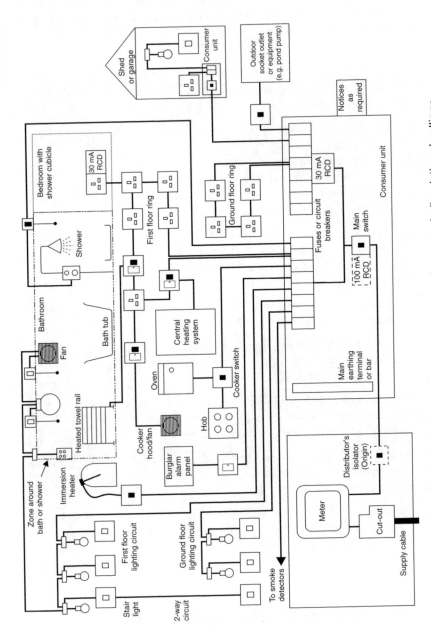

Figure 4.24 Typical fixed installations that might be encountered in new (or upgraded) existing dwellings

conductive parts) via the earth conductor to provide an earth fault current path that ensures dangerous potential differences cannot occur.

Main equipotential bonding conductors connect together the installation earthing system and the metalwork of other services such as gas, electricity and water as close as possible to their point of entry to the building.

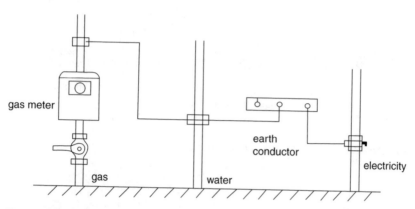

gas meter

earth conductor

electricity

gas   water

Figure 4.25 Main equipotential bonding

Supplementary bonding conductors connect together extraneous conductive parts – that is, metalwork which is not associated with the electrical installation but which may provide a conducting path that could give rise to shock.

4.13.1 Main equipotential bonding conductors

The Regulations require that the main equipotential bonding conductors for every electrical installation is connected to the main earthing terminal of that particular installation and that these shall include the following:

- water service pipes (but see requirements for domestic buildings in Chapter 2)
- gas installation pipes
- other service pipes and ducting
- central heating and air conditioning systems
- exposed metallic structural parts of the building
- the lightning protective system (413-02-02).

Note: Where an installation serves more than one building, the above requirement shall be applied to each building.

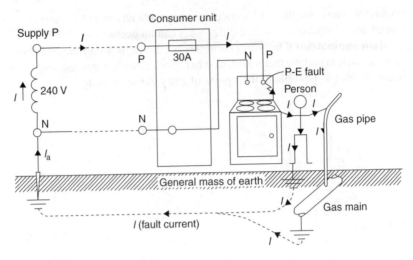

Figure 4.26 Earthed equipotential bonding

Requirements

The cross-sectional area of the main equipotential bonding conductor:	547-02-01

- shall not be less than half the cross-sectional area of the installation's earthing conductor and not less than 6 mm^2
- need not exceed 25 mm^2 if copper.

Where PME (protective multiple earthing) exists, the requirements of Regulation 547-02-01 for the cross-sectional area of a main equipotential bonding conductor shall be selected in accordance with the neutral conductor of the supply and the values shown in Table 54H (p. 120) of the IEE Wiring Regulations.	542-03-01
Simultaneously accessible exposed conductive parts of equipotential bonding conductors shall be connected to the same earthing system individually, in groups or collectively.	413-02-03
Where automatic disconnection of TN, TT and/or IT systems cannot be achieved by using overcurrent protective devices, then either local supplementary equipotential bonding or an RCD shall be used.	413-02-04

The connection of earthing conductors to the earth electrode shall be: 542-03-03

- soundly made
- electrically and mechanically satisfactory
- labelled as per Figure 4.27 (see below)
- suitably protected against corrosion.

The main equipotential bonding connection to any gas, water or other service shall be made: 547-02-02

- as near as practicable to the point of entry of that service into the premises
- to the consumer's hard metal pipework
- before any branch pipework
- within 600 mm of the meter outlet union or at the point of entry to the building if the meter is external.

💡 Equipotential bonding may also be applied to any metallic sheath of a telecommunications cable.

A permanent label (with the words as shown in Figure 4.27) shall be permanently fixed at or near: 514-13-01

- the connection point of every earthing conductor to an earth electrode
- the connection point of every bonding conductor to an extraneous conductive part
- the main earth terminal (when separated from the main switchgear).

Safety Electrical Connection – Do Not Remove

Figure 4.27 Earthing and bonding notice

4.13.2 Supplementary bonding conductors

💡 Supplementary bonding to a fixed appliance **shall** be provided by a supplementary conductor or a conductive part of a permanent and reliable nature.   547-03-04

For installations and locations where there is an increased risk of shock (such as agricultural and horticultural premises and building sites etc.) additional measures may be required, such as reduction of maximum fault clearance time and supplementary equipotential bonding are used.

Where supplementary equipotential bonding is necessary, it shall connect together the exposed conductive parts of equipment including the earthing contacts of socket outlets and extraneous conductive parts. — 413-02-27

Supplementary bonding conductors that connect:

- an exposed conductive part to an extraneous conductive part shall have a conductance (if sheathed or mechanically protected) that is not less than the protective conductor connected to an exposed conductive part — 547-03-02
- two exposed conductive parts shall have a conductance (if sheathed or mechanically protected) not less than the smaller protective conductor that is connected to an exposed conductive part. — 547-03-01

 Note: If mechanical protection is not provided, then the cross-sectional area shall be not less than 4 mm².

- Two extraneous conductive parts shall have a cross-sectional area not less than 2.5 mm² if sheathed or mechanically protected. — 547-03-03

Locations which contain a bath or shower, and where body resistance is lowered as a result of water, are potentially very hazardous environments and it is important to ensure that no dangerous potentials exist between exposed and extraneous conductive parts. For this reason:

local supplementary equipotential bonding shall be provided to connect together the terminals of the protective conductors of each circuit supplying Class I and Class II equipment in zones 1, 2 or 3 with extraneous conductive parts in those zones, such as: — 601-04-01

- metallic pipes supplying services and metallic waste pipes (e.g. water, gas)
- metallic central heating pipes
- air conditioning systems
- accessible metallic structural parts of the building
- metallic baths and shower basins.

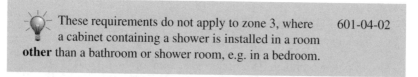

These requirements do not apply to zone 3, where   601-04-02
a cabinet containing a shower is installed in a room
other than a bathroom or shower room, e.g. in a bedroom.

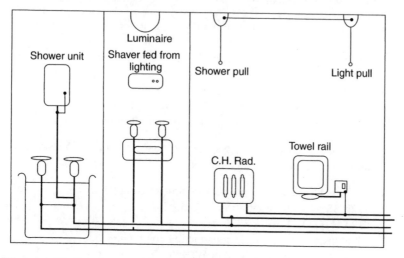

Figure 4.28 Supplementary equipotential bonding

4.14 Protection against impact and vibration

Buried cables, conduits and ducts **shall** be at a   522-06-03
sufficient depth to avoid being damaged.

Care shall be taken when selecting and using fixings and   522-07-01
connections for cables for a wiring system that are fixed to
(or supported by) structure equipment is subject to
vibration.

The Regulations state that *All electrical joints and connections shall meet stipulated requirements concerning conductance, insulation, mechanical strength and
protection* (133-01-04) and requires that:

unless installed in a conduit or duct: 522-06-03

• cables buried in the ground shall either be of an
insulated concentric construction or be protected by
earthed armour or metal sheath

- buried cables shall be marked by cable covers or marking tape.

Cables passing through floor timber joist used for floors and ceilings shall: 522-06-05

- be at least 50 mm from the top, or bottom, of the joist, or
- use an earthed armour or metal sheath as a protective conductor, or
- be of insulated concentric construction, or
- be protected by an earthed, steel conduit enclosure, or
- be protected by some form of mechanical protection.

Cables that are concealed in a wall or partition less than 50 mm from the surface shall either be: 522-06-06

- protected by an earthed metallic conductor complying with BS 5467, BS 6346, BS 6724, BS 7846, BS EN 60702-1 or BS 8436, or
- of an insulated concentric construction complying with BS 4553-1, BS 4553-2 or BS 4553-3, or
- enclosed in earthed conduit, trunking or ducting, or
- mechanically protected against penetration of the cable by nails, screws etc., or
- installed within 150 mm from the top of the wall or partition or
- within 150 mm of an angle formed by two adjoining walls or partitions.

4.15 Protective conductors

A **protective conductor** is a conductor that provides a measure of protection against electric shock and is used to connect together any of the following parts:

- exposed conductive parts
- extraneous conductive parts
- the main earthing terminal
- earth electrode(s)
- the earthed point of the source, or an artificial neutral.

A **circuit protective conductor**, on the other hand, is an arrangement of conductors that join all of the exposed conductive parts together and connect them to the main earthing terminal. There are many types of circuit protective conductor such as:

- a separate conductor
- a conductor included in a sheathed cable with other conductors
- the metal sheath and/or armouring of a cable
- a conducting cable enclosure (such as conduit or trunking)
- exposed conductive parts (such as the conducting cases of equipment).

The main requirements for protective conductors are that:

if a protective conductor is formed by conduit (trunking, ducting or the metal sheath and/or armour of a cable) then the earthing terminal of each accessory, shall be connected by a **separate** protective conductor to an earthing terminal that is incorporated in the associated box or enclosure.	543-02-07
Unless all of the conductors in a ring are protected by a metal covering (or enclosure), the circuit protective conductor of every ring final circuit shall also be run in the form of a ring which has both ends connected to the earthing terminal at the origin of the circuit.	543-02-09

4.15.1 Preservation of electrical continuity of protective conductors

No switching device shall be inserted in a protective conductor except multipole linked switching or plug-in devices.	543-03-04

If electrical earth monitoring is used, then the operating coil shall be connected in the pilot conductor and not in the protective earthing conductor.	543-03-05
Screw or mechanical clamps may be used to ensure that the joints in metallic conduits are mechanically and electrically continuous.	543-03-06
Plain slip or pin-grip sockets shall **not** be used.	

4.15.2 Accessibility of connections

Connections and joints shall be accessible for inspection,   526-04-01
testing and maintenance, unless:

- they are in a compound-filled or encapsulated joint
- the connection is between a cold tail and a heating element
- the joint is made by welding, soldering, brazing or compression tool.

4.15.3 Identification of conductors by colour

For safety, it is important that the following colour conventions are observed.

Neutral or mid-point conductor

 Neutral or mid-point conductors **shall** be coloured blue.   514-04-01

Protective conductor

The protective conductor shall be a bi-colour combination   514-04-02
of green and yellow. Neither colour shall cover more
than 70% of the surface being coloured.

 This combination of colours shall not be used for any other purpose.

Single-core cables that are used as a protective conductor   514-04-02
shall be coloured green-and-yellow.

 Note: Other than PEN conductors, these may not be over-marked at their terminations.

Bare conductors or busbars that are used as protective   514-04-02
conductors shall be identified by equal green and
yellow stripes that are 15 mm to 100 mm wide.

 If adhesive tape is used, it shall be bi-coloured.

PEN conductor

 Note: A PEN conductor is a *conductor that combines the functions of both protective conductor and neutral conductor.*

A PEN conductor shall (when insulated) either be: 514-04-03

- green-and-yellow throughout its length with blue markings at the terminations, or
- blue throughout its length with green-and-yellow markings at the terminations.

Other conductors

All other conductors shall be coloured as shown in accordance with Table 4.4.   514-04-04

 Green shall not be used on its own.   514-04-05

Bare conductors

Bare conductors shall be painted or be identified by a coloured tape, sleeve or disc as per Table 4.4.   514-04-06

Note: For your assistance, a pictorial representation of these colours is shown on the inside back cover of this book.

Table 4.4 Identification of conductors

Function	Alphanumeric	Colour
Protective conductors		Green-and-yellow
Functional earthing conductor		Cream
a.c. power circuit (see note 1)		
Phase of single-phase circuit	L	Brown
Neutral of single- or three-phase circuit	N	Blue
Phase 1 of three-phase a.c. circuit	L1	Brown
Phase 2 of three-phase a.c. circuit	L2	Black
Phase 3 of three-phase a.c. circuit	L3	Grey

(continued)

Table 4.4 (*continued*)

Function	Alphanumeric	Colour
Two-wire unearthed d.c. power circuit		
Positive of two-wire circuit	L+	Brown
Negative of two-wire circuit	L−	Grey
Two-wire earthed d.c. power circuit		
Positive (of negative earthed) circuit	L+	Brown
Negative (of negative earthed) circuit (see note 2)	M	Blue
Positive (of positive earthed) circuit (see note 2)	M	Blue
Negative (of positive earthed) circuit	L−	Grey
Three-wire d.c. power circuit		
Outer positive of two-wire circuit derived from three-wire system	L+	Brown
Outer negative of two-wire circuit derived from three-wire system	L−	Grey
Positive of three-wire circuit	L+	Brown
Midwire of three-wire circuit (see notes 2 & 3)	M	Blue
Negative of three-wire circuit	L−	Grey
Control circuits, ELV and other applications		
Phase conductor	L	Brown, black, red, orange, yellow, violet, grey, white, pink or turquoise
Neutral or midwire (see note 4)	N or M	Blue

Notes:
(1) Power circuits include lighting circuits.
(2) M identifies either the midwire of a three-wire d.c. circuit, or the earthed conductor of a two-wire earthed d.c. circuit.
(3) Only the middle wire of three-wire circuits may be earthed.
(4) An earthed PELV conductor is blue.

4.15.4 Identification of conductors by letters and/or numbers

M shall identify either the midwire of a three-wire d.c. circuit, or the earthed conductor of a two-wire earthed d.c. circuit.	514-06-01

Protective conductor

Conductors coloured green-and-yellow shall not be numbered other than for the purpose of circuit identification.	514-05-02

Numeric

Conductors may be identified by numbers.   514-05-04

The number 0 **shall** indicate the neutral or mid-point conductor.

Omission of identification by colour or marking

Colour or marking is not required for: 514-06-01

- concentric conductors of cables
- metal sheath or armour of cables (when used as a protective conductor)
- bare conductors (where permanent identification is not practicable)
- extraneous conductive parts used as a protective conductor
- exposed conductive parts used as a protective conductor.

Identification of a protective device

Protective devices shall be arranged and identified so   514-08-01
that the circuit protected is easily recognisable.

4.15.5 Warning notice – earthing and bonding connections

A permanent label with the words shown in Figure 4.27   514-13-01
shall be permanently fixed at or near:

- the connection point of every earthing conductor to an earth electrode
- the connection point of every bonding conductor to an extraneous conductive part
- the main earth terminal (when separated from the main switchgear).

4.16 Protective equipment (devices and switches)

The type of protective equipment chosen will depend on the type of protection that it is required to provide (e.g. whether overcurrent, earth fault current, overvoltage or undervoltage).

Requirements

The design of protective equipment shall include protection against the effects of earth fault current.	131-08-01
Switches, circuit breakers (except where linked) and/or fuses shall **not** be inserted in earthed neutral conductors.	131-13-02
Any linked switch or linked circuit breaker that has been inserted in an earthed neutral conductor shall be capable of breaking all of the related phase conductors.	131-13-02

4.17 Accessories and current-using devices

The following is a brief résumé of the earthing requirements for some of the accessories for electrical systems and current-using devices.

4.17.1 Plugs and socket outlets

All low-voltage plug and socket outlets shall conform with the applicable British Standard listed in Table 4.5.	553-01-03

Table 4.5 Plugs and socket outlets for low-voltage circuits

Type of plug and socket outlet	Rating (amperes)	Applicable British Standard
Fused plugs and shuttered socket outlets, 2-pole and earth, for a.c.	13	BS 1363 (fuses to BS 1362)
Plugs, fused or non-fused, and socket outlets, 2-pole and earth	2, 5, 15, 30	BS 546 (fuses, if any, to BS 646)
Plugs, fused or non-fused, and socket outlets, protected-type, 2-pole with earthing contact	5, 15, 30	BS 196
Plugs and socket outlets (industrial type)	16, 32, 63, 125	BS EN 60309-2

4.17.2 Lampholders

Lampholders with an ignitability characteristic 'P' as specified in BS 476 Part 5 (or where separate overcurrent protection is provided) shall not be connected to any circuit where the rated current of the overcurrent protective device exceeds the appropriate value stated in Table 4.6.

553-03-01

 Unless the wiring is enclosed in earthed metal or insulating material.

Table 4.6 Overcurrent protection of lampholders

Type of lampholder			Maximum rating (amperes) of overcurrent protective device protecting the circuit
Bayonet (BS	B15	SBC	6
EN 61184)	B22	BC	16
Edison screw	E14	SES	6
(BS EN 60238)	E27	ES	16
	E40	GES	16

4.17.3 Electrode water heaters and boilers

The current-carrying capacity of the neutral conductor shall be not less than that of the largest phase conductor connected to the equipment.

554-03-05

Electrode boilers and water heaters shall be earthed.

554-03-03

The shell of electrode boilers and water heaters shall be bonded to the metallic sheath and armour (if any) of the incoming supply cable.

554-03-03

For electrode water heaters or boilers that are connected to a three-phase low-voltage supply, the shell of the heater/ boiler shall be connected to the neutral of the supply as well as to the earthing conductor.

554-03-05

When the supply to an electrode water heater or boiler is connected to a single phase (and one electrode is connected to a neutral conductor earthed by the distributor), the shell of the heater/boiler shall be connected to the neutral of the supply as well as to the earthing conductor.

554-03-06

Provided that heater/boiler: 554-03-07

- is not piped to a water supply
- is not in physical contact with any earthed metal and
- the electrodes and the water in contact with the electrodes are insulated so that they cannot be touched while the electrodes are live

then

- a fuse in the phase conductor may be substituted for a circuit breaker and
- the shell of the heater/boiler need not be connected to the supply neutral.

4.17.4 Water heaters having immersed and uninsulated heating elements

Other than the current-carrying parts of a heater/boiler,   554-05-02
all metal parts that are in contact with the water shall be solidly and metallically connected to the metal water pipe that carries the water supply to that heater/ boiler.

The metal water pipe should be connected to the main earthing terminal by the circuit protective conductor.

4.18 Inspection and testing

The following is a résumé of the Regulations requirements for inspections, tests and measurements in respect of earth and of earthing. (A complete list of inspections and test is provided in Chapter 9.)

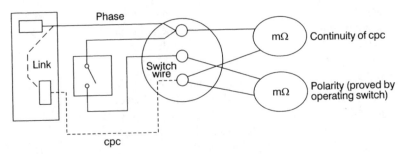

Figure 4.29 Polarity test

Requirements

The insulation resistance between live conductors and   713-04-01
between each live conductor and earth shall be measured
before the installation is connected to the supply.

 Note: The PEN conductor in TN-C systems is considered to be part of the earth.

The insulation resistance (measured with all its final circuits   713-04-02
connected but with current-using equipment disconnected)
shall not be less than that shown in Table 4.7.

Table 4.7

System	Test voltage	Minimum insulation resistance
SELV and PELV	250 V d.c.	0.25 MΩ
LV up to 500 V	500 V d.c.	0.5 MΩ
Over 500 V	1000 V d.c.	1.0 MΩ

Measurements shall be carried out with direct current.   713-04-03

When the circuit includes electronic devices, only a   713-04-04
measurement to protective earth shall be made with
the phase and neutral connected together.

 Precautions may be necessary to avoid damage to electronic devices.

4.18.1 Protection by separation of circuits

The separation of live parts from those of other circuits   713-06-02
and from earth shall be verified by measuring the
insulation resistance in accordance with the requirements
for SELV and electrical separation.

4.18.2 Polarity

A polarity test shall be made to verify that: 713-09-01

- fuses and single-pole control and protective devices are only connected in the phase conductor
- circuits (other than BS EN 60238 E14 and E27 lampholders) which have an earthed neutral conductor centre contact bayonet (and Edison screw lampholders) have their outer or screwed contacts connected to the neutral conductor
- wiring has been correctly connected to socket outlets and similar accessories.

4.18.3 Earth electrode resistance

If the earthing system incorporates an earth electrode   713-10-01
as part of the installation, the electrode resistance to
earth shall be measured.

4.18.4 Earth fault loop impedance

If protective measures are used which require a   713-11-01
knowledge of earth fault loop impedance, the
relevant impedances shall be measured.

4.18.5 Prospective fault current

The prospective short-circuit current and prospective   713-12-01
earth fault current shall be measured at the origin and
at other relevant points in the installation.

5

External influences

Currently Chapter 32 of the Regulations concerning external influences is under development and is, therefore, at too early a stage for adoption as the basis for a national standard. It is anticipated that full details will be included in future revisions of this standard, but in advance of the eventual publication of Chapter 32, a list of external influences and their characteristics have been included as an appendix (i.e. Appendix 5) to the Regulations.

Meanwhile, the BSEN 60721 series of standards on environmental conditions has been available for some time and, in the absence of any 'official' regulations from BSI/IEC, the following notes concerning external influences are offered as guidance. Also included in this chapter are extracts from the current Regulations that have an impact on the environment.

5.1 Environmental factors and influences

The actual environment to which equipment is likely to be exposed is normally complex and comprises a number of environmental conditions. When defining the conditions for a certain application it is, therefore, necessary to consider all environmental influences which may be as a result of:

- conditions from the surrounding medium
- conditions caused by the structure in which the equipment is situated or attached
- influences from external sources or activities.

5.1.1 Combined environmental factors

Equipment may, of course, be simultaneously exposed to a large number of environmental factors and corresponding parameters. Some of the parameters are statistically dependent (e.g. low air velocity and low temperature, sun radiation and high temperature). Other parameters are statistically independent (e.g. vibration and temperature). The effect of a combination of environmental factors is, therefore, extremely important and has to be considered during manufacture and operation.

5.1.2 Sequences of environmental factors

Certain effects of exposing electrical equipment and electrical installations to environmental conditions are a direct result of two or more factors, or parameters, happening either simultaneously or after each other (e.g. thermal shock caused when exposing equipment to a high temperature immediately after exposing it to a low temperature). These possibilities must always be taken into account when designing and installing electrical equipment.

5.1.3 Environmental application

Although the conditions affecting electrical equipment mainly consist of the environment (ambient and created), consideration must also be given to where the equipment will be operating from and how it will be used.

For simplicity this can be broken down into two basic categories:

Conditions   The environmental conditions that have been identified as having an effect on equipment (Table 5.1).

Situations   The main uses of electronic equipment (Table 5.2).

5.1.4 Environmental conditions

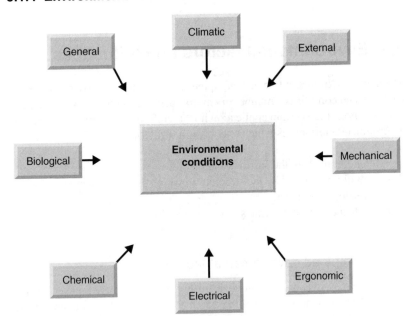

Figure 5.1 Environmental conditions

There are eight basic conditions that have a direct effect on electrical equipment and electrical installations (see Table 5.1).

Table 5.1 Basic conditions

Climatic	Externally generated influences
• ambient temperature • solar radiation • condensation • relative atmospheric humidity • atmospheric pressure • wind • precipitation (i.e. rain, snow and hail) • altitude	• temperature • precipitation (e.g. water spray) • pressure changes (e.g. tunnels) • air movement • dust
Mechanical	**Ergonomic aspects**
• vibration • shock (sinusoidal and random)	• protecting the health of the engineer and end user • the comfort of the operator and end user • achieving maximum task effectiveness
Electrical	**Chemical**
• electromagnetic environment (EMC and EMI) • susceptibility and generation • transients (spikes and surges) • power supplies • earthing and bonding	• pollution and contamination • dangerous substances • corrosion • resistance to solvents
Biological	**General**
• animals • humans (vandalism) • vegetation	• safety • reliability • maintainability • components • waste • earthquakes • flammability and fire hazardous areas • design of equipment

5.1.5 Equipment situations

Obviously not all equipment will be fully operational, all of the time, and so various equipment 'situations' also have to be considered.

Table 5.2 Environmental conditions

Operational	Storage	Transportation
• when installed and operational • when installed and not in use	• when in storage	• when being transported

As a reference on the whole topic of environmental requirements, the following book (by the same author no less!) is recommended. This book has become known as the definitive reference for designers and manufacturers of electrical and electromechanical equipment worldwide.

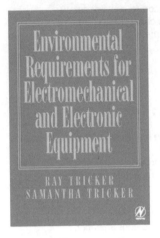

Environmental requirements for the electromechanical and electronic equipment has been written in the form of a reference book, and contains background guidance concerning the environment and environmental requirements. It also provides a case study of typical requirements, values and ranges currently being requested from industry in today's contracts. Test specifications aimed at proving conformance to these requirements are also listed and the most used national, European and international standards and specifications (e.g. BS, EN, CEN, CENELEC, IEC and ISO) are described in relative detail.

ISBN: 0-7506-3902-4

Figure 5.2

 Note: See http://books.elsevier.com/engineering/?isbn=0750639024 for more details of this publication.

5.1.6 Requirements from the Regulations – General

All electrical equipment **shall** be capable of withstanding the stresses, environmental conditions and the characteristics of its location, or be provided with additional protection.	132-01-07

A fault current protective device shall be placed (on the load side) at the point where the current-carrying capacity of the installation's conductors is likely to be lessened owing to inherent environmental conditions.	473-02-02
Conductors and cables shall be capable of withstanding all foreseen electromechanical forces (including fault current) during service.	521-03-01
Equipment shall be capable of withstanding all mechanical, chemical, electrical, thermal influences and stresses normally encountered during service.	412-02-01

Note: Paint, varnish, lacquer or similar products are generally not considered to provide adequate insulation for protection against direct contact in normal service.

Overload protection devices shall be used at all points where due to environmental conditions, the current-carrying capacity of the installation's conductors is reduced.	473-01-01
The insulating enclosure for Class II equipment shall be capable of resisting mechanical, electrical and thermal stresses likely to be encountered.	413-03-05

 Note: A coating of paint, varnish or similar product is generally not considered to comply with these requirements.

5.2 Ambient temperature

Temperature, humidity, rainfall, wind velocity and the duration of sunshine all affect the climate of an area. These elements are in turn the result of the interaction of a number of determining causes, such as latitude, altitude, wind direction, distance from the sea, relief, and vegetation. Elements and their determining causes are similarly interrelated which also contribute to temperature changes; for example, the length of day is a factor which helps to determine temperature. However, the duration of actual sunshine is an element with far-reaching effects on plant and animal life.

Of all the elements that have an effect on man and equipment none is more vital to living organisms than temperature. Temperature has a large influence on where humans live and in the areas where they work. Protective housing or artificial heat sources may overcome low temperatures (and high altitudes); similarly, cooling devices and reflective coatings protect equipment from high temperatures. Temperature is therefore a particularly important aspect of the environment and its accurate measurement and definition requires careful consideration.

The ambient temperature at any given time is the temperature of the air measured under standardised conditions and with certain recognised precautions against errors introduced by radiation from the sun or other heated body. Temperature figures with respect to climate are generally 'shade' temperatures (i.e. the temperature of the air measured with due precautions taken to exclude the influence of the direct rays of the sun) and it is usual for the temperature to be much higher in the direct sunshine. Many mountain areas have air temperatures in the region of zero in winter but the presence of bright sunshine will produce a feeling of warmth and permits the wearing of light clothing.

Figure 5.3 Enjoying the environment

Seasonal fluctuations in temperature do not pass below ground deeper than 60–80 ft. Below that depth, borings and mine-shafts show that the temperature increases (downwards) depending on the geographical position, location and depth.

On average, however, a rise of about 1°F may be taken for each 64 ft of descent. Assuming that this rate of increase is maintained, it stands to reason that

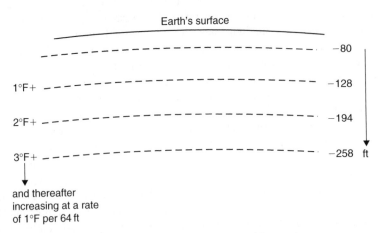

Figure 5.4 Temperature changes below the earth's crust

the interior of the earth must be excessively hot and, therefore, it must warm the surface to some extent. It is not possible to determine the precise influence of this temperature increase but it has an affect on tunnels at a depth greater than 80 ft. As the heat from the core is virtually negligible on the surface of the earth (compared with that of the sun), it has not been considered in this book and the only source of heat that has been taken into account is the sun.

The difference between summer and winter temperatures for any locality is known as the '*annual range of temperature*' or the '*absolute range of temperature*' of that particular locality and it is the difference between the highest and lowest temperatures ever experienced at the place in question. The maximum and minimum temperatures are obviously not the same every year and, should their average over a series of years be taken, it would be known as the '*mean annual extreme range*'.

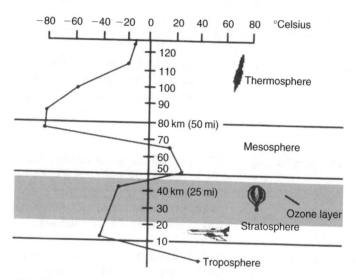

Figure 5.5 Temperature changes above the earth's surface

Although air near the surface (especially at night) may be cooler than the air just above it there is, generally speaking, a gradual falling off of temperature from the ground level up. Over thousands of feet this cooling averages 1°F per 300 ft and thus at approximately 25–55,000 ft (five to ten miles) above the ground the temperature will be down to 55–60°F below zero. Above this the air temperature ceases to fall off regularly, in fact it may even rise for a bit. Usually, however, it remains fairly constant and because of this it is sometimes referred to as the isothermal layer – but to meteorologists and airmen it is known as the stratosphere. The lower layers of the atmosphere (i.e. where temperature falls off with height) is known as the troposphere and the boundary layer between the two is called the tropopause.

5.2.1 Electrical installations

When equipment has been installed without any protection it can be expected to be exposed to more extreme air temperatures and more severe combinations of air temperatures and relative humidities than a similar piece of equipment that has been installed or housed in a temperature-controlled environment.

In addition to open-air temperature, temperature stresses on equipment depend on a number of other environmental parameters (e.g. solar radiation, heating from adjacent equipment and air velocity etc.), and these must be taken into account when designing, manufacturing and installing equipment.

The performance of equipment is also influenced and limited by the internal temperature of the equipment. Internal temperatures, in turn, depend on the external ambient conditions and the heat generated within the device itself. Indeed, whenever a temperature gradient exists within a system formed by a device and its surrounding environment, a process of heat transfer will follow.

Thus, in any generic (or system-specific) specification or standard relating to ambient temperature it is necessary to consider the following.

Operating temperature range	The specified operating temperature for the equipment which must always be the lowest and the highest ambient temperature expected to be experienced by the equipment during its normal operation.
Storage temperature range	The specified storage temperature that is always the lowest and the highest ambient temperature that the equipment is expected to experience (with the power turned off) during storage or from exposure to climatic extremes.

 Note: The equipment is not normally expected to be capable of operating at these extreme temperatures, merely survive them without damage.

5.2.2 Typical requirements – ambient temperature

The following are the most common environmental requirements concerning ambient temperature.

Table 5.3 Typical requirements – ambient temperature

Temperature ranges	Equipment will need to be designed and manufactured to meet the full performance specification requirement for the selected temperature category.
Temperature increases	The design of equipment should always take into account temperature increases within cubicles and equipment cases so as to ensure that the components do not exceed their specified temperature ranges.

(continued)

Table 5.3 (*continued*)

Temperature stresses	In addition to open-air temperature, temperature stresses on equipment caused by other environmental parameters (e.g. solar radiation, air velocity and heating from adjacent equipment) will need to be considered.
Operational requirements	• The specified operating temperatures should be the lowest and the highest ambient temperatures expected to be experienced by equipment during normal operation. • When equipment is turned on it should be expected to operate within the temperature ranges stipulated and be fully operational within a specified time after initial turn on – unless otherwise specified. • The permissible limit temperatures of the operating equipment are not allowed to be exceeded as a result of the temperature rise occurring in operation (including temporary acceleration).
Storage	The specified storage temperatures are normally the lowest and highest ambient temperatures that the sample is expected to experience (with the power turned off) during storage or exposure to climatic extremes. The equipment is not expected to be capable of operating at these extreme temperatures, but to survive them without damage.
Peripheral units	For peripheral units (measuring transducers etc.) or situations where equipment is in a decentralised configuration, ambient temperature ranges are frequently exceeded. In these cases the actual temperature occurring at the location of the equipment concerned needs to be considered during the design and installation.
Installation	When equipment is installed in a controlled climatic environment, provided the equipment is not required to operate outside of those conditions, the temperature range can normally be agreed between purchaser and supplier.

5.2.3 Requirements from the Regulations – ambient temperature

People

Persons, fixed equipment and fixed materials that are adjacent to electrical equipment shall be protected against:

• heat and/or thermal radiation that is emitted by that electrical equipment	130-03-02
• the harmful effects of: – heat developed (and burns caused) by electrical equipment thermal radiation.	421-01-01

Thermal effects

All installations shall comply with the requirements for safety protection in respect of thermal effects.	421-01 01

 Note: 'Installation' in this context is taken to mean either as a whole or in its several parts.

During normal operation, the risk of a person (or livestock) being burnt from electrical equipment shall be minimised.	130-03-01
Electrical equipment shall not become a fire hazard to adjacent materials.	130-03-02 and 421-01-01
Electrical installations shall be organised so that the risk of igniting flammable materials through high temperatures or electric arc is reduced.	130-03-01
In agricultural and horticultural premises:	
• heating appliances shall be fixed so as to minimise the risks of burns to livestock and of fire from combustible material	605-10-01
• radiant heaters shall be fixed not less than 0.5 m from livestock and from combustible material.	605-10-02
Persons, fixed equipment and fixed materials that are adjacent to electrical equipment shall be protected against:	
• heat and/or thermal radiation emitted by that electrical equipment	130-03-02
• the harmful effects of thermal radiation.	421-01-01
The heat generated from fixed electrical equipment shall not damage (or cause danger to) adjacent fixed material.	422-01-01
The thermal rating of generating set conductors shall not be exceeded.	551-05-02
All installations shall comply with the requirements for safety protection in respect of thermal effects.	421-01 01

Earthing arrangements

 Earthing arrangements **shall** be such that they are sufficiently robust (or have additional mechanical protection) to external influences.   542-01-07

Note: This particularly applies with respect to thermal, thermomechanical and electromechanical stresses.

Fixed electrical equipment

 Heat generated from fixed electrical equipment **shall not** damage (or cause danger to) adjacent materials.   422-01-01

All fixed electrical equipment with a surface temperature that is liable to cause a fire or damage adjacent materials shall either be: 422-01-02

- mounted on a support (or placed within an enclosure) that can withstand such temperatures as may be generated or
- screened by material which can withstand the heat emitted by the electrical equipment or
- mounted so that the heat can be safely dissipated.

Fixed equipment that is liable to emit an arc or high temperature particles shall be: 422-01-03

- totally enclosed in arc-resistant material, or
- screened by arc-resistant material from materials that could be effected by emissions or
- mounted so as to allow safe extinction of the emissions at a sufficient distance from material that could be effected by emissions.

Fixed equipment enclosures shall be constructed: 422-01-07

- using heat- and fire-resistant materials according to a relevant product standard or
- (where no product standard exists) be capable of withstanding the highest temperature likely to be produced by the electrical equipment in normal use.

Accessible parts of fixed electrical equipment within arm's reach shall not attain a temperature in excess of the appropriate limit stated in Table 5.4, unless guarded so as to prevent accidental contact.   423-01-01

Table 5.4 The temperature limit under normal load conditions for an accessible part of equipment within arm's reach

Part	Material of accessible surface	Maximum temperature (°C)
A hand-held means of operation	Metallic	55
	Non-metallic	65
A part intended to be touched but not hand-held	Metallic	70
	Non-metallic	80
A part which need not be touched for normal operation	Metallic	80
	Non-metallic	90

Arc-resistant material

Fixed equipment that is liable to emit an arc or high temperature particles shall be: 422-01-03

- totally enclosed in arc-resistant material, or
- screened by arc-resistant material from materials that could be effected by emissions, or
- mounted so as to allow safe extinction of the emissions at a sufficient distance from material that could be effected by emissions.

To provide mechanical stability, arc-resistant material used for this protective measure shall be:

- non-ignitable
- of low thermal conductivity
- of adequate thickness.

Cables and wiring systems

Wiring systems **shall** be capable of withstanding the highest and lowest local ambient temperature likely to be encountered – or shall be provided with additional insulation suitable for those temperatures.   522-01-01 522-02-02

Cables that are run in thermally insulated spaces shall not be covered by the thermal insulation.   523-04-01

The current-carrying capacity of cables that are installed in thermally insulated walls or above thermally insulated ceilings shall be as indicated in Appendix 4 to the Regulations. 523-04-01

Parts of cable or flexible cord within accessories (appliances or luminaire) shall be capable of withstanding the highest and lowest local ambient temperature likely to be encountered – or shall be provided with additional insulation suitable for those temperatures. 522-02-02

Components

Components (including cables and wiring enclosures) shall be installed in accordance with the temperature limits set by the relevant product specification or by the manufacturer. 522-01-02

Forced air heating systems

Electric heating elements of forced air heating systems (other than those of central-storage heaters) shall: 424-01-01

- not be capable of being activated until the prescribed air flow has been established
- deactivate when the air flow is reduced or stopped.

Forced air heating systems shall have two, independent, temperature-limiting devices. 424-01-01

The frames and enclosures of electric heating elements shall be of non-ignitable material. 424-01-02

Electric appliances producing hot water or steam shall be protected against overheating. 424-02-01

Electrical connections

The method of making electrical connections will depend on the: 526-02-01

- conductor's material and insulation
- effectiveness of the insulation of the conductors
- number and shape of the conductor's wires

- number of conductors being connected together
- cross-sectional area of the conductor
- temperature of the terminals during normal service
- type of locking arrangement for situations subject to vibration or thermal cycling.

Motors

Unsupervised motors which are automatically or remotely controlled shall be protected against excessive temperature by a protective device with manual reset.	482-02-11
Motors with star-delta starting shall be protected against excessive temperature in both the star and delta connections.	482-02-11

Enclosures

Surface temperatures of equipment enclosures (such as heaters and resistors) shall be less than: 482-02-18

- 90°C under normal conditions, and
- 115°C under fault conditions.

Switchgear

Switchgear (other than a thermostat or a thermal cut-out) not actually built into the sauna heater: 603-08-01

- shall be installed outside the hot air sauna
- shall only have equipment shown in Table 5.5 installed   603-06-02
  in each of the four temperature zones.

Table 5.5 Permissible equipment in hot air saunas

Temperature zone	Equipment
Zone A	• sauna heaters and equipment directly associated with zone A
Zone B	• no special requirement (concerning heat resistance of equipment)
Zone C	• equipment suitable for an ambient temperature of 125°C
Zone D	• luminaries (and their associated wiring) suitable for an ambient temperature of 125°C
	• control devices for the sauna heater (and their associated wiring) suitable for an ambient temperature of 125°C

5.3 Solar radiation

Of all the factors that control the weather, the sun is by far the most powerful and practically everything that occurs on the earth is controlled, directly or indirectly, by it. The sun affects the places humans inhabit, in the kind of homes that are built, the work that is done, and the equipment that is used.

Less that one-millionth of the energy emitted from the sun's surface travels the ninety-odd million miles to reach this planet. The sun's energy crosses those miles in the form of short electromagnetic radio waves, identical in nature to those used in broadcasting, which pass through the atmosphere and are absorbed by the earth's surface. These waves warm the earth's surface and are then re-radiated back to space. The wavelength of the energy emitted by the earth is very much longer than that emitted by the sun (because the earth is much cooler than the sun) and these longer waves are not able to pass through the atmosphere as freely as short waves. A large proportion of the energy emitted by the earth is absorbed by the water vapour and water droplets in the lower atmosphere which in turn is re-radiated back to earth. Thus the earth plays the part of a receiving station absorbing short electromagnetic waves and converting them into longer electromagnetic waves, while the atmosphere acts as a trap containing most of the longer electromagnetic waves before they are lost to space.

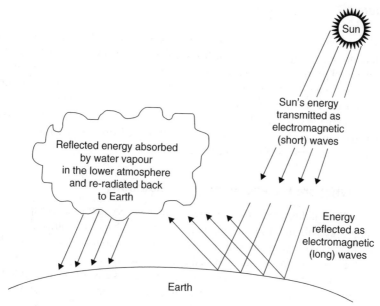

Figure 5.6 Solar radiation – energy

Radiation from the sun consists of rays of three differing wavelengths: heat rays, actinic rays and light rays. Heat rays and actinic rays are intercepted by solid bodies and produce peculiar effects in varying degrees according to the nature of the surface on which they fall. The light rays are responsible for daylight and both light rays and actinic rays are necessary for the life processes of plants. The heat rays' most important aspect is temperature and the amount of sunshine (and therefore the temperature) will depend on latitude and the length of day.

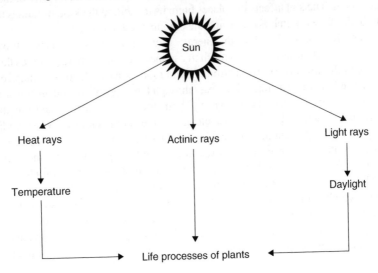

Figure 5.7 Sun's radiation

Radiant energy can be reflected from solid surfaces and intensified by that reflection. For example, reflection from walls is frequently used for ripening peaches and pears. Reflection from bare ground can also assist in the ripening of melons and other creeping plants, while reflection from water surfaces enhances the 'climatic reputation' of waterside resorts. However, radiant energy can also cause damage to equipment as heat rays can warm the material or its environment to dangerous levels and photochemical degradation of materials can be caused by the ultraviolet content of solar radiation.

5.3.1 What are the effects of solar radiation?

On cloudless nights when atmospheric radiation is very low, objects exposed to the night sky will attain surface temperatures below that of the surrounding air temperature. For example (and by experiment), a horizontal disc thermally insulated from the ground and exposed to the night sky during a clear night can attain a temperature of $-14°C$ when the air temperature is $0°C$ and the relative humidity is close to 100% and these values are of assistance when determining the 'under temperature' of components.

The sun's electromagnetic radiation consists of a broad spectrum of light ranging from ultraviolet to near infrared. Owing to the distance of the sun from the earth, solar radiation appears on the earth's surface as a parallel beam and the highest (maximum) level of radiation occurs at noon, on a cloudless day, at a surface perpendicular to the sun.

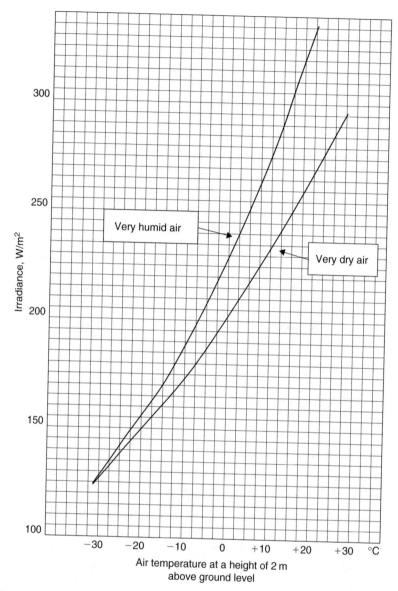

Figure 5.8 Lowest values of atmospheric radiation during clear nights (reproduced from BS 7527 Section 2.4; 1991 by kind permission of the BSI)

Most of the sun's energy reaches the surface of the earth in the 0.3 to 0.4 μm range and the density of the solar radiated power (or irradiance – expressed in watts per square metre) is dependent on the content of aerosol particles, ozone and water vapour in the air. The actual amount of irradiance will vary considerably with geographical latitude and type of climate (i.e. temperature, humidity, air velocity etc.).

Having said that, an object subjected to solar radiation will obtain a temperature depending on the surrounding ambient air temperature, the intensity of radiation, the air velocity, the incidence angle of the radiation on the object, the duration of exposure, the thermal properties of the object itself (e.g. surface reflectance, size, shape, thermal conductance and specific heat), together with other factors such as wind and heat conduction to mountings and surface absorbency etc.

5.3.2 Photochemical degradation of material

One of the biggest problems caused by solar radiation is the photochemical degradation of most organic materials which in turn causes the elasticity and plasticity of certain rubber compounds and plastic materials to be affected and can, in exceptional cases, make optical glass opaque. Although solar radiation can bleach out colours in paints, textiles, paper, etc. (a major consideration when trying to read the colour coding of components), by far the most important effect is the heating of materials.

The combined effect of solar radiation, atmospheric gases, temperature and humidity changes etc., are often termed 'weathering' and result in the 'ageing' and ultimate destruction of most organic materials, e.g. plastics, rubbers, paints, timber etc. Typical defects caused by weathering are:

- rapid deterioration and breakdown of paints
- cracking and disintegration of cable sheathing
- fading of pigments
- bleach out colours in paints, textiles and paper.

5.3.3 Effects of irradiance

To guard against the effects of irradiance, the following guidelines should be considered when locating electrical equipment:

- the sun should be allowed to shine only on the smallest possible casing surfaces
- windows should be avoided on the sunny side of rooms housing electronic equipment
- heat-sensitive parts must be protected by heat shields made, for instance, of polished stainless steel or aluminium plate
- air conditioning plant and cooling fans (when used) in rooms housing electronic equipment should be efficient and reliable
- convection flow should sweep across the largest possible surfaces of materials with good conduction properties.

5.3.4 Heating effects

As previously shown, probably the most important effect of solar radiation heating is mainly caused by the short-term, high-intensity radiation around noon on cloudless days. Typical peak values of irradiance are shown in Table 5.6 below.

Table 5.6 Typical peak values of irradiance in W/m² from a cloudless sky

Area	Large cities	Flat land	Mountainous areas
Subtropical climates and deserts	700	750	1180
Other areas	1050	1120	1180

As equipment (if fully exposed to solar radiation) in an ambient temperature (e.g. 35–40°C) can attain temperatures in excess of 60°C, one has to consider an equipment's outside surface. To a major extent the surface reflectance of an object affects its temperature rise from solar heating and changing the finish from a dark colour to a gloss white can cause a considerable reduction in temperature. On the downside, a pristine finish designed specifically to reduce temperature can be expected to deteriorate in time and result in an increase in temperature.

Another problem found in most of today's materials is that they are also selective reflectors (i.e. their spectral reflectance factor changes with wavelength). For example, paints are, in general, poor infrared reflectors although they may be very efficient as a visible warning. Care should, therefore, be taken when selecting materials and finishes for equipment casings.

5.3.5 Typical requirements – solar radiation

Table 5.7 shows the most common environmental requirements concerning solar radiation.

Table 5.7 Typical requirements – solar radiation

Survivability	Equipment that is exposed to the effect of solar radiation should remain unaffected.
Exposure	The sun should be allowed to shine only on the smallest possible casing surfaces and the convection flow should sweep across the largest possible surfaces of materials with good conduction properties.
Windows	Windows should be avoided on the sunny side of rooms housing electronic equipment.
Heat shields	Heat-sensitive parts shall be protected by heat shields made of (for instance) polished stainless steel or aluminium plate.
Air conditioning	Air conditioning plant and cooling fans (when used) in rooms housing electronic equipment should be efficient and reliable.

5.3.6 Requirements from the Regulations – solar radiation

Wiring systems **shall** withstand solar radiation and/or ultraviolet radiation.   522-11-01

Note: Special precautions may need to be taken for equipment subject to ionising radiation.

> Wiring systems shall be protected for the effects of heat   522-02-01
> from external sources (including solar gain) by:
>
> - shielding
> - placing sufficiently far from the source of heat
> - selecting a system with due regard for the additional temperature rise which may occur
> - local reinforcement or substitution of insulating material.

5.4 Humidity

The atmosphere is normally described as '*a shallow skin or envelope of gases surrounding the surface of the earth which is made up of nitrogen, oxygen and a number of other gases which are present in very small quantities*'. While the ratio of these components shows no appreciable variation either with latitude or with altitude, the water vapour content of the atmosphere is subject to extremely wide fluctuations. The amount of water present in the air is referred to as humidity.

When air and water come into contact, they will exchange particles with each other (i.e. the air particles will pass into the water and water particles will pass into the air in the form of vapour). There is always a certain amount of water vapour present in the air and a certain amount of air present in water and there is always a constant movement between the two mediums. If there is only a small amount of water vapour present in the air, then more particles of water will pass from the water into the air than from the air into the water and so the water will gradually dry up or evaporate. Conversely, if the amount of water vapour in the air is large, then as many particles of vapour will pass from the air into the water, as from the water into the air – and the water will not evaporate. In such a case the air is said to be '*saturated*' – or, to put it another way, it holds as much water vapour as it can contain.

Water vapour is collected in the air above the oceans and is carried by the wind towards the land masses. The amount of water vapour in the air varies greatly depending on the place and season but in general, evaporation is most rapid at high temperature and slower at lower temperatures. We may, therefore, expect to find the greatest amount of water vapour over the oceans near the Equator and the smallest amount over the land in a cold region such as north-eastern Asia in winter. However, even when the surface is covered with snow and ice, water evaporation may take place and occasionally during a long frost, the snow will gradually disappear without melting.

Except for the water vapour being present, the composition of the atmosphere near the surface of the earth up to a height of some 2000 ft is practically uniform throughout the globe. However, at greater altitudes (i.e. above the atmospheric boundary in the tropopause), there is practically no water vapour, or water, in any form.

5.4.1 What is humidity?

Temperature and the relative humidity of air (in varying combinations) are climatic factors which act upon electrical equipment and installations during storage, transportation or operation. Humidity, and the electrolytic damage resulting from moisture, mostly affects plug points, soldered joints (in particular dry joints), bare conductors, relay contacts and switches. Humidity also promotes metal corrosion (see Section 5.7 Pollutants and Contaminants page 237) owing to its electrical conductivity.

In many cases, however, environmental influences such as mechanical and thermal stresses are merely the forerunner of the impending destruction of components by humidity – especially as the majority of electronic component failures are caused by water!

 Note: 'Humidity' (in the context of this book) has been taken to cover relative humidity, absolute humidity, condensation, adsorption, absorption and diffusion and details of these 'subsets' are provided in the following paragraphs.

5.4.2 Relative and absolute humidity and their effect on equipment performance

The performance of virtually all electrical equipment is influenced and limited by its internal temperature which, in turn, is dependent on the external ambient conditions and on the heat generated within the device itself. Fortunately, most electrical and electronic components (especially resistors) will normally remain dry when under load owing to the amount of internal/external heat dissipation. Indeed, many components either have to be de-rated in order too improve their reliability or, for reasons of circuit function, only energised intermittently.

Externally mounted equipment

Equipment and components that are mounted in external cabinets run the risk of coming into contact with water or water vapour (e.g. drifting snow, fog, dew, rain, spray water or water from hoses) and the equipment must, therefore, be adequately protected from such humidity in order to prevent the ingress of vapour into the system within the casing.

Housed equipment

In most locations (e.g. cabinets, equipment rooms, workshops and laboratories) although temperatures above 30°C may often occur, they are normally combined with a lower relative humidity than that found in the open air. In other rooms (e.g. offices), however, where several heat sources are present, temperatures and relative humidities can differ dramatically across the room.

The sun also plays its part because in certain circumstances (such as when equipment is placed in an unventilated enclosure) the intense heat caused through solar radiation can generate relative humidities in excess of 95% when combined with:

- high relative humidity caused by the release of moisture from hygroscopic materials
- the breathing and perspiration of human beings
- open vessels containing water or other sources of moisture.

5.4.3 Condensation

Condensation occurs when the surface temperature of an item is lower than that of the dew point (i.e. the temperature with a relative humidity of 100% at which condensation occurs) a direct relationship which can change electrical characteristics (e.g. decrease surface resistance, increase loss angle) therefore, exists between the absolute point at which atmospheric vapour condenses into droplets (i.e. the dew point), absolute humidity and vapour pressure.

For example. If a piece of equipment has a low thermal time constant, condensation (normally found on the surface of the equipment) will occur only if the temperature of the air increases very rapidly, or if the relative humidity is very close to 100%. Sudden changes in temperature may cause water to condense on parts of equipment and leakage currents can occur.

5.4.4 Adsorption

Adsorption is the amount of humidity that may adhere to the surface of a material and depends on the type of material, the surface structure and the vapour pressure. This layer of water (no matter how small) can cause electrical short circuits and material distortion etc.

5.4.5 Absorption

The quantity of water that can be absorbed by a material depends largely on the water content of the ambient air, and the speed of penetration of the water molecules generally increases with the temperature.

5.4.6 Diffusion

Water vapour can penetrate encapsulations of organic material (e.g. into a capacitor or semiconductor) by way of the sealing compound and into the casing. This factor is frequently overlooked and can become a problem, especially as the moisture absorbed by an insulating material can cause a variation in a number of electrical characteristics (e.g. reduced dielectric strength, reduced insulation resistance, increased loss angle, increased capacitance etc.).

5.4.7 Protection

The effects of humidity mainly depend on temperature, temperature changes and impurities in the air. As shown in Table 5.8, there are three basic methods of protecting the active parts of equipment and components from humidity.

Table 5.8 Protective methods – humidity

Heating the surrounding air so that the relative humidity cannot reach high values	This method normally requires a separate heat source which (especially in the case of equipment mounted in external cabinets) usually means having to have a separate power supply. This method is disadvantaged by the reliability of the circuit being dependent on the efficiency of the heating.
Hermetically sealing components or assemblies using hydroscopic materials	This is an extremely difficult process as the smallest crack or split can allow moisture to penetrate the component, particularly in the area of connecting wire entry points. Metal, glass and ceramic encapsulation do nevertheless produce some very satisfactory results.
Ventilation and the use of moisture-absorbing materials	Most water-retaining materials and paint etc. are suitable for the temporary absorption of excessive high air humidity in the casing, which, because of the risk of pollution and dust penetration, cannot be fully ventilated (i.e. air exchange with the outside temperature is not possible).

5.4.8 Typical requirements – humidity

Table 5.9 shows the most common environmental requirements concerning humidity.

Table 5.9 Typical requirements – humidity

Equipment interoperability	Equipment that is operated adjacent to the sea shore (and, therefore, subject to extreme humidity) must be able to function equally well as the same equipment housed in the low humidity of (for example) the desert.
External humidity levels	Equipment should be designed and manufactured to meet the following external humidity levels (limit values), over the complete range of ambient temperature values anticipated. (See Table 5.10)
Condensation	Operationally caused infrequent and slight moisture condensation should not lead to malfunction or failure of the equipment.
Indoor installations	In all indoor installations, provision must be made for limiting the humidity of the ambient air to a maximum of 75% at $-5°C$.
Equipment in cubicles and cases	The design of equipment should take into account temperature rises within cubicles and equipment cases in order to ensure that the components do not exceed their specified temperature ranges.

(continued)

Table 5.9 (*continued*)

Peripheral units	For peripheral units (e.g. measuring transducers etc.) or equipment employed in a decentralised configuration (i.e. where ambient temperature ranges are exceeded) the actual temperature occurring at the location of the equipment concerned should be utilised when designing equipment.
Product configuration	All proposed and date equipment, components or other articles must be tested in their production configuration without the use of any additional external devices that have been added expressly for the purpose of passing humidity testing.

Table 5.10 Humidity – external humidity levels

Duration	Limit value
Yearly average	75% relative humidity
On 30 days of the year, continuously	95% relative humidity
On the other days, occasionally	100% relative humidity
On the other days, occasionally	30 g/m^3 occurring in tunnels

Note: Meteorological measurements made over many years have shown that, within Europe, a relative humidity greater than 95% combined with a temperature above 30°C does not occur over long periods in free air conditions.

5.4.9 Requirements from the Regulations – humidity

 Wiring systems **shall** be capable of withstanding   522-03-01
condensation and/or ingress of water during installation,
use and maintenance.

Drainage points shall be made available at all points where   522-03-02
water could collect (or condensation could occur) in a
wiring system.

Wiring system subjected to waves shall be protected.   522-03-03

5.5 Air pressure and altitude

Air pressure frequently referred to as atmospheric pressure is '*the force exerted on a surface of a unit area caused by the earth's gravitational attraction on the air vertically above that area*'. Air pressure varies with altitude (i.e. elevation above mean sea level) and location. For instance at the equator where the trade winds of both hemispheres converge, there is a low pressure zone (known as the ITCZ or International Conveyance Zone), which is characterised by high humidity.

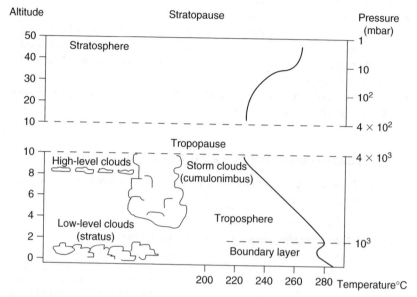

Figure 5.9 Atmospheric structure

It is not widely appreciated that the location of equipment, especially with respect to its altitude above sea level, can affect the working of that equipment. But it is not just the height above sea level that has the most effect! Even air pressure variations at ground level have to be considered.

Table 5.11 Air pressure and altitude (reproduced with permission from ISO 2533)

Air pressure (kPa)	(mbar)	Approximate altitude above sea level (m)
1	10	31,200
2	20	26,600
4	40	22,100
8	80	17,600
15	150	13,600
25	250	10,400
40	400	7,200
55	550	4,850
70	700	3,000

5.5.1 Low air pressure

At altitudes above sea level, low air pressure can cause:

- leakage of gases or fluids from gasket sealed containers
- ruptures of pressurised containers
- change of physical or chemical properties
- erratic breakdown or malfunction of equipment from arcing or corona
- decreased efficiency of heat dissipation by convection and conduction in air, that will affect equipment cooling (e.g. an air pressure decrease of 30% has been found to cause an increase of 12% in temperature)
- acceleration of effects due essentially to temperature (e.g. volatilisation of plasticizers, evaporation of lubricants, etc.).

5.5.2 Typical requirements – air pressure and altitude

Table 5.12 shows the most common environmental requirements concerning air pressure and altitude.

Table 5.12 Typical requirements – air pressure and altitude

High air pressure	High air pressure occurring in natural depressions and mines can have a mechanical effect on sealed containers and should always be borne in mind when designing and installing electrical equipment.
Installations up to 2000 m above sea level	Electrical equipment must be capable of working to an altitude (h) from −120 to 2000 m above sea level – which corresponds to an air pressure range from 110.4 to 74.8 kPa.

5.6 Weather and precipitation

Water is one of the most remarkable substances on the earth. It is the substance that we most often see in all its three states: liquid (water), solid (ice) and gas (steam). No living organism can exist without water and as much as half the weight of plants and animals is made up of water. Water in the oceans makes up approximately eleven times the volume of the solid part of the earth in addition, that is, to water frozen in ice flows, in lakes, rivers, within the ground and in living plants and animals.

Water is a constantly moving cycle. As the sun beats down, some of the surface water is evaporated; this water vapour rises as part of the air and is moved along by the wind. Should it pass over a land mass it may become a cloud and as more moisture is attracted to the cloud or the clouds pass over rising ground the water particles become larger and fall as rain, sleet or snow (see Figures 5.10 and 5.11).

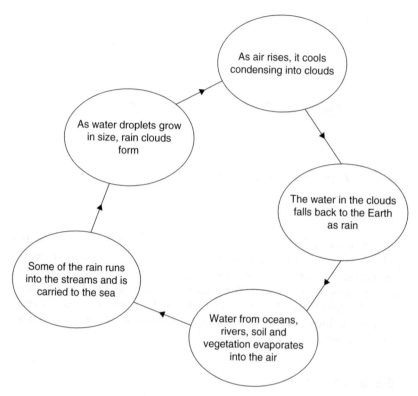

Figure 5.10 The hydrological cycle

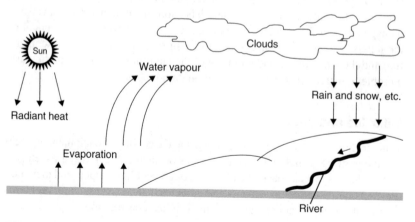

Figure 5.11 Simplified water cycle

As the rain comes into contact with the ground there are several avenues open to it; it may be revaporised and return to the atmosphere, be absorbed by the ground or will remain on the surface of the ground and run downhill, forming streams and rivers which run into the sea and the cycle begins once more.

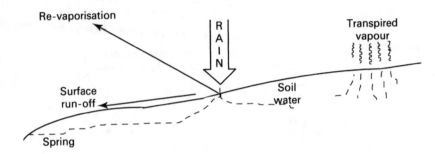

Figure 5.12 The effect of rain

5.6.1 Water

Water (in all of its three forms) is a major cause of failure in every application of electrical and electronic components. The humidity of the air and the possible formation of water particles must always be taken into consideration, especially as humidity possesses (almost without exception) a certain amount of electrical conductivity which increases the possibility of corrosion of metals. Similarly, the ingress of water followed by freezing within electronic equipment can result in malfunction.

5.6.2 Salt water

Salt has an electrochemical effect on metallic materials (i.e. corrosion), which can damage and degrade the performance of equipment and/or parts that have been manufactured from metallic materials. Non-metallic materials can also be damaged by salt through a complex chemical reaction which is dependent on the supply of oxygenated salt solution to the surface of the material, its temperature and the temperature and humidity of the environment. This is particularly a problem in areas close to the sea or mountain ranges.

5.6.3 Ice and snow

Water in the form of ice can cause problems in the cooling of equipment or freezing and thawing, which will result in cracks occurring, breaking cases, etc.

Powdered snow can easily be blown through ventilation ducts and then melt down in equipment compartments and cubicles that can cause damp problems in critical systems if not prevented in the original construction.

5.6.4 Weathering

As shown below, there are several types of 'weathering' (which is the collective term for the processes by which rock at or near the earth's surface is disintegrated and decomposed by the action of atmospheric agents, water, and living things).

Exfoliation

During the day rocks are warmed by the sun and at night the surface can cool more rapidly than the underlying rocks. The outer skin of the rock then becomes tight and cracks, thus layers of the rock peel off and the mountain becomes rounded or dome shaped. This exfoliation can have an effect on equipment that is sunk into rock faces or mounted on the surface of equipment.

Freeze thaw

When water freezes it turns to ice, expanding by about one-twelfth of its volume. If this water is in the joint or a crack in a casing then the space will become enlarged and the casing on either side will be forced apart. When the ice eventually thaws more water will penetrate into the crack and the cycle repeats itself with the crack constantly enlarging.

Chemical weathering

Water can pick up quantities of sulphur dioxide from the atmosphere and will form a weak solution of acid. This acid can attack certain equipment housings and the process whereby the housing is worn away is known as chemical weathering.

Erosion

As the wind blows over dry ground it collects grit and 'throws' it vigorously against the surfaces nearby. This grit, in a similar way as sandpaper, gradually wears away the surface with which it comes into contact.

Mass movement

Once solids, such as sedimentary rocks, have been broken up, there is often a downwards movement of the particles which have been broken off. This 'soil creep' can gather momentum and can, in certain circumstances, submerge equipment.

5.6.5 Typical requirements – weather and precipitation

Table 5.13 shows the most common environmental requirements concerning weather and precipitation.

Table 5.13 Typical requirements – weather and precipitation

Weather protection	Equipment that is operated adjacent to the sea shore or on mountain ranges and therefore subject to water and precipitation must be able to function equally well as the same equipment housed in arid deserts.
Operation	All equipment should be capable of operating during rain, snow, hail and be unaffected by ice, salt and water.
Rain	All equipment should be capable of operating in rain and be capable of preventing the penetration of rainfall at a minimum rate of 13 cm/hour and an accompanying wind rate of 25 m/s.
Snow and hail	Consideration needs to be given to the effect of all forms of snow and/or hail. The maximum diameter of the hailstones is conventionally taken as 15 but larger diameters can occur on occasions.
Salt water	Equipment should be capable of operating in (or be protected from) heavy salt spray, as would be experienced in coastal areas and in the vicinity of salted roadways.

5.6.6 Requirements from the Regulations – weather and precipitation equipment

Equipment likely to be exposed to weather, corrosive atmospheres or other adverse conditions shall be constructed (or protected) to prevent danger arising from such exposure.	131-05-01
Electric appliances producing hot water or steam shall be protected against overheating.	424-02-01
Other than the current-carrying parts of a water heater and/ or water boiler, all metal parts that are in contact with the water shall be solidly and metallically connected to the metal water pipe that carries the water supply to that heater/boiler.	554-05-02

 The metal water pipe should be connected to the main earthing terminal by the circuit protective conductor.

Wiring systems

Wiring systems shall be capable of withstanding condensation and/or ingress of water during installation, use and maintenance.	522-03-01
Drainage points shall be made available at all points where water could collect (or condensation could occur) in a wiring system.	522-03-02

Wiring systems installed near water and subjected to waves shall be protected.	522-03-03
Wiring systems routed near a service liable to cause condensation (e.g. water, steam or gas services) shall be protected.	528-02-03

Construction sites

Reduced low-voltage systems shall use low temperature 300/500 V thermoplastic (PVC) or equivalent flexible cables.	604-10-03
Applications that exceed the reduced low-voltage system shall use H07 RN-F type flexible cable that is resistant to abrasion and water.	604-10-03

5.7 Pollutants and contaminants

Over the last twenty years environmental matters have become an area of widespread public concern, particularly those concerning the issues of pollution and contaminants. Pollutants and contaminants come in many forms and can have an effect on the air, land or water courses and as pollutants move from one medium to another they may be deposited on equipment and equipment housing – they can cause extensive damage.

Pollution of the air can occur in both the troposphere and stratosphere as shown in Figure 5.13 below. In the troposphere, pollutants from chimneys (for example) are carried by the air and can be deposited over time and distance, thus having a limited life span before they are washed out or deposited on the ground. If pollutants are injected straight into the stratosphere (as with a volcanic eruption) they will remain there for some time and result in noticeable effects over the whole region. On the other hand, the roughness of the ground will produce air turbulence which will itself promote the mixing of pollutants and even low wind speeds will result in high pollutant concentrations.

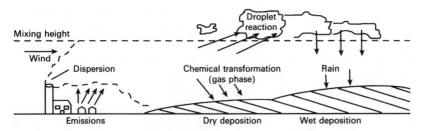

Figure 5.13 Process involved between the emission of air pollutants and being deposited on the ground

Sources of natural pollutants include:

- sulphur – emitted by volcanoes and from biological processes
- nitrogen – from biological processes in soil and lightning and biomass burning
- hydrocarbons – methane from fermentation of rice paddies, fermentation of the digestive tract of ruminants (e.g. cows). Also released by insects, coal mining and gas extraction.

Sources of man-made pollutants include:

- carbon dioxide and carbon monoxide produced during the burning of fossil fuels
- soot formation accompanied by carbon monoxide and generally due to inadequate or poor air supply
- hydrocarbons – most boilers and central heating units burning fossil fuels result in very low emissions of gaseous hydrocarbons or oxygenated hydrocarbons such as aldehydes.

5.7.1 Pollutants

Although pollutant gases are normally only present in low concentrations, they can cause significant corrosion and a marked deterioration in the performance of contacts and connectors. The gases in operating environments which cause corrosion are oxygen, water vapour and the so-called pollutant gases which include sulphur dioxide, hydrogen sulphide, nitrogen oxides and chlorine compounds.

Silver and some of its alloys are particularly susceptible to tarnishing by the minute quantities of hydrogen sulphide that occur in many environments. The tarnish product is dark in colour, consists largely of b-silver sulphide and separable electrical connections using these materials may, therefore, suffer from increased resistance and contact noise as a result.

The amount of tarnishing (of a metal) is dependent upon the amount of humidity present. Less corrosion occurs below 70% Relative Humidity (RH) but above 80% RH the rate of tarnishing increases rapidly. Temperature also has an effect on the amount of tarnishing as the nature of the corrosion mechanism has a tendency to change at temperatures above 30°C.

Sulphur dioxide

Sulphur dioxide is the pollutant gas most commonly found in the atmosphere and is usually present in high concentrations in urban and industrial locations. In combination with other pollutants and moisture (e.g. humidity) it is responsible for the formation of high resistance, visible corrosion layers on all but the most noble metals (e.g. silver and gold) and alloys.

Although sulphur dioxide alone is less corrosive than other gases (such as sulphur trioxide, nitrogen oxides and chlorine compounds), the most extensive corrosion occurs when combustion products are present together with sulphur dioxide.

Hydrogen sulphide

Hydrogen sulphide is caused by bacterial reduction of sulphates in vegetation, soil, stagnant water and animal waste on a worldwide basis. In the atmosphere, hydrogen sulphide is oxidised to sulphur dioxide which, in turn, is brought to the ground by rain. In an aerobic soil, bacteria turns the sulphur dioxide to sulphates. Sulphate-reducing bacteria complete the cycle and turn the sulphates to hydrogen sulphide which is the principal natural sulphur input in the atmosphere and is, therefore, a widespread pollutant of air.

Nitrogen oxides

The production of nitrogen oxides is particularly significant because the rate of corrosion by sulphur dioxide is greatly accelerated in the presence of nitrogen dioxide, although the corrosion products have similar compositions.

5.7.2 Contaminants

Contaminants are composed of dust, sand, smoke and other particles that are contained within the air and these can have an effect on electrical equipment in various ways, especially:

- ingress of dust into enclosures and encapsulations
- deterioration of electrical characteristics (e.g. faulty contact, change of contact resistance)
- seizure or disturbance in motion bearings, axles, shafts and other moving parts
- surface abrasion (erosion and corrosion)
- reduction in thermal conductivity
- clogging of ventilating openings, bushes, pipes, filters and apertures that are necessary for operation etc.

The presence of dust and sand in combination with other environmental factors such as water vapour can also cause corrosion and promote mould growth. Damp heat atmospheres cause corrosion in connection with chemically aggressive dust, and similar effects are caused by salt mist. Effects of ion-conducting and corrosive dusts (e.g. de-icing salts) need also to be considered.

Dust and sand

The concentration of dust and sand in the atmosphere varies widely with geographical locality, local climatic conditions and the type and degree of activity taking place. The amount of dust and sand found in the air is dependent on terrain, wind, temperature, humidity and precipitation. Under certain conditions enormous amounts of dust and sand may be temporarily released and this suspended dust will drift away with the wind (see Table 5.14) depending on its concentration and the size of the particles.

Table 5.14 Concentration of dust and sand (extracted with permission from a paper by Herne European Consultancy)

Region	Dust and sand concentration (μg/m^3)
Rural and suburban	40–110
Urban	100–450
Industrial	500–2000

Particles larger than 150 μm are generally confined to the air layer in the first metre above ground and in this layer, about half of the sand grains move within the first 10 mm above the surface.

The dust and sand appearing in enclosed and sheltered locations is generated by several sources (e.g. quartz, de-icing salts, fertilisers etc., penetrating into locations via ventilating ducts or badly fitting windows). The dust may also come from cloth or carpets in normal use in the working environment.

Dust

Dust may be defined as '*particulate matter of unspecified origin, composition and size ranging from 1 μm to 150 μm originating from quartz, flour, organic fibres etc.*'. Particles of less than 75 μm, because of their low terminal velocity, can remain suspended in the atmosphere for very long periods through the natural turbulence of the air. In sheltered and enclosed locations, the maximum grain size tends to be smaller (e.g. less than 100 μm) than non-weather-protected locations due to the filtering effect of the shelter.

The dust found in and around electrotechnical products may be generated by several different sources. The dust may be quartz, coal, de-icing salts, fertilisers, small fibres from cotton or wool (real or artificial) that has been generated from cloth or carpets by normal use in living rooms and offices, and penetration of dust into a piece of equipment can be:

- carried in by forced air circulation, for example for cooling purposes
- carried in by the thermal motion of the air
- pumped in by variations in the atmospheric pressure caused by temperature changes
- blown in by wind.

Dust itself can act as a physical agent, chemical component (or both) and cause one or more of the following harmful effects:

- seizure of moving parts
- abrasion of moving parts
- adding mass to moving parts thereby causing unbalance
- deterioration of electric insulation
- deterioration of dielectric properties
- clogging of air filters

- reduction of thermal conductivity
- interference with optical characteristics

and also

- corrosion and mould growth
- overheating and fire hazard.

Dust adhering to the surface of materials may contain organic substances that provide a source of food for micro-organisms.

Sand

Sand is the term applied to '*segregated unconsolidated accumulation of detrital sediment, consisting mainly of tiny broken chips of crystalline quartz or other mineral, between 100 μm and 1000 μm in size*'. Particles greater than 150 μm are unable to remain airborne unless continually subjected to strong winds, induced airflows or turbulence. Sand is generally harder than most fused silica glass compositions and can, quite naturally, scratch the surface of most glass optical devices. Pressure applied over trapped grains of sand can also cause fractures to occur in equipment.

The electrostatic charges produced by friction of the particles in sand storms can interfere with the operation of equipment and sometimes be dangerous to personnel. The breakdown of insulators, transformers and lightning arresters and the failure of car ignition systems has also been known to occur as a result of such charges. The electrostatic voltages produced can be very large. Indeed, voltages as high as 150 kV have made telephone and telegraph communications inoperable during sand storms.

Quartz, because of its hardness, can result in rapid wear or damage to products, particularly moving parts. However, erosion of material requires that the presence of dust and sand is combined with a high velocity air stream over an extensive period of time.

Sand and the majority of dusts usually deposited on insulated surfaces are poor conductors in the absence of moisture. The presence of moisture, however, will result in the dissolving of the soluble particles and the formation of conducting electrolytes. For example, the leakage currents flowing over contaminated power line insulators can be of the order of one million times greater than those which flow through clean, dry insulators.

Smoke or fumes

Smoke or fumes are '*dispersive systems in the air consisting of particles below 1 μm*'. As the particles are so small they do not usually effect equipment, provided that the equipment is properly designed.

Fauna and flora

With a few exceptions, fauna (rodents, insects, termites, birds etc.) and flora (plants, trees, seeds, fruit, blossom, mould, bacteria, fungi, etc.) may be present at all locations where equipment is stored, transported or used. Whilst fauna may be the cause of damage inside buildings as well as open-air locations, damage by flora will predominantly occur in open-air conditions. Moulds and bacteria can, however, be present both inside buildings and in open-air conditions.

The frequency of this flora and fauna depends on temperature and humidity. In warm damp climates, fauna and flora, especially insects and micro-organisms such as mould and bacteria, will find favourable conditions of life. Humid or wet rooms in buildings, or rooms in which processes produce humidity, are suitable living spaces for rodents, insects and micro-organisms. The range of temperature in which moulds may grow is from 0°C to 40°C, and the most favourable temperatures for many cultures is between 20°C and 30°C.

If the surfaces of the products carry layers of organic substances (e.g. grease, oil, dust), or animal/vegetable deposits, the surfaces will become ideal locations for the growth of moulds and bacteria.

Effects of flora and fauna

The functioning of equipment and materials can be affected by physical attacks of fauna. Small animals and insects that feed from, gnaw at, eat into and chew at materials are particular problems, as are termites cutting holes into material.

Materials such as wood, paper, leather, textiles, plastics (including elastomers) and even some metals such as tin and lead are all susceptible to attack.

Larger animals can also cause damage by stroke, impact or thrust. These attacks can cause:

- physical breakdown of material, parts, units, devices
- mechanical deformation or compression
- surface deterioration
- electrical failure caused by mechanical deterioration.

Deposits from fauna (especially insects, rodents, birds etc.) can be caused by the presence of the animal itself, nest building, deposited feed stocks and meta-bolic products such as excrement and enzymes etc.

Deposits from flora may consist of detached parts of plants (leaves, blossom, seeds, fruits etc.) and growth layers of cultures of moulds or bacteria. These attacks can lead to:

- deterioration of material
- metallic corrosion
- mechanical failure of moving parts
- electrical failure due to
  - increased conductivity of insulators
  - failure of insulation
  - increased contact resistance
  - electrolytic and ageing effects in the presence of humidity or chemical substances
  - moisture absorption and adsorption
  - decreased heat dissipation.

These in turn can cause interruption of electrical circuits, malfunctioning of mechanical parts and clouding of optical surfaces (including glass).

Mould

Surface contamination in the form of dusts, splashes, condensed volatile nutrients or grease may be deposited on equipment. When that equipment is exposed (in use, storage or transportation) to the atmosphere, and without proper protective covering, mould growth will occur and mould can cause unforeseen damage to equipment, whether constructed from mould-resistant materials or not!

Fungi grows in soil and in, or on, many types of common material. It propagates by producing spores which become detached from the main growth and later germinate to produce further growth. The spores are very small and easily carried by the wind (or moving air). They also adhere to dust particles carried in the air. Contamination can also occur due to handling. Spores may be deposited by the hands or in the film of moisture left by the hands.

Germination and growth

Moisture is essential in allowing the spores to germinate and when a layer of dust or other hydrophilic material (i.e. moisture retaining) is present on the surface, sufficient moisture may be abstracted by it, from the atmosphere. In addition to high humidity, spores require (on the surface of the specimen) a layer of material that will absorb the moisture. Mould growth is also encouraged by stagnant air spaces and lack of ventilation.

When the relative humidity is below 65%, no germination or growth will occur. The higher the relative humidity above this value, the more rapid the growth will be. Spores can survive long periods of very low humidity and even though the main growth may have died, they will germinate and start a new growth as soon as the relative humidity becomes favourable again (i.e. in excess of 65%). The optimum temperature of germination for the majority of moulds is between 20°C and 30°C.

Effects of mould growth

Moulds can live on most organic materials, but some of these materials are much more susceptible to attack than others. Growth normally occurs only on surfaces exposed to the air, and those which absorb or adsorb moisture will generally be more prone to attack.

Even where only a slightly harmful attack on a material occurs, the formation of an electrically conducting path across the surface due to a layer of wet mycelium (i.e. vegetative part of fungus) can drastically lower the insulation resistance between electrical conductors supported by an insulation material. When the wet mycelium grows in a position where it is within the electromagnetic field of a critically adjusted electronic circuit, it can cause a serious variation in the frequency-impedance characteristics of that circuit.

Among the materials that are very susceptible to attack are leather, wood, textiles, cellulose, silk and other natural resources. Most plastic materials, although less susceptible, are also prone to attack as they will probably contain oligomers (i.e. natural or synthetic compounds of usually high molecular weight that consist of millions of linked simple molecules), non-polymerised monomers (i.e. a molecule that can combine with others to form a polymer) and/or additives which may radiate to the surface and be a nutrient for fungi. Some plastic materials depend, for a satisfactory life span, on the presence of a plasticiser (i.e. substances added to plastics to make or keep them soft or pliable) which, if it is readily digested by fungi, will eventually give rise to failure of the main material.

Mould attack on materials usually results in a decrease of mechanical strength and/or changes in other physical properties and the growing mould on the surface of a material can yield acid products and other electrolytes which will cause a secondary attack on the material. This attack can lead to electrolytic or ageing effects, and even glass can lose its transparency due to this process. Oxidation or decomposition may be facilitated by the presence of catalysts secreted by the mould.

Prevention of mould growth

All insulating materials used should be chosen to give as great a resistance to mould growth as possible, thus maximising the time taken for mycelium to grow and minimising any damage to the material consequent upon such growth. The use of lubricants during assembly (e.g. varnishes, finishes etc.) is frequently necessary in order to obtain the required performance or durability of a product. Such materials should be chosen with regard to their ability to resist mould growth for even though it can be shown the lubricants do not support mould growth, they may collect dust which in turn will support mould growth.

Moisture traps which could possibly be formed during the assembly of equipment and in which mould can grow should be avoided. Examples of such less obvious traps are between unsealed mating plugs and sockets, or between printed circuit cards and edge connectors. Other preventatives of mould growth include:

- complete sealing of the equipment in (and with) a dry, clean atmosphere
- continuous heating within an enclosure, which can ensure a sufficiently low humidity
- operation of equipment within a suitable controlled environment
- regularly replaced desiccants (e.g. silica beads)
- periodic and careful cleaning of enclosed equipment.

Where the material and functioning of the equipment allows such treatment, ultraviolet radiation or ozone may be used for sterilisation. Air currents flowing over the parts can retard the development of mould growth and can be used to control the action of acaricides (i.e. mites and ticks).

5.7.3 Typical requirements – pollutants and contaminants

Table 5.15 lists the most common environmental requirements concerning pollutants and contaminants.

Table 5.15 Typical requirements – pollutants and contaminants

Pollutants	• Although the severity of pollution will depend upon the location of the equipment, the effects of pollution must be considered in the design of equipment and components. • Means need to be provided to reduce pollution by the effective use of protective devices. • The requirements of ISO 14001 regarding environmental protection and the prevention of pollution have to be met.
Contaminants	The following should be considered: • chemically active substances • biologically active substances • flora and fauna • dust • sand.
Mould	• In an assembled state, equipment needs to operate when exposed to airborne mould spores and within climates that will be conducive to the growth of moulds. • Insulating materials should be chosen to provide as much resistance to mould growth as possible and all materials used should be chosen with regard to their ability to resist mould growth.

5.7.4 Requirements from the Regulations – pollutants and contaminants

Dust

Precautions shall be taken to prevent the accumulation of dust or other substances which could adversely affect the heat dissipation from a wiring system.	522-04-02
Where materials (e.g. dust and fibres) accumulating on electrical equipment enclosures could be sufficient to cause a fire hazard, the enclosure shall be prevented from exceeding: • 90°C under normal conditions, and • 115°C under fault conditions.	482-02-02
In locations where the presence of dust and fibres could cause a fire hazard, only luminaries with a limited surface temperature in accordance with BS EN 60598-2-24 shall be used.	482-02-12

Where heating and ventilation systems containing heating elements are installed, the dust or fibre content of the air shall not present a fire hazard.	482-02-15
Heat storage appliances shall not ignite combustible dust and/or fibres.	482-02-17
Only luminaries with a limited surface temperature in accordance with BS EN 60598-2-24 shall be used in locations where the presence of dust and fibres causes a fire hazard.	482-02-12

Smoke or fumes

Wiring systems shall **not** be installed near a service which produces smoke or fumes unless protected by shielding.	528-02-02
The location of the electrical source shall be properly and adequately ventilated so that any smoke or fumes from that source cannot penetrate, to a hazardous extent, areas occupied by persons.	562-01-06

Fauna and flora

Wiring systems shall withstand flora and fauna.	522-10-01

Mould

Wiring systems shall withstand mould growth.	522-09-01

5.8 Mechanical

Mechanics is the branch of physics concerned with the motions of objects and their response to forces. Modern descriptions of such behaviour begin with a careful definition of such quantities as displacement (distance moved), time, velocity, acceleration, mass and force.

There is often a tendency to underestimate the effect that the mechanical environment can have on the reliability of equipment, especially the effects of vibration and shock. Mechanical stresses are normally attributed to a moving mass and there is frequently a tendency to underestimate the effect of the mechanical environment on the reliability of static installations. Experience suggests, however, that vibrations and shocks are a significant Reliability, Availability and Maintainability (RAM) factor, not only from the point of view of vehicle-mounted equipment, but also with respect to permanent installations.

If a spurious vibration acts on a Printed Circuit Board (PCB), module or equipment, resonant oscillations will be induced in all components at their natural frequency. If, however, the frequency spectrum has some more or less distinctive frequency bands, the elements will perform forced oscillations at the cyclic frequency of the interference and generally depend on both the characteristics of the oscillator (i.e. the component) and on the interference.

5.8.1 Shock

'Shock' is generally defined as *'an impact shock characterised by a simple acceleration and free impact on a firm base'* and is usually the result of a violent collision, or a heavy blow. Whilst it is difficult to design and install electrical equipment, components and systems so that they are completely immune to shock, precautions should, nevertheless, be taken to guard against potential problem areas.

5.8.2 Vibration

Components, equipment and other articles during transportation or in service may be subjected to conditions involving vibration of a harmonic pattern, generated primarily by rotating, pulsating or oscillating forces caused by machinery and seismic incidents.

5.8.3 Acceleration

Equipment, components and electrotechnical products that are likely to be installed in moving bodies (e.g. rotating machinery) will be subjected to forces caused by steady accelerations. In general the accelerations encountered in service will have different values along each of the major axes of the moving body, and, in addition, usually have different values in the opposite senses of each axis.

5.8.4 Protection

- The resonant frequency of components is greatly influenced by the length of their connecting wires and the actual length of their connecting wires may well be the decisive factor as to whether a component fails or remains functioning under given vibration and impact conditions.
- The amplitude of shocks on the equipment can be reduced by use of special mounting devices.

- Shock absorption is based on storing the impact and releasing it at a retarded rate. The peak acceleration is reduced and the high frequencies damped, thus providing protection for the components with their relatively higher natural frequencies (of some hundred hertz).
- Vibration dampers and shock absorbers are often used as a form of protection against mechanical stresses. The basic difference between vibration dampers and shock absorbers is that with the former the natural frequency lies below the interference frequency, whilst the latter is above.

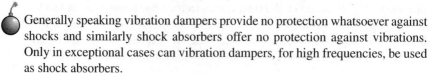 Generally speaking vibration dampers provide no protection whatsoever against shocks and similarly shock absorbers offer no protection against vibrations. Only in exceptional cases can vibration dampers, for high frequencies, be used as shock absorbers.

- Elastic suspension of equipment can cause a critical increase in amplitude at certain frequencies and, where translatory and rotary displacements greater than six degrees of freedom are possible, very complex motions may arise. Wherever possible, therefore, one should try to ensure that none of the (possible) resonant frequencies fall within the range of the induced displacements.
- Sheathing circuits by means of cast resins can, in most cases, be a very effective means of counteracting mechanical stresses combined with temperature humidity.

5.8.5 Typical requirements – mechanical

Table 5.16 lists the most common environmental requirements concerning solar radiation.

Table 5.16 Typical requirements – mechanical

Vibrations and shocks	Any dampers or anti-vibration mountings must be integral with the equipment to prevent the unit being accidentally installed without them.
Mechanical shock	Equipment should be capable of withstanding shock pulses (e.g. a minimum of 20,000 shocks at a shock level of 20 g).
On or near the roadside	Equipment located on or near the roadside must be capable of withstanding vibrations and shocks.
Long-term exposure	Equipment must be capable of withstanding long-term exposure to shocks.
Encapsulated outdoor installations	Equipment contained in encapsulated outdoor installations must be capable of withstanding vibrations and shocks.
Closed rooms	Equipment located in closed room installations must be capable of withstanding self-induced vibrations.
In service	Equipment should be capable of withstanding without deterioration or malfunction all mechanical stresses that occur in service.
Random vibration	Equipment should be capable of withstanding random vibration.

5.8.6 Requirements from the Regulations – mechanical

Conductors

Conductors **shall not** be subjected to excessive mechanical stress.	522-12-01

All parts of a flexible cable (or cord) that is liable to mechanical damage shall be visible throughout its length.	413-06-03
Flexible cables and cords shall be suitably protected against mechanical damage.	482-02-10
When determining the cross-sectional area of conductors, account shall be taken of the mechanical stresses that conductors are likely to be exposed to.	131-06-01
Where a conductor or a cable is not continuously supported it shall be supported by suitable means at appropriate intervals and in such a manner that the conductor or cable does not suffer mechanical damage by its own weight.	522-08-04
Cables and/or conductors used as fixed wiring shall be supported and not be exposed to undue mechanical strain, particularly at terminations.	522-08-05

Earthing arrangements

Earthing arrangements shall be sufficiently robust (or have additional mechanical protection) to external influences.	542-01-07

Enclosures

The insulating enclosure shall be capable of resisting mechanical, electrical and thermal stresses likely to be encountered.	413-03-05

Equipment

Equipment **shall** be capable of withstanding all mechanical stresses normally encountered during service.   412-02-01

Fixed equipment that is liable to emit an arc or high temperature particles shall be: 422-01-03

- totally enclosed in arc-resistant material, or
- screened by arc-resistant material from materials that could be effected by emissions.

To provide mechanical stability, arc-resistant material used for this protective measure shall be:

- non-ignitable
- of low thermal conductivity
- of adequate thickness.

Live parts

Live parts shall be completely covered with insulation which is capable of strongly withstanding mechanical stresses normally encountered during service.   412-02-01

Exposed conductive parts which might attain different potentials through failure of the basic insulation of live parts shall be arranged so that a person will not come into simultaneous contact with: 413-04-02

- two exposed conductive parts, or
- an exposed conductive part and any extraneous conductive part.

 This may be achieved if the location has an insulating floor and insulating walls the insulation is of acceptable electrical and mechanical strength.

Maintenance

Equipment **shall not** be capable of becoming unintentionally or inadvertently reactivated during mechanical maintenance.   462-01-03

If there is a risk of burns or injury from mechanical movement   462-01-01
during maintenance, a switching-off device shall be provided.

This device shall be suitably located and identified.   462-01-02

All installations shall be provided with the facility of being   476-01-01
able to be switched off (i.e. electrically) for mechanical
maintenance or an emergency switching device should
be provided.

Where a soldered connection is used, the design shall take   526-02-01
account of the potential mechanical stress under fault
current conditions.

Mechanical strength

All electrical joints and connections shall meet stipulated   133-01-04
requirements concerning mechanical strength.

Protective devices

Circuit protective devices shall break any fault current   434-01-01
flowing before such current causes danger due to
mechanical effects produced in circuit conductors or
associated connections.

An emergency stopping facility shall be provided where   463-01-05
mechanical movement of electrically actuated equipment
is considered dangerous.

In locations where the operation of a single protective   473-02-05
device is ineffectual, conductors should be protected
against mechanical damage and conductors kept away
from combustible material.

RCDs that are used to provide protection against indirect   531-02-08
contact and that are designed to be used with an overcurrent
protective device, shall be capable of withstanding the
mechanical stresses that it is likely to be subjected to if
fault occurred.

Wiring systems

The type of wiring system and method of installation will   131-07-01
depend (amongst other affects) on the mechanical stresses

that the wiring is likely to be exposed to during the erection of the electrical installation or in service.

Wiring system shall be selected and erected so as to minimise (during installation, use and maintenance) mechanical damage to the sheath and insulation of cables and insulated conductors and their terminations.   522-08-01

To avoid the effects of mechanical stresses: 522-08-02

- where the wiring system is designed to be withdrawable, there shall be adequate means of access for drawing cable in or out
- if buried in the structure, a conduit or cable ducting system for each circuit shall be completely erected before cable is drawn in.

The radius of every bend in a wiring system shall be such that conductors and cables shall not suffer mechanical damage.   522-08-03

Flexible wiring systems shall be installed so that excessive tensile and torsional stresses to the conductors and connections are avoided.   522-08-06

5.9 Electromagnetic compatibility

Most car owners normally accept that when they drive near electric pylons, their listening pleasure may be interrupted by loud crackles and/or buzzing noises but with the increased use of electronic equipment, however, the problem of interference has become one of our prime concerns. Although most forms of interference are usually tolerated as being one of those things *that you cannot do much about*, the design of modern sophisticated equipment has become so susceptible to electromagnetic interference that some form of regulation has had to be agreed.

Within Europe, this regulation is contained in the Electromagnetic Compatibility Directive 89/336/EEC, which clearly states that all electronic equipment shall be constructed so that:

- the electromagnetic disturbance it generates does not exceed a level allowing radio and telecommunications equipment and other apparatus to operate as intended
- the apparatus has an adequate level of intrinsic immunity to electromagnetic disturbance.

5.9.1 Typical requirements – electromagnetic compatibility

Table 5.17 lists the most common environmental requirements concerning electromagnetic compatibility:

Table 5.17 Typical requirements – electromagnetic compatibility

Equipment	<ul><li>The use of electronic equipment shall not interfere with the operation of other equipment.</li><li>All active electronic devices shall comply with the EMC Directive.</li></ul>
CE marking	Only CE (Conformity Europe) marked equipment may be offered for sale and all active equipment connected to an electrical installation shall carry a CE mark.
Apparatus cases	Input/output connections from apparatus cases should always be of non-screened non-balanced signalling cable and are normally restricted in length.
Atmospheric disturbances	To counteract the affects of storms, it is generally recommended that all equipment should be capable of withstanding (as a minimum) the following overvoltages.

Atmospheric disturbances:

Magnitude	2000 V
Rise time	$1.2\,\mu s$
Middle voltage time	$50\,\mu s$

Equipment immunity levels	<ul><li>Equipment should be immune to induced common-mode voltages.</li><li>Equipment should not experience a permanent loss of availability or suffer component damage for any induced common-mode voltage within this range.</li></ul>
Magnetic field	As low-frequency fields can influence cathode ray tubes, equipment should be capable of withstanding the following intensities:

Hz	A/m
5	0.8
50	3.0
250	1.5

Power supply lines	Equipment should be immune to the following high-frequency bursts:

Initial peak to peak voltage	1 kV
Burst repetition rate	5 kHz

Transients	All electronic equipment should be capable of withstanding:
	<ul><li>transients (either directly induced or indirectly coupled) so that no damage or failure occurs during operation.</li><li>without damage or abnormal operation, transient non-repetitive surges.</li></ul>

5.9.2 Requirements from the Regulations – electromagnetic compatibility

Electrical installations

 Electrical installations **shall** be arranged so that they do not mutually interfere (including electromagnetic interference) with other electrical installations and non-electrical installations in a building.   131-11-01

Equipment

Equipment chosen:

- shall be immune to the electromagnetic influences of other equipment and installations   515-02-01
- shall not cause electromagnetic influences to other equipment and/or installations.   515-02-02

The characteristics of electrical equipment (e.g. harmonic currents, fluctuating loads, high-frequency oscillations etc.) that might have a harmful effect on other electrical equipment and/or services shall be assessed.   331-01-01

Note: If the energy is being provided by a distributor then the distributor will supply this information.

All equipment shall comply with the appropriate EN or HD or implementing National Standard or (in certain circumstances) the appropriate IEC Standard or National Standard of another country.   132-01-01

If no applicable standard is available, the item of equipment concerned shall be selected by special agreement between the person specifying the installation and the installer.   132-01-02

The degree of protection of selected equipment shall be suitable for its intended use and mode of installation. Where this is not possible, additional protection shall be provided.   512-06-01

Electrical equipment shall be selected and erected so as to avoid any harmful influence between the electrical installation and any non-electrical installations envisaged.   515-01-01

Where equipment is carrying current of different types   515-01-02
(or at different voltages) and is grouped in a common
assembly (such as a switchboard, a cubicle or a control
desk or box), all equipment belonging to any one particular
type of current, or any one type of voltage, shall
be effectively segregated (wherever necessary) so as to
avoid mutual detrimental influence.

External influences

The mutual effect of a number of different external influences   512-06-02
occurring simultaneously shall be taken into account.

Property, persons and livestock

Property shall be protected from being damaged by   130-04-01
electromagnetic stresses caused by overcurrents occurring
in live conductors.

Persons and livestock shall be protected from being injured   130-04-01
by excessive temperatures and/or electromagnetic stresses
that are caused by overcurrents occurring in live conductors.

Cables

Single-core cables armoured with steel wire or tape shall   521-02-01
not be used for a.c. circuits.

Conductors

Conductors of a.c. circuits installed in ferromagnetic   521-02-01
enclosures shall ensure that the conductors of all phases
plus the neutral conductor (if any) and the appropriate
protective conductor of each circuit are contained in the
same enclosure.

Conductors entering a ferrous enclosure shall be arranged   521-02-01
so that the conductors are not individually surrounded by
a ferrous material, or other provision shall be made to
prevent eddy (induced) currents.

5.10 Fire

A fire will normally start when sufficient thermal energy from, for example, a burning cigarette or an electric short circuit is supplied to a combustible material. Following ignition, the fire will then produce its own thermal energy, some of which will be used as feedback to maintain combustion and some transferred via radiation and convection to other materials. These materials may also ignite and spread the fire.

The environmental conditions relating to the occurrence, development and spread of fire within a building and its effect on electrotechnical products exposed to fire is primarily covered by Section 8 (Fire Exposure) of IEC 721.2. This section provides background information for selecting the appropriate parameters and severities related to exposure of products to fire. More detailed information on fire condition characteristics and fire hazard testing is contained in specialist documentation.

The development of the fire generally consists of three processes:

- thermal
- aerodynamic
- chemical.

As a rule, radiation, convection and flame spread are the dominant physical factors.

5.10.1 Fire growth

Once a fire has started in a space (e.g. a room) its growth and spread is determined by:

- site
- volume
- arrangement of the fuel or fire load, its distribution, continuity, porosity and combustion properties
- aerodynamic conditions of the space
- shape and size of the space
- thermal properties of the space.

During the growth of a fire, a hot layer of gas builds up under the ceiling of the space. Under certain conditions, this gas layer can give rise to a rapid-fire growth and flashover might occur.

5.10.2 Flashover

One normally defines flashover as the time when flames begin to emerge from the openings of the space, which correlates with a temperature of 500°C to 600°C in the upper gas layer.

Flashover marks the transition from the growing fire (pre-flashover) to the fully developed fire (post-flashover).

Pre-flashover

A pre-flashover fire primarily concerns the operation and function of products (e.g. detectors, alarm systems, associated cables and sprinklers etc.) that are vital to maintaining the level of safety required for escape and/or the rescue of people caught in a fire.

Characteristics of pre-flashover fire

The ignitability properties of exposed material will depend on:

- the heat supplied
- the exposure time
- the presence or not of flames
- the geometrical location
- the thermal data

together with time variations such as:

- rate of heat release
- rate of flame spread
- gas temperature.

Fire hazard of a pre-flashover

The fire hazard of a pre-flashover situation is normally considered in terms of a series of probabilities, which depend on:

- the presence of ignition sources
- the presence of products
- the product fire performance properties
- the environmental factors
- the presence of people
- the presence/operation of detection and suppression devices
- the availability of escape.

Post-flashover

Whilst most standards are normally concerned with conditions during the pre-flashover stage of a fire, conditions following flashover must also be considered. A post-flashover fire can seriously damage some of the structural and load-bearing elements of a building and the fire can then, quite easily, spread from one fire space to another via partitions and ventilation systems. This can, of course, seriously damage electrical equipment located in these voids. For example, in a large space it is quite possible that a fire, small in relation to that space, could be large enough to damage some of the structural elements in the post-flashover state. An important factor of the post-flashover fire, which is often overlooked, is the amount of smoke and toxic gases that can affect people in escape routes and remote safety areas in a building. Smoke and toxic gases can also significantly affect equipment.

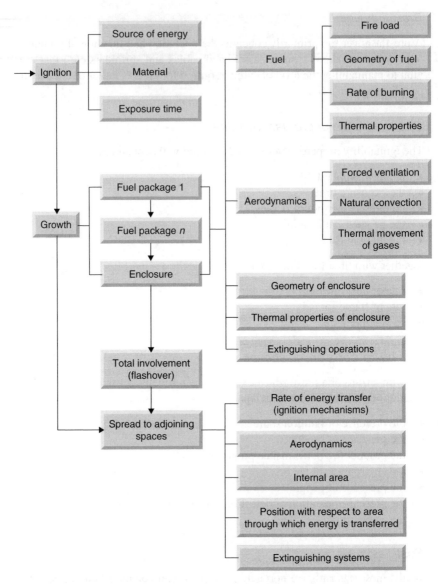

Figure 5.14 Factors affecting ignition, growth and spread of fire in a building

Characteristics of a post-flashover fire

The main characteristics of a post-flashover fire are:

- the rate of heat release
- the gas temperature
- the geometrical and thermal data for external flames

- the smoke and its optical properties
- the composition of the combustion products, particularly corrosive and toxic gases.

The possibility of a large external fire spreading from one storey to another in the same building (and eventually from one building to another) must also be considered. For these cases the first three characteristics – i.e. primarily gas temperature, geometrical and thermal data for the flames emerging from the window openings – are the most relevant.

5.10.3 Characteristics of smoke and gases as a fire product

Smoke is a mixture of heated gases, small liquid drops and solid particles from the combustion. During a fire (pre- and post-flashover), smoke will be distributed within the building through the airflow between rooms and via ventilation ducts etc. In most circumstances this can have disastrous effects because smoke can not only damage and in some cases even destroy property, it can also prevent the functioning of critical equipment. Most of the effects of smoke are of a chemical nature and the most prevalent is destruction or damage to electrotechnical products; in particular, corrosion caused by hydrogen chloride, which is a substance in smoke.

Metal surfaces exposed to air under normal (non-fire) conditions often have a chloride deposit up to $10 \, \text{mg/m}^2$. Such an amount is normally not harmful. However, after exposure to smoke from a fire involving PolyVinyl Chloride (PVC), a surface contamination of up to thousands of milligrams per square metre can be found, often causing significant damage. Chloride contamination of electrotechnical equipment can be removed by, for instance, detergents, solvents, neutralising agents, ultrasonic vibrations, and clean air jets, but the procedures are not always effective, sometimes giving a temporary but not permanent cure.

Experiments involving PVC-coated electrical wires and carried out on a scale large enough to be representative of real fires are currently in hand.

5.10.4 Building designs

In the design of buildings, the fire design of load-bearing structural elements and partitions is normally considered as a national problem and directly related to the results of standard national and (when available) international fire resistance tests. In such tests, the specimen is exposed, in a furnace, to a temperature rise, which is varied with time and within specified limits, according to the particular test being used.

Over the last decades, rapid progress has been made in the development of analytical and computational methods for determining the fire design of load-bearing and separating structures and structural elements. In the long term, it is foreseeable that this will develop into an analytical and/or computational design, directly based on a natural fire exposure. These will be specified with regard to the combustion characteristics of a fire load and the geometrical, ventilation and thermal properties of the fire space.

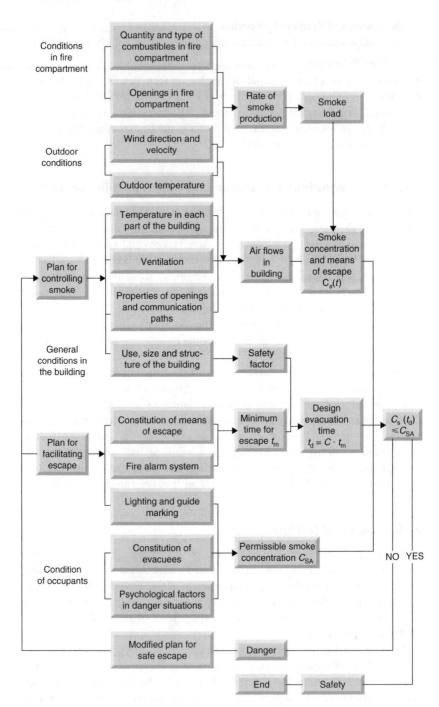

Figure 5.15 Flow diagram of a smoke control design system in a building

5.10.5 Test standards

Fire tests on building materials, components and structures normally focus on the characteristics of pre-flashover fire. Simplified full-scale (i.e. room) tests for surface products against smoke and in particular toxic combustion products are already available, but considerable development work needs to be completed before a useful small-scale test is available.

If no mathematical model of a small-scale test is available, the test results should be statistically correlated directly to full-scale test data. If a validated mathematical model of a small-scale test exists, important material characteristics controlling the space fire growth can be given quantitative values which can then be used as input data in mathematical models of full-scale pre-flashover space fire for specified scenarios.

With a view to practical, long-term use the results of small-scale reaction to fire tests to predict fire hazard should be based on a fundamental and scientific approach.

Figure 5.16 outlines the structure of such an approach.

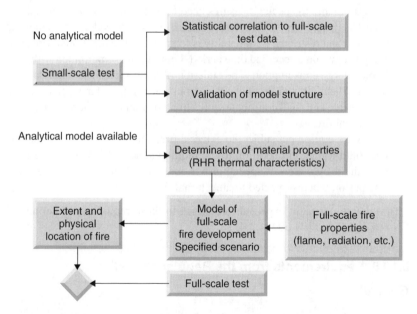

Figure 5.16 Combination of basic property tests and mathematical models for assessing the contribution of a tested material or product to the overall fire safety

5.10.6 Other related standards and specifications

IEC 60695 Series   Fire hazard testing – Guidance, tests and specifications for assessing fire hazard of electrotechnical products

ISO 5657   Fire tests – Reaction to fire – Ignitability of building products

ISO 5658	Reaction to fire tests – Spread of flame on building products and vertical configuration
ISO 5660	Fire tests – Reaction to fire – Rate of heat release from building products
ISO 9705	Fire tests – Full-scale room test for surface products
ISO TR 5924	Fire tests – Reaction to fire – Smoke generated by building products (dual-chamber test)
ISO TR9112.1	Toxicity testing of fire effluents – General.

5.10.7 Typical contract requirements – fire

In most contracts, reference is made to the IEC 60695 series of standards which cover the assessment of electrotechnical products against a nominated fire hazard.

CENELEC, on the other hand, show the requirement for equipment to operate in fire hazardous areas as three distinct clauses, as follows:

- Class FO – no special fire hazard envisaged. This is considered a normal service condition and except for the characteristics inherent to the design of the equipment, no special measures need to be taken to limit flammability.
- Class F1 – equipment subject to fire hazard. This is considered an abnormal condition and restricted flammability is required. Self-extinction of fire shall take place within a specified time period. Poor burning is permitted with negligible energy consumption. The emission of toxic substances shall be minimised. Materials and products of combustion shall, as far as possible, be halogen-free and shall contribute with a limited quantity of thermal energy to an external fire.
- Class F2 – equipment subject to external fire. This is considered an abnormal condition and in addition to the requirements of Class F1, the equipment shall (by means of special provisions), be able to operate for a given time period when subjected to an external fire.

Materials are normally expected to confirm to those requirements defined in EN 60721.3.3 and EN 60721.3.4.

5.10.8 Requirements from the Regulations – fire

General

All fixed electrical equipment with a surface   422-01-02
temperature that is liable to cause a fire or damage
adjacent materials **shall** either be:

- mounted on a support (or placed within an enclosure) that
  can withstand such temperatures as may be generated or

- screened by material which can withstand the heat emitted by the electrical equipment or
- mounted so that the heat can be safely dissipated.

All fixed equipment that focuses or concentrates heat **shall** be isolated from other fixed objects or building elements that could be effected by high temperatures.   422-01-06

Cables **shall** comply with the requirements of BS EN 50265-1 for flame propagation.   527-01-03

Fire alarm and emergency lighting circuits **shall** be segregated from all other cables and from each other in accordance with BS 5839 and BS 5266.   528-01-04

Fixed equipment enclosures **shall** be constructed:

- using heat and fire resistant materials or   422-01-07
- be capable of withstanding the highest temperature likely to be produced by the electrical equipment in normal use.

Fixed equipment that is liable to emit an arc or high temperature particles **shall** be: 422-01-03

- totally enclosed in arc-resistant material, or
- screened by arc-resistant material from materials that could be effected by emissions, or
- mounted so as to allow safe extinction of the emissions at a sufficient distance from material that could be effected by emissions.

 Note: To provide mechanical stability, arc-resistant material used for this protective measure shall be:

- non-ignitable
- of low thermal conductivity
- of adequate thickness.

Live conductors (or joints between them) shall be terminated within an enclosure.   422-01-04

The heat generated from fixed electrical equipment shall not damage (or cause danger to) adjacent fixed material.   422-01-01

The risk of spread of fire shall be minimised by selection of an appropriate material and erection.   527-01-01

The wiring system shall not reduce the structural performance and fire safety of a building.	527-01-02
Where the risk of fire is high special precautions shall be taken.	527-01-03

Burns

The temperature of accessible parts of fixed electrical equipment within arm's reach shall be limited, unless guarded so as to prevent accidental contact.	423-01-01

Fire precautions

The risk of spread of fire shall be minimised by selection of an appropriate material and erection.	527-01-01
The wiring system shall not reduce the structural performance and fire safety of a building.	527-01-02
Cables shall comply with the requirements of BS EN 50265-1 for flame propagation.	527-01-03
Wiring systems that pass through floors, walls, roofs, ceilings, partitions or cavity barriers shall be sealed according to the degree of fire resistance required of the element concerned (if any).	527-02-01

 Note: See BS 476-23 for details concerning the sealing of wiring systems.

Wiring systems (such as conduit, cable ducting, cable trunking, busbar or busbar trunking) that penetrate a building construction that has specified fire resistance, shall be internally sealed to maintain the degree of fire resistance.	527-02-02
Temporary sealing arrangements shall be provided during the erection of a wiring system.	527-03-01
Sealing that has been disturbed during alterations work shall be reinstated as soon as practicable.	527-03-02

Sealing arrangement shall be visually inspected during   527-04-01
erection, to verify that it conforms to the manufacturer's
erection instructions. Details shall be recorded.

Risk of fire owing to manufacturing process

Where there is a risk of fire: 482-01-01

* owing to the nature of the manufacturing process
* owing to the type of processing, storage and use of
  combustible materials or
* where the location has been constructed out
  of combustible material

electrical equipment shall be selected, constructed and
installed so that its temperature (and possible temperature
rise in the event of a fault) is unlikely to cause a fire.

Locations with risks of fire due to the nature of processed or stored materials

 Exposed bare and live conductors shall not be used.   482-02-09

Electrical equipment shall be restricted to only that   482-02-01
necessary for use in the location.

Electrical equipment shall be chosen according to its   482-02-03
suitability for a particular location.

Electrical equipment enclosures shall provide a degree   482-02-03
of protection of at least 1P5X.

Cables that are not completely embedded in non-combustible   482-02-04
material (such as plaster or concrete), or otherwise
protected from fire, shall meet the flame propagation
characteristic specified in BS EN 50265-2-1 or 2-2.

All circuits shall be capable of being isolated from live   482-02-08
supply conductors by a linked switch or linked circuit
breaker.

Where possible, groups of circuits shall be isolated by a common means.   482-02-08

Flexible cables and flexible cords shall be:

- of a heavy-duty type with a voltage rating of not less than 45 or 50 V, or suitably protected against mechanical damage.   482-02-10

Heating appliances shall be fixed.   482-02-16

Where a heating appliance is mounted close to combustible materials it shall be protected by barriers to prevent the ignition of such materials.   482-02-16

Enclosures of equipment such as heaters and resistors shall not attain surface temperatures higher than: 482-02-02

- 90°C under normal conditions,
- and 115°C under fault conditions

particularly where the accumulation of materials (i.e. such as dust and fibres) accumulating on electrical equipment enclosures could be sufficient to cause a fire hazard.

Locations with combustible constructional materials

Electrical equipment such as installation boxes and distribution boards that are installed on or in a combustible wall shall comply with the relevant standard for enclosure temperature rise.   482-03-01

Overheating

Electric appliances producing hot water or steam **shall** be protected against overheating.   424-02-01

Electric heating elements of forced air heating systems (other than those of central-storage heaters) shall: 424-01-01

- not be capable of being activated until the prescribed airflow has been established
- deactivate when the airflow is reduced or stopped.

Forced air heating systems shall have two, independent, temperature-limiting devices.   424-01-01

The frames and enclosures of electric heating elements shall be of non-ignitable material.   424-01-02

Fire precautions

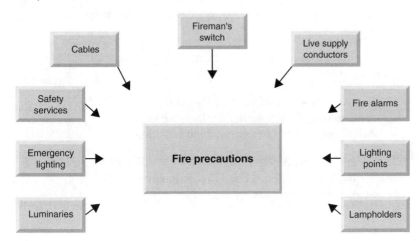

Figure 5.17 Fire precautions

Cables

Cables **shall not** be run in a lift (or hoist) shaft unless it forms part of the lift installation.   528-02-06

Cable locations containing long vertical runs or bunched cables (and where the risk of flame propagation is high) shall meet the requirements specified in BS 4066-3.   482-02-04

Cables shall comply with the requirements of BS EN 50265-1 for flame propagation – if not, they shall:

- be limited to short lengths for connection of appliances to the permanent wiring system   527-01-03
- not pass from one fire-segregated compartment to another.   527-01-04

Cables that are not completely embedded in a non-   482-02-04
combustible material (such as plaster or concrete) or
otherwise protected from fire, shall meet the flame
propagation characteristic specified in BS EN 50265-2-1
or 2-2.

Flexible cables and flexible cords shall be: 482-02-10

• of a heavy-duty type with a voltage rating of not less
  than 45 or 50 V, or suitably protected against mechanical
  damage.

Heating cable passing through (or be in close proximity to)   554-06-01
a fire hazard:

• shall be enclosed in material with an ignitability
  characteristic 'P' as specified in BS 476 Part 5
• shall be protected from any mechanical damage.

The cables and connections from circuits with different   528-01-07
voltage bands in Band I and Band II voltage circuits shall
be segregated by a partition.

 Note: If metal, this segregation shall be earthed.

Fire alarm and emergency lighting

Fire alarm and emergency lighting circuits **shall** be   528-01-04
segregated from all other cables and from each other in
accordance with BS 5839 and BS 5266.

Fireman's switch

All exterior installations that are located on single   476-03-06
premises **shall** (wherever practicable) be controlled by a
single fireman's switch.

All firemen's switches **shall** meet the requirements of the   476-03-07
Regulations plus any requirements of the local fire authority:

Firemen's switches shall:

- be coloured red   537-04-06
- be fixed on a permanent nameplate (minimum size 150 mm by 100 mm) with not less than 36-point lettering stating 'FIREMAN'S SWITCH'
- have its ON and OFF positions clearly indicated by lettering legible to a person standing on the ground
- have the OFF position at the top
- be provided with a device to prevent the switch being inadvertently returned to the ON position
- be positioned so as to allow trouble-free operation by a fireman.

High-voltage discharge lighting installations

High-voltage electric signs and high-voltage luminous   554-02-01
discharge tube installations shall meet the requirements
of BS 559.

Lampholders

Except for E14 and E27 lampholders complying with   553-03-04
BS EN 60238, the outer contact of all Edison screw
or single centre bayonet cap type lampholders in
TN or TT system circuits shall be connected to the
neutral conductor.

 This regulation also applies to track-mounted systems.

Lampholders with an ignitability characteristic 'P' as   553-03-01
specified in BS 476 Part 5 or where separate overcurrent
protection is provided, shall not be connected to any
circuit where the rated current of the overcurrent
protective device exceeds the appropriate value stated
in Table 5.18.

Unless the wiring is enclosed in earthed metal or insulating material.

Table 5.18 Plugs and socket outlets for low-voltage circuits

Type of plug and socket outlet	Rating (amperes)	Applicable British Standard
Fused plugs and shuttered socket outlets, 2-pole and earth, for a.c.	13	BS 1363 (fuses to BS 1362)
Plugs, fused or non-fused, and socket outlets, 2-pole and earth	2, 5, 15, 30	BS 546 (fuses, if any, to BS 646)
Plugs, fused or non-fused, and socket outlets, protected-type, 2-pole with earthing contact	5, 15, 30	BS 196
Plugs and socket outlets (industrial type)	16, 32, 63, 125	BS EN 60309-2

Lighting points

Ceiling roses shall not be installed in circuits whose operating voltage is greater than 250 volts.   553-04-02

Fixed lighting points shall use one of the following: 553-04-01

- a batten lampholder to BS 7895, BS EN 60238 or BS EN 61184
- a ceiling rose to BS 67
- a connection unit to BS 5733 or BS 1363-4
- a luminarie designed to be connected directly to the circuit wiring
- a luminarie supporting coupler to BS 6972 or BS 7001
- a suitable socket outlet.

Lighting installations shall be controlled by: 553-04-01

- a switch (or combination of switches) to BS 3676 and/or BS 5518, or a suitable automatic control system.

Unless specifically designed for multiple pendants, ceiling roses shall only have one outgoing flexible cord.   553-04-03

Live supply conductors

Circuits **shall** be capable of being isolated from all live supply conductors by a linked switch or linked circuit breaker. 482-02-08

Luminaries

Extra-low-voltage luminaries that do not use a protective conductor shall only be installed as part of a SELV system. 554-01-02

Luminaire supporting couplers shall only be used to hold up one luminaire electrical connection. 553-04-04

Safety services

Alarm and control devices **shall** be clearly indicated and identified. 563-01-06

Protection against electric shock and fault current **shall** be met for each source. 565-01-02

Safety service circuits **shall** have adequate fire resistance for the locations through which it passes. 563-01-02

Mechanical or electrical interlocking shall be used to avoid the paralleling of sources. 565-01-01

Protection against short circuit and against electric shock shall be met (regardless of whether the installation is supplied by either of the two sources, or by both, in parallel). 566-01-01

Safety service circuits shall be independent of any other circuit so that an electrical fault or an intervention or modification in one system shall not affect the correct functioning of the other. 563-01-01

Safety service equipment operating in fire conditions shall be provided with protection against fire risk. 561-01-02

Safety service sources shall be: 562-01-03

• accessible only to skilled or instructed persons   562-01-02
  installed as fixed equipment.

Safety service sources shall be one of the following: 562-01-01

• a primary cell or cells
• a storage battery
• a generating set capable of independent operation
• a separate feeder effectively independent of the
  normal feeder.

Switchgear and control gear shall be clearly identified   563-01-05
and grouped in locations that are only accessible to
skilled or instructed persons.

The location of the source shall be properly and sufficiently 562-01-06
ventilated so that any exhaust gases, smoke or fumes from
that source cannot penetrate, to a hazardous extent, areas
that are occupied by persons.

Sealing arrangements

Wiring systems (such as conduit, cable ducting,   527-02-02
cable trunking, busbar or busbar trunking) that penetrate
a building construction that has a specified fire resistance,
shall be internally sealed to maintain the degree of fire
resistance.

Sealing of wiring system arrangements shall: 527-02-03

• be compatible with wiring system's material
• permit thermal movement of the wiring system
• be removable without damage to existing cables
• resist relevant external influences to the same degree
  as the wiring system with which it is used.

Special installations and locations

The Regulations list the following specific requirements for wiring systems in
special installations and locations (see Table 5.19).

Table 5.19 Requirements for fire precautions in special installations and locations

Installation or location	Requirement	
Caravans and motor caravans	The following wiring systems shall be used: • **flexible** single-core insulated conductors in non-metallic conduits • stranded insulated conductors (with a minimum of seven strands) in non-metallic conduits • sheathed flexible cables. **Flame** propagating wiring systems shall not be used.	608-06-01

Wiring systems

Wiring systems installed near non-electrical services **shall** be arranged so as not to cause damage to each other.   528-02-04

Wiring systems routed near a service liable to cause condensation (e.g. water, steam or gas services) **shall** be protected.   528-02-03

In locations where there is a risk of fire owing to the nature of processed or stored materials: 482-02-06

• wiring systems (less those using mineral insulated cables and busbar trunking systems) shall be protected against earth insulation faults:
  – in TN and TT systems, by RCDs having a rated residual operating current (IAn) not exceeding 300 mA in accordance with Regulation 531-02-04 in IT systems, by insulation monitoring devices with audible and visible signals.

Wiring systems that pass through floors, walls, roofs, ceilings, partitions or cavity barriers shall be sealed according to the degree of fire resistance required of the element concerned (if any).   527-02-01

 Note: See BS 476-23 for details concerning the sealing of wiring systems.

Wiring systems (such as conduit, cable ducting, cable   527-02-02
trunking, busbar or busbar trunking) that penetrate a
building construction that has specified fire resistance
shall be internally sealed to maintain the degree of
fire resistance.

Wiring systems located in close proximity to a non-electrical   528-02-01
service shall be protected against:

- the hazards likely to arise from that other service indirect
  contact with that other service.

6

Electrical equipment, components, accessories and supplies

The amount of different types of equipment, components, accessories and supplies for electrical installations currently available is enormous and any attempt to cover every type, model and/or manufacture would prove an impossible task for a book such as this. The intention of this chapter, therefore, is to provide a catalogue of all the different types identified and referred to in the Wiring Regulations (e.g. luminaries, RCDs, plugs and sockets etc.) and then make a list of the specific requirements that are sprinkled throughout the Regulations. For your convenience this catalogue has been compiled in alphabetical order.

 Similar to other chapters, please remember that these lists of requirements are **only** the author's impression of the most important aspects of the Wiring Regulations and electricians should **always** consult BS 7671 to satisfy compliance!

6.1 Equipment, components and accessories

Components, cables and wiring enclosures shall be installed in accordance with the temperatures limits set by the relevant product specification or by the manufacturer.	522-01-02
Accessories shall not be connected to conductors intended to operate at a temperature exceeding 70°C.	512-02-01

6.1.1 Ceiling roses

Ceiling roses shall **not** be installed in circuits with an operating voltage greater than 250 volts.   553-04-02

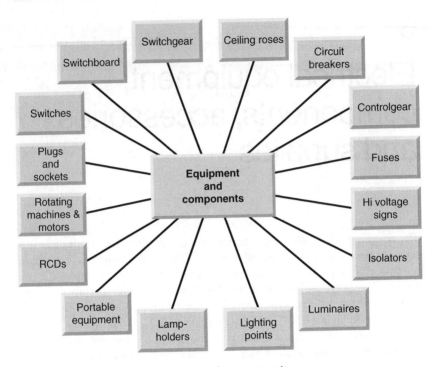

Figure 6.1 Electrical equipment and components

Fixed lighting points shall use a ceiling rose to BS 67.	553-04-01
Unless specifically designed for multiple pendants, ceiling roses shall only have one outgoing flexible cord.	553-04-03

6.1.2 Circuit breakers

Non-skilled personnel **shall** be prevented from changing the setting or calibration of the overcurrent release of a circuit breaker unless they use a key or a tool and this results in a visible indication of its setting (or calibration).	533-01-05

Any linked circuit breaker that has been inserted in an earthed neutral conductor shall be capable of breaking all of the related phase conductors.	131-13-02

Circuit breakers (except where linked) shall be inserted in an earthed neutral conductor.	131-13-02
Circuit breakers shall be capable of breaking and making any overcurrent up to and including the prospective fault current at the point where the device is installed (see Table 6.1).	432-02-01 and 433-02-01 to 433-02-04

Table 6.1 Coordination between conductor and protective device (courtesy of StingRay)

	Design current (I_B) of the circuit	Lowest current-carrying capacity (I_Z) of any conductor in a circuit	Operating current of any protective device (I_2)
Requirements	shall be greater than the nominal current or current setting (I_n) of a protective device.	shall be greater than the nominal current or current setting (I_n) of a protective device.	shall not exceed 1.45 times the lowest of the current-carrying capacities (I_Z) of any of the conductors of the circuit.
BS EN 60898 circuit breakers		Yes	Yes
BS EN 60947-2 circuit breakers		Yes	Yes
Circuit breakers to (or residual current circuit breaker with) integral overcurrent protection (RCBO) to BS EN 61009-1		Yes	Yes

Circuit breakers shall clearly indicate their intended nominal current.	533-01-01
Circuit breakers used as:	
• fault current protection devices shall be capable of making any fault current up to and including the prospective fault current	432-04-01

- emergency switching devices that are operated by   537-04-03
  remote control shall open on de-energisation of
  the coil.

Single-pole circuit breakers shall only be inserted in   131-13-01
the phase conductor.

Construction site, agricultural and horticultural installations

If a circuit breaker is used in a construction, agricultural or horticultural site to disconnect a circuit supplying intentionally moveable installations and/or equipment either directly or through socket outlets, the maximum values of earth fault loop impedance are given in Table 604B2 of the Regulations. As an example, the following table provides an indication of the maximum earth fault loop impedance for Type B circuit breakers (with a nominal voltage of 230 V) corresponding to a disconnection time of 0.2 s.

Table 6.2 Maximum earth fault loop impedance for circuit breakers (courtesy of StingRay)

Type B circuit breakers to BS EN 60898 and RCBOs to BS EN 61009

Rating (amperes)	6	10	20	40	60	80	100	125
Z_s (ohms)	8.00	4.80	2.40	1.20	0.76	0.60	0.48	0.38

 Note: The circuit loop impedances given in the table above should not be exceeded when the conductors are at their normal operating temperature. If the conductors are at a different temperature when tested, then the reading should be adjusted accordingly.

6.1.3 Control gear

Control gear **shall** be clearly identified and grouped   563-01-05
in locations accessible only to skilled or instructed persons.

Control gear assemblies shall be functionally tested to   713-13-02
show that they are properly mounted, adjusted and
installed in accordance with the Regulations.

Protection by use of Class II equipment may be provided by low-voltage control gear assemblies having total insulation (see BS EN 60439).

413-03-01

When control gear assemblies are used as protective conductors, any associated metal enclosure/frame shall:

543-02-04

- be protected against mechanical, chemical or electrochemical deterioration and have a cross-sectional area not less than:
  - 2.5 mm^2 copper equivalent if protection against mechanical damage is provided or
  - 4 mm^2 copper equivalent if mechanical protection is not provided
- allow other protective conductors to be connected at predetermined tap-off points.

Special installations and locations

The Regulations list the following specific requirements for control gear in special installations and locations (see Table 6.3).

Table 6.3 Requirements for switchgear in special installations and locations

Installation or location	Requirement	
Locations containing a bath or a shower	Control gear shall not be installed in zone O.	601-08-01
Swimming pools	In zones A and B, control gear shall **not** be installed, unless socket outlets comply with BS EN 60309-2 and they are installed: • more than 1.25 m outside of zone A at least 0.3 m above the floor.	602-07-01
Hot air saunas	All control gear (other than a thermostat or a thermal cut-out) not actually built into the sauna heater shall be installed outside the hot air sauna.	603-08-01
Agricultural and horticultural premises	Emergency stopping devices shall: • be inaccessible to livestock • not be impeded by livestock in the event of them panicking.	605-13-01

6.1.4 Fuses

 Off-load fuses and links shall **not** be used for functional switching.   537-05-03

Fuses shall **not** be inserted in the neutral conductor of TN or TT systems.   530-01-02

Fuses shall **not** be inserted in an earthed neutral conductor.   131-13-02

Single-pole fuses shall only be inserted in the phase conductor.   131-13-01

A polarity test shall be made to verify that fuses are only connected in the phase conductor.   713-09-01

Where a fuse is used to disconnect a circuit supplying socket outlets and other final circuits which supply portable equipment that is intended for manual movement during use, or hand-held Class I equipment:

the operating current of any protective device (I_2) shall not exceed 1.45 times the lowest of the current-carrying capacities (I_z) of any of the conductors of the circuit.   433-02-01

The maximum values of earth fault loop impedance corresponding to a disconnection time of 0.4 s shall be in accordance with Table 41B1 of the Regulations for a nominal voltage to earth (U_0) of 230 V.   413-02-10

 Note: Table 6.4 below is an indication of some of these values.

Where the device is a general purpose type fuse to BS 88-2.1, a fuse to BS 88-6, a fuse to BS 1361 or a semi-enclosed fuse to BS 3036, the conditions shown in Table 6.5 shall apply (433-02-02 to 433-02-04).

Note: The circuit loop impedances given in Table 6.5 should not be exceeded when the conductors are at their normal operating temperature. If the conductors are at a different temperature when tested, then the reading should be adjusted accordingly.

Table 6.4 British Standards for fuse links

	Standard	Current rating	Voltage rating	Breaking capacity	Notes
1	BS 2950	Range 0.05–25 A	Range 1000 V (0.05 A) to 32 V (25 A) a.c. and d.c.	Two or three times current rating	Cartridge fuse links for telecommunication and light electrical apparatus. Very low breaking capacity
2	BS 646	1, 2, 3 and 5 A	Up to 250 V a.c. and d.c.	1000 A	Cartridge fuse intended for fused plugs and adapters to BS 546: 'round-pin' plugs
3	BS 1362 cartridge	1, 2, 3, 5, 7, 10 and 13 A	Up to 250 V a.c.	6000 A	Cartridge fuse primarily intended for BS 1363: 'flat pin' plugs
4	BS 1361 HRC cut-out fuses	5, 15, 20, 30, 45 and 60 A	Up to 250 V a.c.	16,500 A 33,000 A	Cartridge fuse intended for use in domestic consumer units. The dimensions prevent interchangeability of fuse links which are not of the same current rating
5	BS 88 motors	Four ranges, 2–1200 A	Up to 660 V, but normally 250 or 415 V a.c. and 250 or 500 V d.c.	Ranges from 10,000 to 80,000 A in four a.c. and three d.c. categories	Part 1 of Standard gives performance and dimensions of cartridge fuse links, whilst Part 2 gives performance and requirements of fuse carriers and fuse bases designed to accommodate fuse links complying with Part 1
6	BS 2692	Main range from 5 to 200 A; 0.5 to 3 A for voltage transformer protective fuses	Range from 2.2 to 132 kV	Ranges from 25 to 750 MVA (main range) 50 to 2500 MVA (VT fuses)	Fuses for a.c. power circuits above 660 V
7	BS 3036 rewirable	5, 15, 20, 30, 45, 60, 100, 150 and 200 A	Up to 250 V to earth	Ranges from 1000 to 12,000 A	Semi-enclosed fuses (the element is a replacement wire) for a.c. and d.c. circuits
8	BS 4265	500 mA to 6.3 A 32 mA to 2 A	Up to 250 V a.c.	1500 A (high breaking capacity) 35 A (low breaking capacity)	Miniature fuse links for protection of appliances of up to 250 V (metric standard)

Table 6.5 Coordination between conductor and protective device (courtesy of StingRay)

	Design current (I_B) of the circuit	Lowest current-carrying capacity (I_Z) of any conductor in a circuit	Operating current of any protective device (I_2)
Requirements	Shall be greater than the nominal current or current setting (I_n) of a protective device	Shall be greater than the nominal current or current setting (I_n) of a protective device	Shall not exceed 1.45 times the lowest of the current-carrying capacities (I_Z) of any of the conductors of the circuit
BS 88-2.1 fuses		Yes	Yes
BS 88-6 fuses		Yes	Yes
BS 1361 fuses		Yes	Yes
BS 3036 (semi-enclosed) fuse			Yes provided (I_n) does not exceed 0.725 (I_Z)

All fuses shall clearly indicate their intended nominal current.   533-01-01

Any fuse that can be removed or replaced by an unskilled person shall: 533-01-02

- be capable of being removed or replaced without danger
- clearly indicate the type of fuse link that should be used
- preferably be of a type that cannot be replaced inadvertently by one having a higher fusing factor.

Fuses shall ideally be of the cartridge type.   533-01-04

Fuses that could be removed and replaced whilst the supply is connected shall be of a type where this removal and replacement can be done without incurring any danger – either to the person or the electrical installation.   533-01-03

Semi-enclosed fuses shall be fitted with an element in accordance with the manufacturer's instructions or (in the absence of any instructions) they shall be fitted with a single element of tinned copper wire as specified in Table 6.6.   533-01-04

Table 6.6 Sizes of tinned copper wire for use in semi-enclosed fuses

Nominal current (A)	Nominal diameter of wire (mm)
3	0.15
5	0.2
10	0.35
15	0.5
20	0.6
25	0.75
30	0.85
45	1.25
60	1.53
80	1.8
100	2.0

All low-voltage fused plug and socket outlets shall conform with the applicable British Standard listed in Table 6.7.

553-01-03

Table 6.7 Plugs and socket outlets for low-voltage circuits

Type of plug and socket outlet	Rating (amperes)	Applicable British Standard
Fused plugs and shuttered socket outlets, 2-pole and earth, for a.c.	13	BS 1363 (fuses to BS 1362)
Plugs, fused or non-fused, and socket outlets, 2-pole and earth	2, 5, 15, 30	BS 546 (fuses, if any, to BS 646)
Plugs, fused or non-fused, and socket outlets, protected-type, 2-pole with earthing contact	5, 15, 30	BS 196
Plugs and socket outlets (industrial type)	16, 32, 63, 125	BS EN 60309-2

Semi-enclosed fuses shall clearly indicate their intended nominal current.

533-01-01

Where a fuse is used, the maximum value of earth fault loop impedance shall permit a disconnection time of less than 5 s.

471-15-06

Construction site installations and agricultural/horticultural installations

Where a fuse is used to disconnect a circuit supplying socket outlets and other final circuits supplying portable equipment (that is intended for manual movement during use), or general purpose hand-held Class I equipment (being used in construction, agricultural and/or horticultural sites), the maximum values of earth fault loop impedance are shown in Table 604B1 of the Regulations. As an example, Table 6.8 lists the maximum earth fault loop impedance for a disconnection time of 0.2 s for a nominal voltage of 230 V (604-04-03 and 605-05-03).

Table 6.8 Maximum earth fault loop impedence (Z_s) for fuses, for 0.2 s disconnection time with U_o of 230 V (Reproduced courtesy of BSI).

General purpose fuses to BS 88-2.1 and BS 88-6

Rating (amperes)	6	10	16	20	25	32	40	50
Z_s (ohms)	7.74	4.71	2.53	1.60	1.33	0.92	0.71	0.53

Highway power supplies (street furniture and street-located equipment)

For highway power supplies where it is intended that isolation and switching will **only** be carried out by instructed persons, a suitably rated fuse carrier may be used.	611-03-01

6.1.5 High-voltage signs

High-voltage electric signs and high-voltage luminous discharge tube installations shall meet the requirements of BS 559.	554-02-01

6.1.6 Isolators

Off-load isolators shall **not** be used for functional switching.	537-05-03
Isolators shall **not** break a protective conductor or a PEN conductor.	460-01-03

If an isolating device for a particular circuit is some distance away from the equipment to be isolated, then that isolating device shall be capable of being secured in the open position.

476-02-02

 Note: When this is achieved by a lock or removable handle, the key or handle shall be non-interchangeable with any others used for a similar purpose within the premises.

Isolators, when used with a circuit breaker for isolating main switchgear for maintenance, shall be interlocked with that circuit breaker or placed and/or guarded so that it can only be operated by skilled persons.

476-02-01

The isolating distance between contacts when in the open position shall be not less than that determined for an isolator (disconnector).

537-02-02

The location of each isolator shall be indicated unless there is no possibility of confusion.

514-11-01

6.1.7 Lampholders

Filament lampholders shall **not** be installed in circuits operating in excess of 250 V.

553-03-02

Lampholders with an ignitability characteristic 'P' as specified in BS 476 Part 5 (or where separate overcurrent protection is provided) shall not be connected to any circuit where the rated current of the overcurrent protective device exceeds the appropriate value stated in Table 6.9.

553-03-01

Unless the wiring is enclosed in earthed metal or insulating material.

Table 6.9 Overcurrent protection of lampholders

Type of lampholder			Maximum rating (amperes) of overcurrent protective device protecting the circuit
Bayonet (BS	B15	SBC	6
EN 61184)	B22	BC	16
Edison screw (BS	E14	SES	6
EN 60238)	E27	ES	16
	E40	GES	16

Bayonet lampholders B15 and B22 shall:

- comply with BS EN 61184
- have a temperature rating T2 as described in BS EN 61184.   553-03-03

Dual-voltage luminaries shall be fitted with separate   608-08-07
lampholders for each voltage.

Except for E14 and E27 lampholders complying with   553-03-04
BS EN 60238, the outer contact of all Edison screw
or single centre bayonet cap type lampholders in
TN or TT system circuits shall be connected to the
neutral conductor.

 This regulation also applies to track-mounted systems.

Except for suspended lampholders which have no exposed   471-08-08
conductive parts, installations that provide protection
against indirect contact by automatically disconnecting
the supply shall have a circuit protective conductor run
to (and terminated at) each point in the wiring and at
each accessory.

Except for suspended lampholders which have no exposed   471-09-02
conductive parts, circuits supplying Class II equipment shall
have a circuit protective conductor run to, and terminated
at, each point in the wiring and at each accessory.

6.1.8 Luminaries

Figure 6.2 An example of a luminaire used for street lighting

A luminaire is defined as '*an equipment which distributes, filters or transforms the light from one or more lamps and which includes any parts necessary for supporting, fixing and protecting the lamps (but not the lamps themselves) and, where necessary, circuit auxiliaries together with the means for connecting them to the supply*'. For the purposes of the Regulations a lampholder, however supported, is also deemed to be a luminaire and the following requirements must be met.

Construction

Self-contained luminaries (or circuits supplying luminaries) with an open circuit voltage exceeding low voltage shall have:

- an interlock (in addition to the switch normally used for controlling the circuit) that automatically disconnects the live parts of the supply before access to these live parts is permitted   476-02-02
- some form of isolatator (in addition to the switch normally used for controlling the circuit) to isolate the circuit from the supply
- a switch (with a lock or removable handle) or a distribution board which can be locked.

Lighting points

Fixed lighting points shall use one of the following: • a batten lampholder to BS 7895, BS EN 60238 or BS EN 61184 • a ceiling rose to BS 67 • a connection unit to BS 5733 or BS 1363-4 • a luminaire designed to be connected directly to the circuit wiring • a luminaire-supporting coupler to BS 6972 or BS 7001 • a suitable socket outlet.	553-04-01
Lighting installations shall be controlled by: • a switch (or combination of switches) to BS 3676 and/or BS 5518, or • a suitable automatic control system.	553-04-01
Lighting track systems shall comply with BS EN 60570.	521-06-01
Supporting couplers shall only be used for the mechanical and electrical connection of luminaries.	553-04-04
Where a pendant luminaire is installed, the associated accessory shall be suitable for the mass suspended.	554-01-01

SELV systems

Luminaire supporting couplers with a protective conductor contact shall **not** be installed in a SELV system.	411-02-11
Extra-low-voltage luminaries that do not allow the connection of a protective conductor shall only be installed as part of a SELV system.	554-01-02

Non-conducting locations

Luminaire-supporting couplers which have a protective conductor contact shall **not** be installed in a non-conducting location.	413-04-03

Locations with risk of fire

Lumaires shall be of a type that prevents lamp components   482-02-14
from falling from the luminaire – particularly in locations
where there is a possible risk of fire owing to the nature
of processed or stored materials.

Special installations and locations

The Regulations list the following specific requirements for luminaries in special installations and locations (see Table 6.10).

Table 6.10 Requirements for luminaries in special installations and locations

Installation or location	Requirement	
Hot air saunas	Luminaries shall be mounted so as to prevent overheating.	603-09-01
Construction sites	Luminaire-supporting couplers shall not be used.	604-12-03
Caravans and motor caravans	Luminaries shall (preferably) be fixed directly to the structure or lining of the caravan or motor caravan.	608-08-06
	Pendant luminaires shall be securely installed to prevent damage when the caravan or motor caravan is moved.	608-08-06
	Dual-voltage luminaries shall: • be fitted with separate lampholders for each voltage • clearly indicate the lamp wattage • not be damaged if both lamps are lit at the same time • provide adequate separation between LV and ELV circuits • have lamps that cannot be fitted into lampholders intended for lamps of other voltages.	608-08-07
Highway power supplies (street furniture and street-located equipment)	Access to the light source for luminaries located less than 2.80 m above ground level shall only be possible after removing a barrier or enclosure that requires the use of a tool.	611-02-02

6.1.9 Portable equipment

Permanent protection **shall** be provided in all   413-04-05
areas where the use of mobile and/or portable
equipment is envisaged.

The maximum disconnection times to a circuit supplying socket outlets and final circuits that are the source for portable equipment (intended for manual movement during use) or hand-held Class I equipment shall not be exceeded.	413-02-09
Where a fuse or a circuit breaker is used to disconnect a circuit supplying socket outlets and final circuits which supply portable equipment (intended for manual movement during use) or hand-held Class I equipment, the maximum values of earth fault loop impedance shall not be exceeded.	413-02-10 and 413-02-11

Supplies for portable equipment outdoors

The following are additional requirements for installations and locations where the risk of electric shock is increased by a reduction in body resistance and/or by contact with earth potential.

All socket outlets and permanently connected hand-held equipment (with a current rating of up to and including 32 A) shall be protected.	604-08-03
Circuits supplying portable equipment for use outdoors and not connected to a socket by means of flexible cable or cord (having a current-carrying capacity of 32 A or less) shall be protected by an RCD.	471-16-02
Socket outlets rated at 32 A or less supplying portable equipment for use outdoors shall be protected by an RCD.	471-16-01

 This regulation does not apply to socket outlets supplied by a circuit protected by:

- SELV
- electrical separation
- automatic disconnection and reduced low-voltage systems.

Where portable equipment is likely to be used, a conveniently accessible socket outlet shall be provided.	553-01-07

6.1.10 Residual Current Devices

Automatic disconnection using an RCD shall **not** be applied to a circuit incorporating a PEN conductor. — 471-08-07

RCDs **shall** be capable of disconnecting all of the phase conductors of a circuit at the same time. — 531-02-01

Although RCDs reduce the risk of electric shock, they may **not** be used as the sole means of protection against direct contact. — 412-06-01 and 412-06-02

RCDs **shall** be located so that their operation will not be harmed by magnetic fields from other equipment. — 531-02-07

RCDs that are used to provide protection against indirect contact **shall** be capable of withstanding the likely thermal and mechanical stresses that it will probably be subjected to if fault occurred. — 531-02-08

RCDs that can be operated by a non-skilled person **shall** be designed and installed so that its operating current and/or time delay mechanism cannot be altered without using a key or a tool. — 531-02-10

Socket outlets rated at 32 A or less that supply portable equipment for use outdoors shall be provided with an RCD. — 471-16-01

The magnetic circuit of the transformer of an RCD **shall** enclose all the live conductors of the protected circuit. — 531-02-02

The protective conductor of an RCD **shall** be outside the magnetic circuit. — 531-02-02

Two or more RCDs connected in series **shall** be capable of operating separately. — 531-02-09

If an installation includes an RCD then it shall have a notice (fixed in a prominent position) that reads as Figure 6.3.

This installation, or part of it, is protected by a device which automatically switches off the supply if an earth fault develops. Test quarterly by pressing the button marked 'T' or 'Test'. The device should switch off the supply and should then be switched on to restore the supply. If the device does not switch off the supply when the button is pressed, seek expert advice.

Figure 6.3 Inspection and testing notice

Other requirements from the Regulations include:

Circuits supplying portable equipment for use outdoors and which are not connected to a socket by means of flexible cable or cord (having a current-carrying capacity of 32 A or less) shall be provided with supplementary protection in the form of an RCD.	471-16-02
Electrode water heaters or boilers that are directly connected to a supply that exceeds low voltage shall include an RCD capable of disconnecting the supply from the electrodes when the sustained protective conductor exceeds 10% of the rated current of the heater or boiler.	554-03-04

Note: A time delay may be incorporated in the device to prevent unnecessary operation in the event of imbalance of short duration.

Protection against indirect contact by automatic disconnection may be provided by an RCD.	471-15-06
RCDs associated with a circuit that would normally be expected to have a protective conductor shall not be considered sufficient protection against indirect contact if there is no such conductor.	531-02-05
RCDs shall ensure that any protective conductor current which may occur will be unlikely to cause unnecessary tripping of the device.	531-02-04
RCDs which are powered from an auxiliary source (and which do not operate automatically in the case of failure of that auxiliary source) shall only be used if: • protection against indirect contact is ensured • the device is part of a supervised (tested and inspected) installation.	531-02-06
RCDs that can be operated by a non-skilled person shall be designed and installed so that its operating current and/or time delay mechanism cannot be altered without using a key or a tool.	531-02-10

All alterations that have been made to the setting or calibration must be visibly indicated.

Socket outlets rated at 32 A or less that supply portable | 471-16-01
equipment for use outdoors shall be provided with
an RCD.

This regulation does not apply to socket outlets supplied by a circuit protected by:

- SELV
- electrical separation
- automatic disconnection and reduced low voltage.

The effectiveness of an RCD providing protection | 713-13-01
against indirect contact or supplementary protection
against direct contact shall be verified by a test
simulating an appropriate fault condition.

Where an RCD is used, the product of the rated residual | 471-15-06
operating current (in amperes) and the earth fault loop
impedance shall not exceed 50 Ω.

Where automatic disconnection of TN, TT and/or IT | 413-02-04
systems cannot be achieved by using overcurrent
protective devices, then an RCD shall be used.

Wiring systems (less those using mineral insulated | 482-02-06
cables and busbar trunking systems) shall be protected
against earth insulation faults:

- in TN and TT systems, by RCDs having a rated
  residual operating current ($I_{\Delta n}$) not exceeding 300 mA
- in IT systems, by insulation monitoring devices with
  audible and visible signals.

Note: The disconnection time of the overcurrent protective device, in the event of a second fault, shall not exceed 5 s.

RCDs in a TN system

For TN-S systems, an RCD shall be positioned to avoid | 551-06-02
incorrect operation due to the existence of any parallel
neutral earth path.

If an RCD is used for automatic disconnection of a   413-02-17
circuit which extends beyond the conductive parts,
then exposed conductive parts do not need to be
connected to the TN system protective earthed
equipotential zone, provided that they are connected
to an earth electrode which has a resistance appropriate
to the operating current of the RCD.

If protection is provided by an RCD then:

$$Z_s \times I_{An} \leq 50\,V$$ 413-02-16

where Z_s is the earth fault loop impedance in ohms and I_{An}
is the rated residual operating current of the protective
device in amperes.

Where an RCD is used in a TN-C-S system, the protective   413-02-07
conductor to the PEN conductor shall be on the source
side of the RCD.

RCDs in a TT system

If the installation is part of a TT system, all socket outlet   471-08-06
circuits shall be protected by an RCD.

Parts of a TT system that are protected by a single RCD   531-04-01
shall be placed at the origin of the installation unless that
part between that origin and the device is protected by a
Class II equipment or an equivalent insulation.

 Where there is more than one origin this requirement applies to each origin.

RCDs in an IT system

For installations and locations where there is an increased   471-08-01
risk of shock (such as agricultural and horticultural
premises, building sites, bathrooms, swimming pools etc.)
automatic disconnection of the supply by means of an
RCD with a rated residual operating current (I_{sn}) not
exceeding 30 mA may be required, as an additional
protective measure.

Where an IT system is protected against indirect contact   413-02-25
by an RCD, each final circuit shall be separately protected.

Where protection is provided by an RCD (and   531-05-01
disconnection following a first fault is not envisaged)
the non-operating residual current of the device
shall be at least equal to the current which circulates
on the first fault to earth.

Special installations and locations

The Regulations list the following specific requirements for luminaries in special installations and locations (Table 6.11).

Table 6.11   Requirements for luminaries in special installations and locations

Installation or location	Requirements	
Locations containing a bath or a shower	Socket outlets (other than a SELV socket outlet or shaver supply unit) located in shower cubicles that are installed in rooms other than a bathroom or shower room, and that is outside zones 0, 1, 2 or 3, shall be protected by an RCD.	601-08-02
	In zone 3, current-using equipment other than fixed current-using equipment shall be protected by an RCD.	601-09-03
Swimming pools	In zones A and B, switchgear, control gear and accessories shall **not** be installed, unless socket outlets comply with BS EN 60309-2 and they are protected by an RCD and: • installed more than 1.25 m outside of zone A installed at least 0.3 m above the floor.	602-07-01
	In zone C, socket outlets, switches or accessories are only permitted if they are protected by an RCD or: • protected individually by electrical separation protected by SELV, or are a shaver socket complying with BS 3535. ☀ This requirement does not apply to the insulating cords of cord-operated switches complying with BS 3676.	602-07-02
	In zone C, equipment may be protected by an RCD. ☀ This requirement does not apply to instantaneous water heaters.	602-08-03
Restrictive conductive locations	Circuits (less SELV) supplying a socket outlet shall be protected by an RCD.	605-03-01
	For fire protection purposes an RCD (with a residual operating current not exceeding 0.5 A) shall be installed on equipment supplies other than that essential to the welfare of livestock.	605-10-01

(continued)

Table 6.11 (*continued*)

Installation or location	Requirements	
Locations with equipment having high protective conductor currents	If more than one item of equipment has a protective conductor current exceeding 3.5 mA and is going to be supplied from an installation incorporating an RCD, the circuit used shall ensure that the residual current (including switch-on surges), will not trip the device.	607-07-01
Caravan parks	When protection by automatic disconnection of supply is used, an RCD complying with BS 4293, BS EN 61008-1 or BS EN 61009-1 that breaks all live conductors shall be provided and the wiring system shall include a circuit-protective conductor that is connected to the • inlet protective contact, and • electrical equipment's exposed conductive parts, and • socket outlets' protective contacts.	608-03-02
	Socket outlets shall be protected (individually or in groups of not more than three) by an RCD complying with BS 4293, BS EN 61008-1 or 61009-1.	608-13-05

6.1.11 Rotating machines and motors

Emergency switching devices **shall** be capable of cutting off the full load current of the relevant part of the installation.   537-04-01

Fixed electric motors **shall** be provided with a readily accessible, efficient means of switching off, that is easily operated and safely positioned.   131-14-02

Unsupervised motors which are automatically (or remotely) controlled **shall** be protected against excessive temperature by a protective device with manual reset.   482-02-11

Other requirements include:

All equipment (including the cables of circuits carrying the starting, accelerating and load currents of a motor) shall be suitable for a current of at least equal to the full-load current rating of the motor.   552-01-01

All motor circuits shall be provided with a disconnecter to disconnect the motor and equipment.   476-02-03

Electric motors with a rating of more than 0.37 kW shall be protected against overload of the motor.	552-01-02
Emergency switching devices used to interrupt the supply shall be capable of cutting off the full load current of the relevant part of the installation.	537-04-01

 Note: Where appropriate, consideration of stalled motor conditions should be taken into account.

Motors shall be provided with means to prevent automatic restarting after a stoppage due to a drop in voltage or failure of supply – particularly where an unexpected restarting of the motor might cause danger.	552-01-03

 This requirement does not apply where failure to start after a brief interruption would be likely to cause greater danger.

Motors that are intended for intermittent duty (and/or for frequent starting and stopping) shall take into account the potential temperature rise of the equipment of the circuit caused by the cumulative effects of the starting or braking currents.	552-01-01
Motors with star-delta starting shall be protected against excessive temperature in both the star and the delta connections.	482-02-11
Overload protection devices need not be used for circuits supplying exciter circuits of rotating machines where unexpected disconnection of the circuit could cause danger.	473-01-03

 In such situations consideration shall be given to the provision of an overload alarm.

The voltage drop between the supply terminals of an installation and a socket outlet (or the terminals of a fixed current-using equipment) shall not exceed 4% of the nominal voltage of the supply.	525-01-02

 Note: A greater voltage drop is acceptable during motor starting periods and for equipment with high inrush currents provided that it is verified that the voltage variations are within the limits specified in the relevant British Standards.

Where reverse-current braking is provided, motors shall be prevented from rotating in a reverse direction at the end of braking.	552-01-04
Where safety depends on the direction of rotation of a motor, provision shall be made to prevent reverse operation.	552-01-05

6.1.12 Plugs and socket outlets

Plug and socket outlet shall **not** be used as a device for emergency switching.	537-04-02
Plug and socket outlets with a rating of more than 16 A may **not** be used as a switching device for d.c.	537-05-05
All potentially hazardous fixed or stationary appliances that are not connected to the supply by means of a plug and socket outlet (rated at less than 16 A) **shall** be provided with a means of interrupting the supply on load.	476-03-04
In a non-conducting location, socket outlets shall **not** incorporate an earthing contact.	413-04-03
The voltage drop between the supply terminals of an installation and a socket outlet **shall** not exceed 4% of the nominal voltage of the supply.	525-01-02

Other requirements include:

A plug and socket outlet rated at not more than 16 A may be used for functional switching.	464-01-04

 Except for d.c., where this purpose is specifically excluded.

All plugs and socket outlets of a reduced low-voltage system: • shall have a protective conductor contact and shall not be dimensionally compatible with any plug, socket outlet or cable coupler being used at any other voltage or frequency in the same installation.	471-15-07

All socket outlets shall use a plug which is dimensionally different with those being used by any other system in the same premises.	471-14-06
Circuits supplying portable equipment for use outdoors (and not connected to a socket by means of flexible cable or cord) and which have a current-carrying capacity of 32 A or less, shall be provided with an RCD as supplementary protection.	471-16-02
If the installation is part of a TF system, all socket outlet circuits shall be protected by an RCD.	471-08-06

Plug and socket outlets with a rating not exceeding 16 A may be used:

- as a means for switching off for mechanical maintenance (537-03-05)   537-05-05 and
- as a switching device, provided that the plug and socket outlet has a breaking capacity appropriate to the use intended regarding emergency switching.   537-04-02

Plug and socket outlets may be used:

- as a means of isolation   537-02-10
- as a switching device providing their rating does not exceed 16 A.   537-05-04

Socket outlets rated at 32 A or less supplying portable equipment for use outdoors shall be provided with an RCD as supplementary protection.	471-16-01

 This regulation does not apply to socket outlets supplied by a circuit protected by:

- SELV
- electrical separation
- automatic disconnection and reduced low-voltage systems.

Source supplies may supply more than one item of equipment provided that all socket outlets are provided with a protective conductor contact that is connected to the equipotential bonding conductor.	413-06-05

The earth fault loop impedance at all points of utilisation (including socket outlets) shall be such that the disconnection time does not exceed 5 s.   471-15-06

The maximum disconnection times to a circuit supplying socket outlets which supply portable equipment intended for manual movement during use, or hand-held Class I equipment, shall not exceed those given in Table 41A (p. 44) of the IEE Wiring Regulations.   413-02-09

 Note: This requirement does not apply to a final circuit supplying an item of stationary equipment connected by means of a plug and socket outlet where precautions are already taken to prevent the use of the socket outlet for supplying hand-held equipment, nor to the reduced low-voltage circuits.

When Class II equipment is used as the sole means of protection against indirect contact, the installation or circuit concerned must be under effective supervision whilst in normal use.   471-09-03

 This form of protection shall **not** be used for any circuit that includes socket outlets or where a user can change items of equipment without authorisation.

Where supplementary equipotential bonding is necessary, it shall connect together the exposed conductive parts of equipment in the circuits concerned including the earthing contacts of socket outlets and extraneous conductive parts.   413-02-27

Accessories

Except for SELV circuits, the pin of a plug shall **not** be capable of making contact with any live contact: 553-01-01

- of its associated socket outlet when any other pin of the plug is completely exposed
- of any socket outlet within the same installation other than the type of socket outlet for which the plug is designed.

Wall sockets and outlets shall be mounted above the floor and all working surfaces.	553-01-06

Except for SELV circuits (or a special circuit from Regulation 553-01-05 – see below) all plug and socket outlets shall be non-reversible and capable of including a protective conductor.	553-01-02 and 553-01-05
All low-voltage plug and socket outlets shall conform with the applicable British Standard listed in Table 6.12 – except where Regulation 553-01-05 applies (see below).	553-01-03 and 553-01-05

Table 6.12 Plugs and socket outlets for low-voltage circuits

Type of plug and socket outlet	Rating (amperes)	Applicable British Standard
Fused plugs and shuttered socket outlets, 2-pole and earth, for a.c.	13	BS 1363 (fuses to BS 1362)
Plugs, fused or non-fused, and socket outlets, 2-pole and earth	2, 5, 15, 30	BS 546 (fuses, if any, to BS 646)
Plugs, fused or non-fused, and socket outlets, protected-type, 2-pole with earthing contact	5, 15, 30	BS 196
Plugs and socket outlets (industrial type)	16, 32, 63,125	BS EN 60309-2

All household socket outlets **shall** be of the shuttered type.	553-01-04

Note: For an a.c. installation, they shall preferably comply with BS 1363.

Fixed lighting points shall use a suitable socket outlet.	553-04-01
Plug and socket outlets not complying with BS 1363, BS 546, BS 196 or BS EN 60309-2 may be used in	553-01-05

single-phase a.c. or two-wire d.c. circuits that operate
at a nominal voltage not exceeding 250 volts for:

- an electric clock (provided that the plug and socket
  outlet is specifically designed for that purpose and
  the plug has a fuse (rated no more than 3 amperes)
  which complies with BS 646 or BS 1362)
- an electric shaver (provided that the socket outlet is
  either part of the shaver supply unit that complies with
  BS 3535 or in a room (other than a bathroom) that
  complies with BS 4573)
- circuits with special characteristics (where the
  function of the circuit needs to be distinguished).

Wall sockets and outlets shall be mounted above the floor   553-01-06
and all working surfaces in order to minimise the risk of
mechanical damage during insertion, use or withdrawal
of the plug.

Water heaters and boilers shall be permanently connected   554-05-03
to the electricity supply via a double-pole linked switch
which is either:

- separate from and within easy reach of the heater/
  boiler or
- part of the boiler/heater (provided that the wiring
  from the heater or boiler is directly connected to the
  switch without use of a plug and socket outlet).

Where portable equipment is likely to be used, a   553-01-07
conveniently accessible socket outlet shall be
provided.

SELV

The socket outlet of a SELV system shall:

- be compatible with the plugs used for other systems   411-02-10
  being used, in the same premises
- not have a protective conductor contact.

In locations containing a bath or a shower

In all locations containing a bath or a shower, the following requirements (see
Table 6.13) shall apply.

Table 6.13 Zonal limitation on switchgear and control gear

Zone	Limitation
Zone 0	Switchgear or accessories shall not be installed.
Zone 1	Only switches of SELV circuits (supplied at a nominal voltage not exceeding 12 V a.c. RMS or 30 V ripple-free d.c.) shall be installed.
	Note: The safety source shall be installed outside zones 0, 1 and 2.
Zone 2	Only switches and socket outlets of SELV circuits and shaver supply units complying with BS EN 60742 shall be installed.
	Note: The safety source shall be installed outside zones 0, 1 and 2.
Zone 3	Only SELV socket outlets and shaver supply units complying with BS EN 60742 shall be installed.
	Note: Portable equipment can only be connected as above.

Socket outlets (other than SELV socket outlets and/or shaver supply units complying with BS EN 60742) shall **not** be installed outside zones 0, 1, 2 and 3.   601-08-02

When the heater/ boiler is installed in a room containing a fixed bath, the switch shall comply with the specific requirements for these types of locations.   554-05-03

Socket outlets (other than a SELV socket outlet or shaver supply unit) located in shower cubicles that are installed in rooms other than a bathroom or shower room, and that is outside zones 0, 1, 2 or 3, shall be protected by an RCD.   601-08-02

Swimming pools

In zones A and B:

- the only protective measure against electric shock that may be used is SELV and the safety source shall be installed outside zones A, B and C, unless socket outlets are protected by an RCD.   602-04-01

- switchgear, control gear and accessories shall **not**   602-07-01
  be installed, unless socket outlets comply with
  BS EN 60309-2 and they are:
  – installed more than 1.25 m outside of zone A
  – installed at least 0.3 m above the floor
  – protected by either an RCD or electrical
    separation (with the safety isolating transformer
    placed outside zones A, B and C).

In zone C, socket outlets are only permitted if they comply with BS EN 60309-2 and are:

- protected individually by electrical separation or   602-07-02
- protected by SELV or   and
- protected by an RCD or   602-08-01
- a shaver socket complying with BS 3535.

This requirement does not apply to the insulating cords of cord-operated switches complying with BS 3676.

On construction site installations

The following are additional requirements for installations and locations where the risk of an electric shock is increased by a reduction in body resistance and/or by contact with earth potential.

Circuits supplying portable equipment for use   471-16-02
outdoors **shall** be provided with an RCD as
supplementary protection.

Plug and socket outlets **shall** comply with   604-12-02
BS EN 60309-2.

Other requirements from the Regulations include:

If required, circuits supplying current using equipment   604-11-05
shall be fed from a distribution assembly that includes
socket outlets.

Socket outlets rated at 32 A or less supplying portable equipment for use outdoors shall be provided with an RCD as supplementary protection.	471-16-01

 This regulation does not apply to socket outlets supplied by a circuit protected by:

- SELV
- electrical separation
- automatic disconnection and reduced low-voltage systems.

Socket outlets shall be part of an assembly complying with BS 4363 and BS EN 60439-4.	604-12-01
Where socket outlets are protected by electrical separation they shall be supplied from a separate isolating transformer.	604-08-04
Circuits supplying portable equipment for use outdoors and not connected to a socket by means of flexible cable or cord (having a current-carrying capacity of 32 A or less) shall be provided with an RCD as supplementary protection.	471-16-02
RCDs shall be used as supplementary protection for all portable equipment being used outdoors which is not connected to a socket by means of flexible cable or cord with a current-carrying capacity of at least 32 A.	471-16-02

In agricultural and horticultural premises

Circuits (less SELV) supplying a socket outlet **shall** be protected by an RCD.	605-03-01

Except for socket outlets protect by an RCD, the maximum disconnection times to a circuit supplying socket outlets which supply portable equipment intended for manual movement during use (or hand-held Class I equipment) shall not exceed those shown in Table 6.14.	605-05-02

Note: This requirement does not apply to a final circuit supplying an item of stationary equipment connected by means of a plug and socket outlet where precautions are already taken to prevent the use of the socket outlet for supplying hand-held equipment, nor to the reduced low-voltage circuits.

Table 6.14 Maximum disconnection times for TN systems

Installation nominal voltage U_o (volts)	Maximum disconnection time t (seconds)
120	0.35
220 to 277	0.2
400, 480	0.05
580	0.02

In restrictive conductive locations

The supply to socket outlets used by a:

- hand lamp, shall be protected by SELV   606-04-02
- hand-held tool, shall be protected by SELV or by   606-04-04
  electrical separation.

Where earthing requirements for the installation of equipment having high protective conductor currents is a requirement

Equipment with a protective conductor current of   607-02-02
between 3.5 mA and 10 mA shall be connected by a plug
and socket complying with BS EN 60309-2.

Equipment with a protective conductor current   607-02-03
exceeding 10 mA shall be connected to the supply
via a flexible cable with a plug and socket outlet
complying with BS EN 60309-2.

When two protective conductors are used, the ends of   607-02-05
the protective conductors shall be terminated
independently of each other at all socket outlets.

Electrical installations in caravans and motor caravans

Low-voltage socket outlets **shall** include a protective   608-08-02
contact (unless supplied by an individual winding of
an isolating transformer).

Plugs designed for extra-low-voltage socket outlets **shall** be incompatible with low-voltage socket outlets.   608-08-03

Other requirements from the Regulations include:

Extra-low-voltage socket outlets shall: 608-08-03

- clearly show their voltage
- prevent the insertion of a low-voltage plug.

Electrical installations in caravan parks

Socket outlets must **not** be bonded to the PME terminal.   608-13-05

Socket outlets **shall** be protected (individually or in groups of no more than three) by an RCD complying with BS 4293, BS EN 61008-1 or BS EN 61009-1.   608-13-05

At least one socket outlet **shall** be provided for each pitch.   608-13-02

A polarity test **shall** be made to verify that wiring has been correctly connected to socket outlets and similar accessories.   713-09-01

Other requirements from the Regulations include:

Grouped socket outlets shall be on the same phase.   608-13-06

Socket outlets and enclosures forming part of the caravan pitch supply equipment shall: 608-13-02

- comply with BS EN 60309-2
- be protected to at least IPX4
- be placed between 0.80 m and 1.50 m above the ground (to the lowest part of the socket outlet)
- have a current rating of not less than 16 A.

Socket outlets shall be protected (individually) by an overcurrent device.   608-13-04

The protective conductor of each socket outlet circuit used for the supply from a distributor's multiple earthed network shall be connected to an earth electrode.	608-13-05
When protection by automatic disconnection of supply is used, the wiring system shall include a circuit-protective conductor that is connected to the socket outlets' protective contacts.	608-03-02

6.1.13 Switches

Protective switches

Switches (unless linked) shall **not** be inserted in an earthed neutral conductor.	131-13-02
Single-pole switches **shall only** be inserted in the phase conductor.	131-13-01

Linked switches that have been inserted in an earthed neutral conductor shall be capable of breaking all of the related phase conductors.	131-13-02

Auxiliary switches

Electrical separation between live parts of a SELV system (including auxiliary switches) and any other system shall be maintained.	411-02-06

Isolation and switching

A main switch that is intended to be operated by an unskilled person (e.g. householder) **shall** interrupt both live conductors of a single-phase supply.	476-01-03
Switches (unless supplied from more than one energy source) **shall** not break a protective conductor or a PEN conductor.	460-01-03

A main linked switch (or linked circuit breaker) **shall** be provided as near as practicable to the origin of every installation.   460-01-02

Note: For d.c. systems, all poles shall be provided with a means of isolation.

Other requirements from the Regulations include:

If an installation is supplied from more than one energy source and one of these sources needs to be independently earthed (and it is necessary to ensure that not more than one means of earthing is applied at any time) then a linked switch may be inserted in the connection between the neutral point and the means of earthing.   460-01-05

If an installation is supplied from more than one source:

- a main switch shall be provided for each source of supply
- a notice shall be prominently placed warning operators that more than one switch needs to be operated.   460-01-02

Non-automatic switching facilities shall be used (i.e. to prevent or remove hazards) associated with electrical installations and electrically powered equipment.   460-01.01

Functional switching (control)

Off-load isolators, fuses and links shall **not** be used for functional switching.   537-05-03

Plug and socket outlets with a rating of more than 16 A may **not** be used as a switching device for d.c.   537-05-05

Functional switching devices:

- shall be provided for all circuits that need to be independently controlled   464-01-01
- may control several items of simultaneously operating equipment   464-01-03

- shall be suitable for the most difficult duty intended   537-05-01
- that are used to change over alternative supply   464-01-05
  sources shall:
  - switch all live conductors
  - not be capable of connecting the sources in parallel
- need not necessarily control all live conductors   464-01-02
  of a circuit
- shall not be placed solely in the neutral conductor.

Plug and socket outlets: 537-05-04
and
- may be used as a switching device providing their   464-01-04
  rating does not exceed 16 A.

💣 Except for d.c. where this purpose is specifically
excluded.

With a rating of more than 16 A, they may be used as   537-05-05
a switching device, provided that the plug and socket
outlet has a breaking capacity appropriate to the use
intended.

Semiconductor functional switching devices (such as   537-05-02
semiconductor switching devices) may control the current
without necessarily opening the corresponding poles.

Emergency switching

💣 Plug and socket outlet shall **not** be used as a device   537-04-02
for emergency switching.

A readily accessible, easily operated, emergency stopping   476-03-02
facility **shall** be provided in every place where an
electrically driven machine may give rise to danger.

Resetting the emergency switching device shall **not**   537-04-05
re-energise the equipment concerned.

Emergency switching devices **shall** use a latching   537-04-05
type operation and be capable of being restrained
in the OFF or STOP position.

Where practicable, emergency switching devices that directly   537-04-03
interrupting the main circuit **shall** be manually operated.

An emergency stopping facility shall be provided:

- where mechanical movement of electrically actuated equipment is considered dangerous   463-01-05
- where it is necessary to quickly cut off the supply in order to prevent or remove danger (except where a risk of electric shock is involved where the means shall interrupt all live conductors).   463-01-01

Circuit breakers shall open on de-energisation of the coil.   537-04-03

Emergency switching devices may automatically reset, provided that both the operating means and the means of re-energising are under the control of one and the same person.   537-04-05

Emergency switching devices used to interrupt the supply shall be capable of cutting off the full load current of the relevant part of the installation.   537-04-01

Emergency switching shall consist of: 537-04-02

- a single switching device that directly cuts off the incoming supply, or
- a combination of several items of equipment that are operated by a single action that cuts off the appropriate supply.

Emergency switching shall: 463-01-02 to 463-01-04

- act as directly as possible on the appropriate supply conductors
- require only one initiative action
- not introduce a further hazard when activated
- be readily accessible and durably marked.

The operating means (e.g. handle or pushbutton) for emergency switching devices shall be: 537-04-04

- clearly identifiable and preferably coloured red
- installed in a readily accessible position where the hazard might occur.

When installing an emergency switching device, account shall be taken of the intended use of the premises and the type of emergency foreseen.   476-03-01

Where inappropriate operation of an emergency switching device could cause a dangerous condition, then that switching device shall be organized so that it can only be operated by skilled personnel.   476-03-03

Potentially hazardous fixed or stationary appliances

Potentially hazardous fixed and/or stationary appliances that are not connected to the supply by means of a plug and socket outlet shall be provided with a means of interrupting the supply on load which:	476-03-04

- shall be located in a position that does not put the operator in any danger
- may be incorporated in the appliance or, if separate from the appliance, is in a readily accessible position.

Where two or more potentially hazardous appliances are installed in the same room, one interrupting device may be used to control all the appliances.	476-03-04

Operational conditions

Switchgear shall **not** be connected to conductors intended to operate at a temperature exceeding 70°C.	512-02-01

Diagrams

Diagrams, charts, tables or schedules shall be used to indicate the classification (and location) of all protection, isolation and switching devices.	514-09-01

Live conductors

The neutral conductor of switchgear that is used to disconnect live conductors shall not be capable of being:	530-01-01

- disconnected before the phase conductors
- reconnected before (or at the same time as) the phase conductors.

Mechanical maintenance

All devices and/or switches (such as plug and socket outlets) with a rating not exceeding 16 A that are used for switching off for mechanical maintenance, shall:

• be manually operated	537-03-02
• have an externally visible contact gap or clearly indicate the OFF or OPEN position	
• prevent unintentional reclosure, caused by mechanical shock or vibration	537-03-03
• be capable of cutting off the full load current of the relevant part of the installation	537-03-04
• either be inserted in the main supply circuit (which is always the preferred option) or in the control circuit.	537-03-01

Devices for functional switching

Off-load isolators, fuses and links shall **not** be used for functional switching.	537-05-03
Plug and socket outlets with a rating of more than 16 A may **not** be used as a switching device for d.c.	537-05-05

Functional switching devices shall be suitable for the most difficult duty intended.	537-05-01
Plug and socket outlets may be used as a functional switching device providing their rating does not exceed 16 A.	537-05-04
Plug and socket outlets with a rating of more than 16 A may be used as a functional switching device, provided that the plug and socket outlet has a breaking capacity appropriate to the use intended (also see Regulation 537-04-02 regarding emergency switching).	537-05-05
Semiconductor functional switching devices may control the current without necessarily opening the corresponding poles.	537-05-02

PEN conductors

Isolation devices or switching shall **not** be inserted in the outer conductor of a concentric cable.   546-02-06

Lighting points

Lighting installations shall be controlled by: 553-04-01

- a switch (or combination of switches) to BS 3676 and/or BS 5518, or
- a suitable automatic control system.

Water heaters having immersed and uninsulated heating elements

When the heater/ boiler is installed in a room containing a fixed bath, the switch shall comply with the requirements for locations containing a bath or shower (see Section 601 of the IEE Wiring Regulations).   554-05-03

No single-pole switch, non-linked circuit breaker or fuse shall be fitted in the neutral conductor between the heater/boiler and the origin of the installation.   554-05-04

The heater or boiler shall be permanently connected to the electricity supply via a double-pole linked switch.   554-05-03

This linked switch shall either be: 554-05-03

- separate from and within easy reach of the heater/boiler or
- part of the boiler/heater (provided that the wiring from the heater or boiler is directly connected to the switch without use of a plug and socket outlet).

Autotransformers and step-up transformers

If a step-up transformer is used, then a linked switch shall be provided for disconnecting the transformer from the supply's live conductors.	555-01-03

Circuits

Switchgear and control gear shall be clearly identified and grouped in locations accessible only to skilled or instructed persons.	563-01-05

Functional testing

Assemblies (such as switchgear, control gear, drives, controls and interlocks) shall be functionally tested to show that they are properly mounted, adjusted and installed in accordance with the Regulations.	713-13-02

Locations with risks of fire due to the nature of processed or stored materials

All circuits shall be capable of being isolated from all live supply conductors by a linked switch or linked circuit breaker.	482-02-08

Fireman's switch

All fireman's switches **shall** meet the requirements of the Regulations **plus** any requirements of the local fire authority.	476-03-07

A fireman's switch shall be provided in the low-voltage circuit supplying all: • exterior electrical installations operating at a voltage exceeding low voltage	476-03-05

> • interior discharge lighting installations operating
>   at a voltage exceeding low voltage.

 This requirement does not apply to a portable discharge lighting luminaire or to a sign (not exceeding 100 W) that is fed from a readily accessible socket outlet.

Firemen's switches shall: 537-04-06

- be coloured red
- have fixed on it a permanent nameplate (minimum
  size 150 mm by 100 mm) with not less than
  36-point lettering stating 'FIREMAN'S SWITCH'
- have its ON and OFF positions clearly indicated
  by lettering legible to a person standing on the
  ground
- have the OFF position at the top
- be provided with a device to prevent the switch
  being inadvertently returned to the ON position
- be positioned so as to allow trouble-free operation
  by a fireman.

For an interior installation, the switch shall (unless   476-03-07
otherwise agreed with the local fire authority):

- be in the main entrance to the building
- be placed in a conspicuous position
- be reasonably accessible to firemen
- be not more than 2.75 m from the ground or the
  standing beneath the switch.

For any exterior installation, the switch shall be placed   476-03-07
outside the building, adjacent to the equipment.

 Note: Where this is not possible, a notice showing the position of the switch shall be placed adjacent to the equipment and a notice fixed near the switch shall indicate its use.

Where more than one switch is installed on any one   476-03-07
building, each switch shall be clearly marked, to
indicate the installation or part of the installation
which it controls.

6.1.14 Switchboards

Figure 6.4 Certainly NOT the right sort of installation! (courtesy of StingRay)

Passageway and working platforms which have access to an open type switchboard or an equipment that has dangerous exposed live parts, needs to allow persons, without hazard, to:

471-13-02

- operate and maintain the equipment
- pass one another as necessary with ease, and
- back away from the equipment.

Equipment carrying different types of current (or at different voltages) that is grouped in a common assembly (e.g. a switchboard) shall ensure that all of the equipment belonging to any one type of current or any one type of voltage is effectively segregated, wherever necessary, so as to avoid mutual detrimental influence.

515-01-02

The identification of switchboard busbars and conductors shall be in accordance with Table 6.15.

514-03-03

Identification of conductors by colour

Table 6.15 Identification of conductors

Function	Alphanumeric	Colour
Protective conductors		Green-and-yellow
Functional earthing conductor		Cream
a.c. power circuit (Note 1)		
Phase of single-phase circuit	L	Brown
Neutral of single- or three-phase circuit	N	Blue
Phase 1 of three-phase a.c. circuit	L1	Brown
Phase 2 of three-phase a.c. circuit	L2	Black
Phase 3 of three-phase a.c. circuit	L3	Grey
Two-wire unearthed d.c. power circuit		
Positive of two-wire circuit	L+	Brown
Negative of two-wire circuit	L−	Grey
Two-wire earthed d.c. power circuit		
Positive (of negative earthed) circuit	L+	Brown
Negative (of negative earthed) circuit (Note 2)	M	Blue
Positive (of positive earthed) circuit (Note 2)	M	Blue
Negative (of positive earthed) circuit	L−	Grey
Three-wire d.c. power circuit		
Outer positive of two-wire circuit derived from three-wire system	L+	Brown
Outer negative of two-wire circuit derived from three-wire system	L−	Grey
Positive of three-wire circuit	L+	Brown
Midwire of three-wire circuit (Notes 2 & 3)	M	Blue
Negative of three-wire circuit	L−	Grey
Control circuits, ELV and other applications		
Phase conductor	L	Brown, black, red, orange, yellow, violet, grey, white, pink or turquoise
Neutral or midwire (Note 4)	N or M	Blue

Notes:
(1) Power circuits include lighting circuits.
(2) M identifies either the midwire of a three-wire d.c. circuit, or the earthed conductor of a two-wire earthed d.c. circuit.
(3) Only the middle wire of three-wire circuits may be earthed.
(4) An earthed PELV conductor is blue.

6.1.15 Switchgear

All circuits and final circuits **shall** be provided with a   476-01-02
means of switching for interrupting the supply on load.

Notes:

(1) This regulation does not apply to short connections between the origin of the installation and the consumer's main switchgear.
(2) A group of circuits may be switched by a common device.

Protection by low-voltage switchgear having total insulation (see BS EN 60439) may be used as Class II equipment.   413-03-01

Switchgear that is isolated by circuit breakers for maintenance shall be interlocked with that circuit breaker or placed and/or guarded so that it can only be operated by skilled persons.   476-02-01

Switchgear shall not be connected to conductors that are intended to operate above 70°C.   512-02-01

Labels (or other means of identification) shall indicate the purpose of each item of switchgear.   514-01-01

Where different nominal voltages exist, access to all live parts of switchgear and other fixed live parts shall be marked to indicate the voltages present.   514-10-01

The neutral conductor of switchgear used to disconnect live conductors shall not be capable of being: 530-01-01

- disconnected before the phase conductors
- reconnected before (or at the same time as) the phase conductors.

When low-voltage switchgear is used as a protective conductor, any associated metal enclosure/frame shall: 543-02-04

- be protected against mechanical, chemical or electrochemical deterioration
- have a cross-sectional area not less than:
  - 2.5 mm^2 copper equivalent if protection against mechanical damage is provided or
  - 4 mm^2 copper equivalent if mechanical protection is not provided
- shall allow other protective conductors to be connected at predetermined tap-off points.

Switchgear shall be clearly identified and grouped in locations accessible only to skilled or instructed persons.   563-01-05

Special installations and locations

The Regulations list the following specific requirements for switches and switchgear in special installations and locations:

Table 6.16 Requirements for switches and switchgear in special installations and locations

Installation or location	Requirement		
Locations containing a bath or a shower	Zone 0	Switchgear or accessories shall **not** be installed.	601-08-01
	Zone 1	Only switches of SELV circuits (supplied at a nominal voltage not exceeding 12 V a.c. RMS or 30 V ripple-free d.c.) **shall** be installed.	
		Note: The safety source shall be installed outside zones 0, 1 and 2.	
	Zone 2	Only switches and socket outlets of SELV circuits and shaver supply units complying with BS EN 60742 shall be installed as accessories to switchgear.	
		Note: The safety source shall be installed outside zones 0, 1 and 2.	
		Only insulating pull cords of cord-operated switches meeting the requirements of BS 3676 are permitted in zones 1 and 2.	
	Zone 3	Only SELV socket outlets and shaver supply units complying with BS EN 60742 shall be installed.	
		Note: Portable equipment can only be connected as above.	
		When a heater/ boiler is installed in a room containing a fixed bath, the switch shall comply with the specific requirements for these types of locations.	554-05-03

Swimming pools		In zones A and B, switchgear, control gear and accessories shall **not** be installed, unless socket outlets comply with BS EN 60309-2 and they are:
		• installed more than 1.25 m outside of zone A
		• installed at least 0.3 m above the floor protected by either:
		  – a residual current device complying with Regulation 412-06-02, or
		  – electrical separation (in accordance with Regulation 413-06) with the safety isolating transformer placed outside zones A, B and C.
			602-07-01

		In zone C, socket outlets, switches or accessories are only permitted if they are: • protected individually by electrical separation (Regulation 413-06), or	602-07-02

(continued)

Table 6.16 (*continued*)

Installation or location	Requirement	
	• protected by SELV (Regulation 411-02), or • protected by a residual current device complying Regulation 412-06-02, or • a shaver socket complying with BS 3535. This requirement does not apply to the insulating cords of cord-operated switches complying with BS 3676.	
Hot air saunas	All switchgear (other than a thermostat or a thermal cut-out) not actually built into the sauna heater shall be installed outside the hot air sauna.	603-08-01
	Accessories (other than a thermostat or a thermal cut-out) shall **not** be installed within the hot air sauna.	603-08-02
Construction sites	Emergency switching shall be provided for supplies that need to have live conductors disconnected in order to remove a hazard.	604-11-03
Agricultural and horticultural premises	Emergency switching and emergency stopping devices shall: • be inaccessible to livestock • not be impeded by livestock in the event of them panicking.	605-13-01
Caravans and motor caravans	All installations shall have a main isolating switch that disconnects all live conductors.	608-07-04
	The isolating switch for installations consisting of only one final circuit may be the overcurrent protection device.	608-07-04
	Appliances not connected to the supply by a plug and socket outlet shall be controlled by a switch.	608-08-05
Highway power supplies (street furniture and street located equipment)	Where it is intended that isolation and switching shall only be carried out by instructed persons, isolation may be provided via a suitably rated fuse carrier.	611-03-01
	If the distributor's cut-out is used for isolating a highway power supply, then the approval of the distributor shall be obtained.	611-03-02

6.2 Supplies

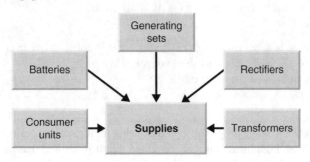

Figure 6.5 Types of supplies

6.2.1 Consumer units

A consumer unit (sometimes known as a consumer control unit or electricity control unit) is a particular type of distribution board for the control and distribution of electrical energy, primarily in domestic premises. It includes a manually operated method for isolation of both poles of the incoming circuit(s) and assemblies of one or more fuses, circuit breakers, RCDs, signalling and other devices that have been purposely manufactured for this purpose.

 All circuits and final circuits **shall** be provided with a   476-01-02
means of switching for interrupting the supply on load.

 Note: This regulation particularly applies to circuits and parts of an installation that (for safety reasons) need to be switched independently of other circuits and/or installations. It does not apply to short connections between the origin of the installation and the consumer's main switchgear.

A group of circuits may be switched by a common device.

Voltage drop in consumers' installations

The voltage drop between the supply terminals of an   525-01-02
installation and a socket outlet (or the terminals of a fixed
current-using equipment) shall **not** exceed 4% of the
nominal voltage of the supply.

The voltage at the terminals of a fixed current-using   525-01-01
equipment shall be greater than the lower limit shown
in the relevant British Standard for that particular
equipment.

The voltage at the terminals of equipment that is not   525-01-01
subject to a British Standard shall not prejudice the
safe functioning of that equipment.

 Note: A greater voltage drop is acceptable during motor starting periods and for equipment with high inrush currents provided that it is verified that the voltage variations are within the limits specified in the relevant British Standards.

Earthing arrangements

Earthing arrangements shall be such that: 542-01-07

- they are sufficiently robust (or have additional mechanical protection) to external influences
- earth fault currents and protective conductor currents that may occur are carried without danger
- the value of impedance from the consumer's main earthing terminal to the earthed point of the supply:
  - for TN systems, meets the protective and functional requirements of the installation
  - for TT and IT systems, meets the protective and functional requirements of the installation.

 Note: This particularly applies with respect to thermal, thermomechanical and electromechanical stresses.

The main equipotential bonding connection to any gas or   547-02-02
other service shall be made:

- as near as practicable to the point of entry of that service into the premises
- to the consumer's hard metal pipework
- before any branch pipework
- within 600 mm of the meter outlet union or at the point of entry to the building if the meter is external.

6.2.2 Batteries

A SELV source may be from a battery or from another   411-02-02
form of electrochemical source.

Provided that the unprotected conductor has been   473-02-04
installed so as to minimise the risk of fault current and
the risk of fire (or danger to persons), then there is no
need to have a fault current protective device for a
conductor connecting a battery with its control panel.

Safety service sources shall include: 562-01-01

- a primary cell or cells
- a storage battery.

6.2.3 Generating sets

The safety and proper functioning of all other sources of supply shall **not** be weakened by the generating set.

551-02-01

Protection against indirect contact **shall** be provided for each supply source which is capable of operating independently of other sources.

551-04-01

When a generator set is provided with overcurrent detection, this **shall** be located as near as practicable to the generator terminals.

551-05-01

Where a generating set is intended to operate in parallel with a distributor's, circulating harmonic currents **shall** be limited so that the thermal rating of conductors is not exceeded.

551-05-02

Potential short-circuit and earth fault currents shall be assessed for each supply source.

551-02-02

The requirements of the distributor shall be ascertained before a generating set is installed in any installation that is connected to the distributor's network.

551-01-01

The short-circuit rating of protective devices connected to the distributor's network shall not be exceeded.

551-02-02

Generating sets that provide a supply for an installation that is not connected to the distributor's network (or to provide a supply as a switched alternative to the distributor's network) shall not endanger or damage equipment after the connection (or disconnection) of any intended load as a result of a deviation in voltage or frequency from the intended operating range.

551-02-03

When required, methods shall be provided to automatically disconnect parts of the installation if the capacity of the generating set is exceeded.

551-02-03

Extra-low-voltage systems supplied from more than one source

Where a SELV or PELV system is supplied by more than one source, the following requirements shall apply to each source.

The SELV source shall be from one of the following: 551-03-01
and
- a safety isolating transformer complying with   411-02-02
  BS 3535
- a motor-generator with windings providing electrical
  separation equivalent to that of the safety isolating
  transformer specified in (i) above
- a battery or other form of electrochemical source
- a source independent of a higher voltage circuit
  (e.g. an engine driven generator)
- electronic devices which (even in the case of an
  internal fault) restrict the voltage at the output.

Where one or more of the sources is earthed:

Protection against direct contact shall be provided by   471-14-02
either:

- barriers or enclosures with a degree of protection of at
  least 1P2X or IPXXB, or
- insulation capable of withstanding 500 V a.c. RMS
  for 60 seconds.

 This form of protection against direct contact is **not** required if the equipment is
within a building in which main equipotential bonding is applied and the voltage
does not exceed:

- 25 V a.c. RMS or 60 V ripple-free d.c. when the equipment is normally
  only used in dry locations and large-area contact of live parts with the
  human body is not to be expected.
- 6 V a.c. RMS or 15 V ripple-free d.c. in all other cases.

If one or more of the circuits does not meet the requirements for SELV in some
respect, then the system shall be treated as a FELV system and the following
will apply:

Socket outlets and luminaire-supporting couplers shall   471-14-06
use a plug which is dimensionally different from those
used for any other system in the same premises.

Exposed conductive parts of equipment in that FELV   471-14-04
system shall be connected to the protective conductor

of the primary circuit provided that the primary circuit of
the FELV source is protected by automatic disconnection.

If the primary circuit of the FELV source is protected   471-14-05
by electrical separation, exposed conductive parts of
equipment in that FELV voltage system shall be
connected to the non-earthed protective conductor
of the primary circuit.

In order to maintain the supply to an extra-low-voltage   551-03-02
system following the loss of one or more sources of
supply, each source of supply that is capable of operating
independently shall be capable of supplying the
intended load of the extra-low-voltage system.

Protection against direct contact shall be provided by   471-14-03
one or more of the following:

• barriers or enclosures
• insulation corresponding to the minimum voltage
  required for the primary circuit.

The insulation of equipment shall be reinforced to   471-14-03
withstanding a voltage of 1500 V a.c. RMS for
60 seconds when an extra-low-voltage circuit is
used to supply equipment whose insulation does
not comply with the minimum test voltage required
for the primary circuit.

The loss of low-voltage supply to an extra-low-voltage   551-03-02
source shall not lead to danger or damage for other
extra-low-voltage equipment.

Protection against indirect contact

Protection against indirect contact shall be provided for   551-04-01
each supply source which is capable of operating
independently of other sources.

Protection by automatic disconnection of supply

Protection by automatic disconnection of supply shall   551-04-02
be applied in accordance with the requirements for the   and
type of system earthing in use (i.e. TN, TT or IT).   413-02-01

Additional requirements for installations where the generating set provides a switched alternative supply to the distributor's network (stand-by systems)

When the generator is operating as a switched alternative to a TN system, protection by automatic disconnection of supply shall not rely on the connection to the earthed point of the distributor's network.   551-04-03

Additional requirements for installations incorporating static inverters

Supplementary equipotential bonding shall be provided on the load side of the static converter when automatic disconnection cannot be achieved   551-04-04

The operation of protective devices shall not be damaged: 551-04-05

- by direct current generated by a static inverter or
- by the presence of filters.

Additional requirements for protection by automatic disconnection where the installation and generating set are not permanently fixed

Installations which are not permanently fixed (e.g. portable generating sets) shall be provided with protective conductors that are part of the cord or cable, between separate items of equipment.   551-04-06

In TN, TT and IT systems an RCD (with a rated residual operating current not exceeding 30 mA) shall be installed.   551-04-06

Protection against overcurrent

The effects of circulating harmonic currents shall be limited by the: 551-05-02

- selection of generating sets
- provision of a suitable impedance in the connection to generator star points

- provision of interlocked switches that ensure
  protection against indirect contact provision of
  filtering equipment.

When a generator set is provided with overcurrent   551-05-01
detection, this shall be located as near as practicable
to the generator terminals.

Where a generating set is intended to operate in parallel with   551-05-02
a distributor's, circulating harmonic currents shall be limited
so that the thermal rating of conductors is not exceeded.

Fault current protective device

Provided that the unprotected conductor has been installed   473-02-04
so as to minimise the risk of fault current (and to
minimise the risk of fire or danger to persons), then there
is no need to have a fault current protective device for a
conductor connecting a generator with its control panel.

Additional requirements for installations where the generating set provides a supply as a switched alternative to the distributor's network (stand-by systems)

Precautions **shall** be taken to ensure that the generator   551-06-01
cannot operate in parallel with the distributor's network.

For TN-S systems, RCDs shall be positioned to avoid incorrect operation owing
to the existence of any parallel neutral earth path. Precautions may include:

- an electrical, mechanical or electromechanical   551-06-01
  interlock between the operating mechanisms or
  control circuits of the changeover switching devices
- a system of locks with a single transferable key
- a three-position break-before-make changeover switch
- an automatic changeover switching device with a
  suitable interlock.

Additional requirements for installations where the generating set may operate in parallel with the distributor's network

The generating set **shall** be capable of being isolated from the distributor's network.   551-07-04

When using a generating set in parallel with a distributor's network, care shall be taken to avoid adverse effects in respect of: 551-07-01

- power factor
- voltage changes
- harmonic distortion
- unbalance
- starting
- synchronising, or
- voltage fluctuation effects.

Protection shall be provided to disconnect the generating set from the distributor's network.   551-07-02

6.2.4 Rectifiers

Provided that any unprotected conductor has been installed so as to minimise the risk of fault current (as well as to minimise the risk of fire or danger to persons) then there are no requirements to have a fault current protective device for a conductor connecting a rectifier with its control panel.   473-02-04

6.2.5 Transformers

The neutral (i.e. star) point of the secondary windings of three-phase transformers (or the mid-point of the secondary windings of single-phase transformers) **shall** be connected to earth.   471-15-04

A system which does **not** use a device such as an   411-02-03
autotransformer, potentiometer, semiconductor device
etc., to provide electrical separation, shall **not** be
deemed to be a SELV system.

Overload protection devices need not be used for   473-01-03
secondary circuits supplying current where unexpected
disconnection of the circuit could cause danger.

 In such situations consideration shall be given to the provision of an overload
alarm.

Protection by electrical separation may be applied to the   471-12-01
supply of any individual item of equipment by means of
a transformer complying with BS 3535 the secondary of
which is not earthed, or a source affording equivalent safety.

Provided that the unprotected conductor has been installed   473-02-04
so as to minimise the risk of fault current and to minimise
the risk of fire or danger to persons, then there is no need
to have a fault current protective device:

• for a conductor connecting a transformer with its
  control panel
• in a measuring circuit (such as a secondary circuit
  of a current transformer)
• where an unexpected disconnection could cause a
  greater danger than a fault current condition.

The magnetic circuit of the transformer of an RCD shall   531-02-02
enclose all the live conductors of the protected circuit.

The SELV source may be from a safety isolating   411-02-02
transformer complying with BS 3535.

The supply source to a reduced low-voltage circuit   471-15-03
may be:

• a double wound isolating transformer complying
  with BS 3535-2, or
• a motor generator whose windings provide isolation
  equivalent to an isolating transformer.

The supply source to the circuit shall be either: 413-06-02

• an isolating transformer complying with BS 3535 or
• a motor generator.

Autotransformers and step-up transformers

Step-up autotransformers **shall** not be connected to an IT system.	555-01-02
If a step-up transformer is used, then a linked switch shall be provided for disconnecting the transformer from the live conductors of the supply.	555-01-03
The common terminal of the winding of all autotransformers shall be connected to the neutral conductor.	555-01-01

Special installations and locations

The Regulations list the following specific requirements for transformers in special installations and locations (Table 6.17).

Table 6.17 Requirements for transformers in special installations and locations

Installation or location	Requirement	
Swimming pools	The safety source shall be installed outside zones A, B and C, except if floodlights are installed, then each floodlight shall be supplied from its own transformer (or an individual secondary winding of a multi-secondary transformer), having an open circuit voltage not exceeding 18 V.	602-04-01
Construction sites	Where socket outlets and permanently connected hand-held equipment socket outlets are protected by electrical separation, they shall be supplied from a separate isolating transformer.	604-08-04
Locations with equipment having high protective conductor currents	The wiring of final and distribution circuits to equipment with a protective conductor current exceeding 10 mA shall have a high-integrity protective connection (e.g. to the supply by means of a double-wound transformer which has its secondary winding connected to the protective conductor of the incoming supply and the exposed conductive parts).	607-02-04
Caravans and motor caravans	Low-voltage socket outlets shall include a protective contact (unless supplied by an individual winding of an isolating transformer).	608-08-02

7

Cables and conductors

Within the Wiring Regulations there is frequent reference to different types of cables (e.g. single-core, multicore, fixed, flexible etc.), conductors (e.g. live supply, protective, bonding etc.) and conduits, cable ducting, cable trunking and so on. Unfortunately, as is the case for equipment and components, the requirements for these items are liberally sprinkled throughout the Standard. The aim of this chapter, therefore, is to provide a catalogue of all the different types identified and referred to in the Wiring Regulations under three main headings (cables, conductors and conduits etc.) and then make a list of their essential requirements.

 Similar to other chapters, please remember that these lists of requirements are **only** the author's impression of the most important aspects of the Wiring Regulations and electricians should **always** consult BS 7671 to satisfy compliance!

7.1 Cables

An electric power cable is defined as an assembly of two or more electrical conductors consisting of a core protected by twisted wire strands held together with (and typically covered by) an overall sheath. The conductors may be of the same or different sizes, each with their own insulation. A bare conductor is normally used for the equipment safety earth.

There are five main types of cables found in electrical installations. These are:

1. single-core cables
2. multicore cables
3. flexible cables
4. heating cables
5. fixed wiring.

7.1.1 General

> Cables and cords shall comply with the requirements of
> BS EN 50265-2-1 or 2-2.   482-03-03
>
> Cables with a non-metallic sheath or non-metallic
> enclosure are not considered to be a Class II construction.   471-09-04

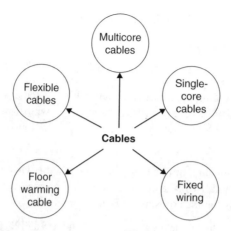

Figure 7.1 Cables

7.1.2 Single-core cables

Owing to possible electromagnetic effects, single-core cables that are armoured with steel wire or tape shall **not** be used for a.c. circuits.   521-02-01

Metallic sheaths and/or the non-magnetic armour of single-core cables that are in the same circuit shall either be bonded together: 523-05-01

- at both ends of their run (solid bonding) or
- at one point in their run (single point bonding)

provided that, at full load, voltages from sheaths and/or armour to earth:

- do not exceed 25 volts, and
- do not cause corrosion and
- do not cause danger or damage to property.

Single-core cables may be used as protective conductors but shall be coloured green-and-yellow throughout their length.   543-02-02 and 514-04-02

The conductance of the outer conductor of a concentric single-core cable shall not be less than the internal conductor.   546-02-04

When non-twisted single-core cables with a cross-sectional area greater than 50 mm^2 in copper (or 70 mm^2 in aluminium) are connected in parallel, the load current shall be shared equally between them.   523-02-01

This ruling does not apply to final ring circuits.

7.1.3 Multicore cables

 A Band I circuit shall **not** be contained in the same wiring system as a Band II voltage circuit except as shown below.

A Band I circuit shall not be contained in the same wiring   528-01-02
system as a Band II voltage circuit unless it is in a multicore
cable (or cord) and the cores of the Band I circuit are:

- insulated for the highest voltage present in the Band II circuit
- separated from the cores of the Band II circuit by an earthed metal screen

and the cables:

- are insulated for their system voltage
- are installed in a separate compartment of a cable ducting or trunking system
- are installed on a tray or ladder separated by a partition
- use a separate conduit, trunking or ducting system.

Note: For telecommunication circuits, data transfer circuits and similar, consideration shall also be given to electrical interference, both electromagnetic and electrostatic (see BS EN 50081 and BS EN 50082).

SELV circuit conductors that are contained in a multicore   411-02-06
cable with other circuits shall be insulated for the highest
voltage present in that cable.

Separated circuits shall, preferably, use a separate wiring   413-06-03
system. If this is not feasible, multicore cables (without a
metallic sheath or insulated conductors) may be used.

The conductance of the outer conductor of a concentric
cable for a multicore cable:

- serving a number of points contained within one final   546-02-04
  circuit (or where the internal conductors are connected
  in parallel) shall not be less than that of the internal
  conductors
- in a multiphase or multipole circuit shall not be   546-02-04
  less than that of one internal conductor.

7.1.4 Flexible cables

Flexible cables and flexible cords:	482-02-10
• shall be of a heavy-duty type with a voltage rating of not less than 45 or 50 V, or	
• shall be suitably protected against mechanical damage	
• that are liable to mechanical damage shall be visible throughout its length	413-06-03
• that are used as an overhead low-voltage line shall comply with the relevant British or harmonised Standard	521-01-01 and 521-01-03
• shall only be used for fixed wiring where the relevant Regulations permit.	521-01-04
Insulated flexible cables and cords may include a flexible metallic armour, braid or screen.	521-01-01
Non-flexible cables (and flexible cords not forming part of a portable appliance or luminaire) that are sheathed with lead, PVC or an elastomeric material may include a catenary wire (or hard-drawn copper conductor) for aerial use or when suspended.	521-01-01
Provided that all flexible equipment cables (other than Class II equipment) have a protective conductor for use as an equipotential bonding conductor, then source supplies may supply more than one item of equipment.	413-06-05

Special installations and locations

The Regulations list the following specific requirements for flexible cables in special installations and locations (Table 7.1).

Table 7.1 Requirements for flexible cables in special installations and locations

Installation or location	Requirement	
Construction sites	Reduced low-voltage systems shall use low-temperature 300/500 V thermoplastic (PVC) or equivalent flexible cables.	604-10-03
Locations with equipment having high protective conductor currents	Equipment with a protective conductor current exceeding 10 mA shall be connected to the supply by either: • a permanent connection to the wiring of the installation (which is the preferred option) or • a flexible cable with a plug and socket outlet complying with BS EN 60309-2.	607-02-03

(continued)

Table 7.1 (*continued*)

Installation or location	Requirement	
Caravans and motor caravans	Wiring systems may use flexible single-core insulated conductors in non-metallic conduits.	608-06-01
	A flexible cord or cable, 25 m (±2 m) in length (HO7RN-F, or HO5VV-F or equivalent) that includes a protective conductor whose cross-section meets the requirements of Table 7.2 may be used to attach the supply to a caravan or motor caravan pitch socket outlet.	608-08-08

Table 7.2 Cross-sectional areas of flexible cords and cables for caravan connectors

Rated current (A)	Cross-sectional area (mm^2)
16	2.5
25	4
32	6
63	16
100	35

7.1.5 Heating cables

Heating cables: 554-06-01

- passing through (or in close proximity to) a fire hazard:
  - shall be enclosed in material with an ignitability characteristic 'P' as specified in BS 476 Part 5
  - shall be protected from any mechanical damage
- that are going to be laid (directly) in soil, concrete, cement screed, or other material used for road and building construction shall be: 554-06-02
  - capable of withstanding mechanical damage
  - constructed of material that will be resistant to damp and/or corrosion
- that are going to be laid (directly) in soil, a road, or the structure of a building, shall be installed so that they are: 554-06-03
  - completely embedded in the substance it is intended to heat
  - not damaged by movement (by the substance in which it is embedded)
  - complies with the makers instructions and recommendations.

The maximum loading of floor-warming cable under operating conditions is shown in Table 7.3 (554-06-04).

Table 7.3 Maximum conductor operating temperatures for a floor-warming cable

Type of cable	Maximum conductor operating temperature (°C)
General-purpose PVC over conductor	70
Enamelled conductor, polychlorophene over enamel, PVC overall	70
Enamelled conductor PVC overall	70
Enamelled conductor, PVC over enamel, lead-alloy 'E' sheath overall	70
Heat-resisting PVC over conductor	85
Nylon over conductor, heat-resisting PVC overall	85
Synthetic rubber or equivalent elastomeric insulation over conductor	85
Mineral insulation over conductor, copper sheath overall	Temperature dependent on type of seal employed, outer covering etc.
Silicone-treated woven-glass sleeve over conductor	180

7.1.6 Cables with thermal insulation

Cables that are run in thermally insulated spaces **shall** not be covered by the thermal insulation.   523-04-01

The current-carrying capacity of cables that are installed in thermally insulated walls (or above a thermally insulated ceiling) shall comply with the requirements shown in Appendix 4 to the Regulations.   523-04-01

The current-carrying capacity of cables that are totally surrounded by thermal insulation for longer than 0.5 m shall be reduced according to the size of cable, length and thermal properties of the insulation.   523-04-01

Sealing arrangements shall permit thermal movement of the wiring system.   527-02-03

7.1.7 Fixed wiring

Cables used as fixed wiring shall be supported as well as not being exposed to undue mechanical strain – particularly at terminations.   522-08-05

7.1.8 Cable couplers

For installations and locations where the risk of electric shock is increased by a reduction in body resistance and/or by contact with earth potential, all cable couplers of a reduced low-voltage system shall:

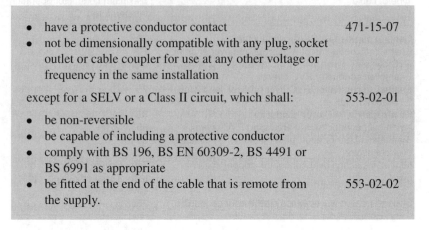

- have a protective conductor contact   471-15-07
- not be dimensionally compatible with any plug, socket outlet or cable coupler for use at any other voltage or frequency in the same installation

except for a SELV or a Class II circuit, which shall: 553-02-01

- be non-reversible
- be capable of including a protective conductor
- comply with BS 196, BS EN 60309-2, BS 4491 or BS 6991 as appropriate
- be fitted at the end of the cable that is remote from   553-02-02
  the supply.

7.2 Conductors

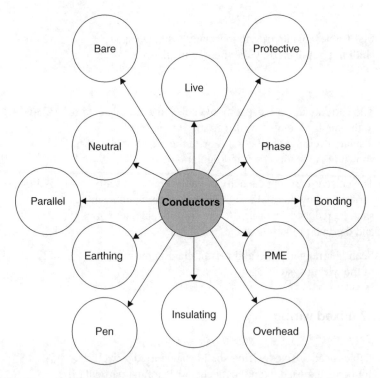

Figure 7.2 Conductors

Conductors intended to operate at temperatures 512-02-01
above 70°C shall **not** be connected to switchgear, protective
devices, accessories or other types of equipment.

Conductors shall **not** be subjected to excessive mechanical 522-12-01
stress.

7.2.1 General

Conductors shall be capable of withstanding all foreseen 521-03-01
electromechanical forces (including fault current)
during service.

Conductors used as fixed wiring shall: 522-08-05

- be supported
- not be exposed to undue mechanical strain (particularly
  at terminations).

Conductors entering a ferrous enclosure shall be arranged 521-02-01
so that they are not individually surrounded by a ferrous
material – unless other provision shall be made to prevent
eddy (induced) currents.

Conductors are considered to be protected against 436-01-01
overcurrent if their current-carrying capacity is greater
than the current supplied by the source.

7.2.2 Colour coding

Conductors may be identified by numbers with the 514-05-04
number 0 indicating the neutral or midpoint conductor.

Colour coding or marking is not required for: 514-06-01

- concentric conductors of cables
- the metal sheath or armour of cables (when used as a
  protective conductor)
- bare conductors (where permanent identification is not
  practicable)
- extraneous conductive parts used as a protective conductor
- exposed conductive parts used as a protective conductor.

7.2.3 Cross-sectional area of conductors

The cross-sectional area of conductors shall be
determined according to the:

    131-06-01

- admissible maximum temperature
- voltage drop limit
- maximum impedance for operation of short-circuit
  and earth fault protection
- electromechanical stresses likely to occur due to
  short-circuit and earth fault currents
- other mechanical stresses to which conductors are
  likely to be exposed.

The minimum cross-sectional area of a phase conductor
in an a.c. circuit and of live conductors in d.c. circuits
shall be as shown in Table 7.4.

    524-01-01

Table 7.4 Minimum nominal cross-sectional area of conductor

Type of wiring system	Use of circuit	Conductor	
		Material	Minimum cross-sectional area (mm)
Cables and insulated conductors	Power and lighting circuits	Copper	1.0
		Aluminium	16.0 (see Note 1)
	Signalling and control circuits	Copper	0.5 (see Note 2)
Bare conductors	Power circuits	Copper	10
		Aluminium	16
	Signalling and control circuits	Copper	4
Flexible connections with insulated conductors and cables	For a specific appliance	Copper	See relevant British Standard
	For any other application	Copper	0.5 (see Note 2)
	Extra-low-voltage circuits for special applications	Copper	0.5

7.2.4 Live conductors

 Bare live conductors **shall** be installed on insulators.   521-07-02

Exposed bare and live conductors shall **not** be used in locations where there is a risk of fire owing to the nature of processed or stored materials.	482-02-09
In an IT system, live conductors shall **not** be directly connected with earth.	413-02-21
The supply to all live conductors **shall** be automatically interrupted in the event of overload current and fault current.	431-01-01
Conductors **shall** be able to carry fault current without overheating.	130-05-01

 If an earth connection is required, then it shall be through a high impedance so that the fault current will be so low that it will not give rise to the risk of an electric shock.

Live conductors (or joints between them) shall be terminated within an enclosure.	422-01-04
The supply to live conductors shall be protected by an automatic interruption device.	431-01-01
Safety protective methods for live conductors shall be in accordance with Part 4 of BS 7671.	312-01-01

 Note: In determining safety protection methods and the selection and erection of equipment, an assessment shall be made of:

- its purpose
- its general structure
- its supplies
- the external influences to which it is to be exposed
- its compatibility
- its maintainability.

The insulation resistance between live conductors and between live conductors and earth shall be measured before the installation is connected to the supply.	713-04-01
The protective conductor of an overcurrent protective device shall be incorporated in the same wiring system as the live conductors **or** be located in their immediate proximity.	544-01-01

The number and type of live conductors (e.g. single-phase   312-02-01
two-wire a.c., three-phase four-wire a.c.) will depend on the:

- type of circuit
- energy source.

 If the energy is being provided by a distributor then the distributor will normally supply this information.

Fire risk locations

In locations where there is a risk of fire owing to the   482-02-09
nature of processed or stored materials:

- exposed live conductors shall not be used
- all circuits shall be capable of being isolated from all   482-02-08
  live supply conductors by a linked switch or linked
  circuit breaker.

Emergency switching

Installations **shall** be provided with an emergency   463-01-01
switching device where there is a risk of an electric
shock occurring.

Emergency switching devices shall:

- act as directly as possible on the appropriate supply   463-01-02
  conductors
- require only one initiative action   463-01-02
- not introduce a further hazard when activated   463-01-03
- be readily accessible and durably marked.   463-01-04

Functional switching

Functional switching devices:

- that are used to change over alternative supply sources   464-01-05
  shall switch all live conductors
- need not necessarily control all live conductors of a circuit.   464-01-02

Main switch

All main switches that are likely to be operated by an unskilled person (e.g. householder) **shall** interrupt both live conductors of a single-phase supply.   476-01-03

Special installations and locations

The Regulations list the following specific requirements for live conductors in special installations and locations (Table 7.5).

Table 7.5   Requirements for live conductors in special installations and locations

Installation or location	Requirement	
Construction sites	Emergency switching shall be provided for supplies that need to have live conductors disconnected in order to remove a hazard.	604-11-03
Caravans and motor caravans	All installations shall have a main isolating switch that disconnects all live conductors.	608-07-04
	Live conductors of all final circuits shall be protected by an overcurrent protective device.	608-04-01

7.2.5 Live supply conductors

Live supply conductors **shall** be capable of being isolated from circuits.   461-01-01

 Whenever possible, groups of circuits shall be isolated from live supply conductors by a common means.

Isolation devices shall isolate all live supply conductors from the circuit concerned.   537-02-01

If a step-up transformer is used, then a linked switch shall be provided to disconnect the transformer from the supply's live conductors.   555-01-03

7.2.6 Bare conductors

 Bare live conductors **shall** be installed on insulators.   521-07-02

Exposed bare live conductors shall **not** be used in   482-02-09
locations where there is a risk of fire owing to the nature
of processed or stored materials.

The minimum cross-sectional area of phase conductors in   524-01-01
a.c. circuits and of live conductors in d.c. circuits shall
be as shown in Table 7.6.

Table 7.6 Minimum cross-sectional area of bare conductors

Type of wiring system	Use of circuit	Conductor	
		Material	Minimum cross-sectional area (mm)
Bare conductors	Power circuits	Copper	10
		Aluminium	16
	Signalling and control circuits	Copper	4

The insulation and/or sheath of cables connected   523-03-01
to a bare conductor (or busbar) shall be capable of
withstanding the maximum operating temperature
of that bare conductor (or busbar).

Bare conductors that are used as protective conductors   514-04-02
shall be identified by equal green and yellow stripes
that are 15 mm to 100 mm wide.

If adhesive tape is used, then it shall be bicoloured.

Bare conductors shall be painted or identified by   514-04-06
a coloured tape, sleeve or disc as per Table 7.7.

Table 7.7   Identification of conductors

Function	Alphanumeric	Colour
Protective conductors		Green-and-yellow
Functional earthing conductor		Cream
a.c. power circuit (Note 1)		
Phase of single-phase circuit	L	Brown
Neutral of single- or three-phase circuit	N	Blue
Phase 1 of three-phase a.c. circuit	L1	Brown
Phase 2 of three-phase a.c. circuit	L2	Black
Phase 3 of three-phase a.c. circuit	L3	Grey
Two-wire unearthed d.c. power circuit		
Positive of two-wire circuit	L+	Brown
Negative of two-wire circuit	L−	Grey
Two-wire earthed d.c. power circuit		
Positive (of negative earthed) circuit	L+	Brown
Negative (of negative earthed) circuit (Note 2)	M	Blue
Positive (of positive earthed) circuit (Note 2)	M	Blue
Negative (of positive earthed) circuit	L−	Grey
Three-wire d.c. power circuit		
Outer positive of two-wire circuit derived from three-wire system	L+	Brown
Outer negative of two-wire circuit derived from three-wire system	L−	Grey
Positive of three-wire circuit	L+	Brown
Midwire of three-wire circuit (Notes 2 & 3)	M	Blue
Negative of three-wire circuit	L−	Grey
Control circuits, ELV and other applications		
Phase conductor	L	Brown, black, red, orange, yellow, violet, grey, white, pink or turquoise
Neutral or midwire (Note 4)	N or M	Blue

Notes:
(1) Power circuits include lighting circuits.
(2) M identifies either the midwire of a three-wire d.c. circuit, or the earthed conductor of a two-wire earthed d.c. circuit.
(3) Only the middle wire of three-wire circuits may be earthed.
(4) An earthed PELV conductor is blue.

Colour or marking is not required for bare conductors (where permanent identification is not practicable) (514-06-01).

7.2.7 Protective conductors

Gas pipes, oil pipes and flexible (or pliable conduit) may **not** be used as a protective conductor. 543-02-01

Exposed conductive parts of equipment shall **not** be used as a protective conductor for other equipment. 543-02-08

In installations and locations where the risk of an electric shock is increased by a reduction in body resistance and/or by contact with earth potential, all plugs, socket outlets and cable couplers of a reduced low-voltage system **shall** have a protective conductor contact. 471-15-07

Types of protective conductor

A protective conductor may consist of one or more of the following:

- a single-core cable 543-02-02
- a conductor in a cable
- an insulated or bare conductor in a common enclosure with insulated live conductors
- a fixed bare or insulated conductor
- a metal covering (for example, the sheath, screen or armouring of a cable)
- a metal conduit or other enclosure or electrically continuous support system for conductors
- an extraneous conductive part (such as a metal enclosure, frame of a low-voltage switchgear, control gear assembly or busbar trunking system) provided that: 543-02-04 and 543-02-06
    - its electrical continuity has been assured
    - it is protected against mechanical, chemical or electrochemical deterioration
    - its cross-sectional area confirms to the Requirements
    - precautions have been taken against its removal
    - it is suitable for its intended use
    - its cross-sectional area has been verified in accordance with BS EN 60439-1
    - it allows other protective conductors to be connected at predetermined tap-off points
- the metal covering including the sheath (bare or insulated) or 543-02-05

- the sheath of a mineral insulated cable (trunking, ducting and metal conduit) provided that:
  - it is protected against mechanical, chemical or electrochemical deterioration
  - its cross-sectional area has been verified in accordance with BS EN 60439-1.

543-02-04
and
543-02-07

Note: If a protective conductor is formed by conduit, trunking, ducting or the metal sheath and/or armour of a cable, the earthing terminal of each accessory shall be connected by a separate protective conductor to an earthing terminal incorporated in the associated box or enclosure.

General

Joints that can be disconnected for test purposes are permitted in a protective conductor circuit.

543-03-04

a.c. circuit conductors that are installed in ferromagnetic enclosures shall be arranged so that the phases, neutral (if any) and the appropriate protective conductor of each circuit are contained in the same enclosure.

521-02-01

If electrical earth monitoring is used, the operating coil shall be connected in the pilot conductor and **not** in the protective earthing conductor.

543-03-05

Unless a protective conductor forms part of a multicore cable (or cable trunking or a conduit is used) then provided that the cross-sectional area is up to 6 mm², it must be protected, throughout, by a covering of at least equivalent to a single-core non-sheathed cable with a voltage rating of at least 450/750 V.

543-03-02

Identification

Bare conductors (or busbars) used as protective conductors shall be identified by equal green and yellow stripes which are 15 mm to 100 mm wide and where no more than 70% of the surface area is covered by one colour.

514-04-02

If an adhesive tape is used, then it shall also be bi-coloured.

Conductors coloured green and yellow shall not be numbered unless required for circuit identification purposes.	514-05-02
Single-core cables used as protective conductors shall be coloured green-and-yellow throughout their length	514-04-02

This combination of green and yellow shall **not** be used for any other purpose.

Cross-sectional areas

If the protective conductor:

- is not an integral part of a cable, or
- is not formed by conduit, ducting or trunking, or
- is not contained in an enclosure formed by a wiring system

then the cross-sectional area shall not be less than:

• 2.5 mm² copper equivalent if protection against mechanical damage is provided or • 4 mm² copper equivalent if mechanical protection is not provided.	543-01-01
The cross-sectional area of all protective conductors shall not be less than:	543-01-03

$$S = \frac{\sqrt{I^2 t}}{K}$$

where S is the nominal cross-sectional area of the conductor in mm²; I is the value in amperes (RMS for a.c.) of fault current for a fault of negligible impedance, which can flow through the associated protective device (also see Regulation 413-02-05); t is the operating time of the disconnecting device in seconds; and k is a factor taking account of the resistivity, temperature coefficient and heat capacity of the conductor material.

 Note: Values of k for protective conductors in various use or service are as given in Tables 54B, 54C, 54D, 54E and 54F of the Regulations.

Earthing arrangements

If the protective conductor forms part of a cable,   542-01-09
then this **shall** only be earthed in the installation containing
the associated protective device.

If a number of installations have separate earthing   542-01-09
arrangements, then protective conductors that are
common to any of these installations:

- shall either be capable of carrying the maximum fault
  current likely to flow through them or
- earth one installation only and be insulated from the earthing
  arrangements of any other installation(s).

Protective conductors buried in the ground shall have a   543-01-01
cross-sectional area not less than the values shown in
Table 54A (p. 113) of the IEE Wiring Regulations.

Final circuits

Unless the circuit protective conductor is formed by   543-02-09
a metal covering or enclosure, the circuit protective
conductor of ring final circuits shall **also** be run in the
form of a ring with both ends connected to the
earthing terminal at the origin of the circuit.

Preservation of electrical continuity

Protective conductors **shall** be suitably covered   543-03-01
against mechanical and chemical deterioration and
electrodynamic effects.

Switching devices shall not be inserted in a protective   543-03-04
conductor unless it is a multipole linked switching
(or a plug-in device) where the protective
conductor circuit is not interrupted before the live
conductors – and re-established not later than when
the live conductors are re-connected.

Note: If an installation is supplied from more than one energy source and one of these sources needs to be independently earthed (and it is necessary to ensure that not more than one means of earthing is applied at any time) then a linked switch may be inserted in the connection between the neutral point and the means of earthing (460-01-05).

Other than joints in metal conduits, ducting, trunking or   543-03-03
support systems, the connection of a protective conductor
shall be accessible for inspection, testing and maintenance,
unless:

- they are in a compound-filled or encapsulated joint
- the connection is between a cold tail and a heating element
- the joint is made by welding, soldering, brazing, or a
  compression tool.

When a cable sheath that includes an uninsulated protective   543-03-02
conductor (and is adjacent to joints and terminations) is
removed, the protective conductor shall be protected by
insulating sleeving complying with BS 2848.

Portable generating sets

Portable generators (and other installations which are not   551-04-06
permanently fixed) shall be provided with protective
conductors that are part of the cord or cable, between
separate items of equipment.

 If a portable generating set is used at construction site installations, earth fault monitoring may be omitted.

Electrode water heaters and boilers

Metal water pipes of water heaters which have   554-05-01
immersed and uninsulated heating elements shall **not** be 554-05-02
connected to the main earthing terminal by the circuit
protective conductor.

The protective conductor shall be connected to the shell   554-03-03
of the heater or boiler.

Electrode water heaters (and/or boilers) that are directly   554-03-04
connected to a supply exceeding low voltage shall include
an RCD supply from the electrodes when the sustained
protective conductor exceeds 10% of the rated current of
the heater or boiler.

Note: A time delay may be incorporated in the device to prevent unnecessary
operation in the event of imbalance of short duration.

Special installations and locations

The Regulations list the following specific requirements for protective conductors in special installations and locations.

Locations containing a bath or shower

Local supplementary equipotential bonding shall be   601-04-01
provided to connect together the terminals of the
protective conductor for each circuit supplying Class I
and Class II equipment in zones 1, 2 or 3 with
extraneous conductive parts in those zones.

Swimming pools

Local, supplementary, equipotential bonding shall   602-03-02
connect all extraneous conductive parts with the
protective conductors of all exposed conductive parts
within zones A, B and C.

Agricultural and horticultural premises

Metallic grids laid in the floor for supplementary   605-08-03
bonding **shall** be connected to the protective conductors
of the installation.

Locations with equipment having high protective conductor currents

No special precautions are necessary for equipment   607-02-01
with a protective conductor current not exceeding 3.5 mA.

Equipment with a protective conductor current of between 3.5 mA and 10 mA shall either be: 607-02-02

- permanently connected to the fixed wiring of the installation (without the use of a plug and socket outlet) or
- connected by a plug and socket complying with BS EN 60309-2.

Equipment with a protective conductor current exceeding 10 mA shall be connected to the supply by either: 607-02-03

- a permanent connection to the wiring of the installation (which is the preferred option) or
- a flexible cable with a plug and socket outlet complying with BS EN 60309-2.

The wiring of final and distribution circuits to equipment with a protective conductor current exceeding 10 mA shall have a protective connection complying with one or more of the following: 607-02-04

- a single protective conductor with a cross-sectional area greater than $10\,\text{mm}^2$
- a single (mechanically protected) copper protective conductor with a cross-sectional area greater than $4\,\text{mm}^2$
- two individual protective conductors
- a BS 4444 earth monitoring system that will automatically disconnect the supply to the equipment in the event of a continuity fault
- connection (i.e. of the equipment) to the supply by means of a double wound transformer which has its secondary winding connected to the protective conductor of the incoming supply and the exposed conductive parts of the equipment.

When two individual protective conductors are used, the ends of the protective conductors shall be terminated independently of each other at all distribution boards, junction boxes and socket outlets.   607-02-05

Socket outlet final circuits

Final circuits whose protective conductor current is likely to exceed 10 mA shall be provided with a high-integrity protective conductor connection.   607-03-01

Cross-sectional area of protective conductors

The cross-sectional area of the protective conductors shall not be less than:

$$S = \frac{\sqrt{I^2 t}}{K}$$

where S is the nominal cross-sectional area of the conductor in mm^2; I is the value in amperes (RMS for a.c.) of fault current for a fault of negligible impedance, which can flow through the associated protective device; t is the operating time of the disconnecting device in seconds; and k is a factor taking account of the resistivity, temperature coefficient and heat capacity of the conductor material (607-04-01).

Requirements for TT systems

Equipment with a protective conductor current exceeding 3.5 mA I shall be supplied from an installation forming part of a TT system.	607-05-01
The product of the total protective conductor current (in amperes) and twice the resistance of the installation earth electrodes (in ohms) shall not exceed 50.	607-05-01

Caravans and motor caravans

All extraneous conductive parts of a caravan or motor caravan that are likely to become live in the event of a fault **shall** be bonded to the circuit protective conductor by a conductor with a minimum cross-sectional area of 4 mm^2.	608-03-04
All protective conductors, regardless of cross-sectional area, **shall** be insulated.	608-06-03

When protection by automatic disconnection of supply is used, the wiring system shall include a circuit protective conductor that is connected to the	608-03-02

- inlet protective contact, and
- the electrical equipment's exposed conductive parts, and
- the socket outlet's protective contacts.

If the cable does not include a protective conductor (as specified above) and the cable is not enclosed in conduit or trunking, then it shall:

608-03-03

- have a minimum cross-sectional area of $4\,mm^2$
- be insulated.

Caravan parks

Overhead conductors

All overhead conductors used in caravan parks shall be completely covered with insulation which is:

608-12-03

- at least 2 m away from the boundary of any caravan pitch
- not less than 6 m in vehicle movement areas and 3.5 m in all other areas.

 Note: Poles and other overhead wiring supports shall be protected against any reasonably foreseeable vehicle movement.

Highway power supplies (street furniture and street-located equipment)

Where protection against indirect contact is provided by Class II equipment or equivalent insulation, no protective conductor shall be provided.

611-02-06

7.2.8 RCD conductors

The magnetic circuit of the transformer of an RCD **shall** enclose all the live conductors of the protected circuit.

531-02-02

The protective conductor of an RCD **shall** be outside the magnetic circuit.

531-02-02

RCDs shall ensure that any protective conductor current which may occur will be unlikely to cause any unnecessary tripping of the device.

531-02-04

RCDs that are associated with a circuit that would normally be expected to have a protective conductor shall not be considered sufficient for protection against indirect contact if there is no such conductor.

531-02-05

7.2.9 Neutral conductors

Functional switching devices shall **not** be placed solely in the neutral conductor.

464-01-02

In a TN or TT system, where the cross-sectional area of the neutral conductor is at least equal or equivalent to that of the phase conductor:

Overcurrent detection and an associated disconnecting device for the neutral conductor need not be provided.

473-03-04

Overcurrent detection shall, however, be provided for the neutral conductor in a polyphase circuit which shall be capable of disconnecting the phase conductors but not necessarily the neutral conductor.

473-03-04

In a TN or TT system, where the cross-sectional area of the neutral conductor is **less than** that of the phase conductors, overcurrent detection shall be provided for the neutral conductor that enables the disconnection of the phase conductors but not necessarily the neutral conductor. Overcurrent detection need **not** be provided for the neutral conductor if:

- the neutral conductor is protected against fault current by the protective device for the phase conductors of the circuit, and
- the maximum current, including harmonics, likely to be carried by the neutral conductor is significantly less than the value of the current-carrying capacity of that conductor.

473-03-05

In addition:

No fuse or unlinked switch or circuit breaker shall be inserted in the neutral conductor of TN or TT systems.	530-01-02
Links that are used as isolation devices and which are inserted in the neutral conductor shall:	537-02-05

- not be capable of being removed without the use of tools
- be accessible to skilled persons only.

Neutral or mid-point conductors shall be coloured blue.	514-04-01
Switches, circuit breakers (except where linked) and/or fuses shall be inserted in an earthed neutral conductor.	131-13-02
The neutral conductor of switchgear used to disconnect live conductors shall not be capable of being:	530-01-01

- disconnected before the phase conductors
- reconnected before (or at the same time as) the phase conductors.

The cross-sectional area of neutral conductors in a single-phase circuit (or a polyphase circuit) shall be the same (or greater) than the phase conductor.	524-02-01 and 524-02-02
The outer contact of all Edison screw or single centre bayonet cap type lampholders in TN or TT system circuits (except for E14 and E27 lampholders complying with BS EN 60238) shall be connected to the neutral conductor.	553-03-04
Unless the neutral conductor is effectively protected:	473-03-06

- against fault current by a protective device placed on the supply side or
- by an RCD with a rated residual operating current (I_{An}) not exceeding 0.15 times the current-carrying capacity of the corresponding neutral conductor

then overcurrent detection shall be provided for the neutral conductor of every circuit.

Where the harmonic content of the phase currents is greater than 10% of the fundamental current, the cross-sectional area of neutral conductors in discharge lighting and polyphase circuits shall be the same (or greater) than the phase conductor(s).	524-02-03

The common terminal of the winding of all auto-   555-01-01
transformers shall be connected to the neutral conductor.

In electrode boilers and water heaters:

- the current-carrying capacity of the neutral conductor   554-03-05
  shall be not less than that of the largest phase
  conductor connected to the equipment
- no single-pole switch, non-linked circuit breaker or   554-05-04
  fuse shall be fitted in the neutral conductor between
  the heater/boiler and the origin of the installation.

7.2.10 Phase conductors

An overcurrent detection device shall be provided for   473-03-01
each phase conductor which shall be capable of only
disconnecting that particular conductor – unless the
disconnection of one phase could cause danger or damage.

In TT systems overcurrent protection need not be provided   473-03-02
for one of the phase conductors, If:

- differential protection exists, in the same circuit or on
  the supply side which can cause the disconnection of
  all phase conductors, and
- the neutral conductor is not distributed from an
  artificial neutral point of the circuit that is located
  on the load side of that differential protective device.

The overload protective device may be omitted in one of   473-03-03
the phase conductors of an IT system that does not have
a neutral conductor, provided that an RCD is installed in
each circuit.

The minimum cross-sectional area of phase conductors   524-01-01
in a.c. circuits shall be as shown in Table 52C (page 99)
of the IEE Wiring Regulations.

The neutral conductor of switchgear that is used to   530-01-01
disconnect live conductors shall not be capable of being:

- disconnected before the phase conductors
- reconnected before (or at the same time as) the
  phase conductors.

RCDs shall be capable of disconnecting all of the phase   531-02-01
conductors of a circuit at the same time.

Provided that a heater/boiler: 554-03-07

- is not piped to a water supply
- is not in physical contact with any earthed metal and
- the electrodes and the water in contact with the
  electrodes are insulated so that they cannot be
  touched while the electrodes are live

then:

- a fuse in the phase conductor may be substituted for
  the circuit breaker.

7.2.11 Earthing conductors

Earthing conductors **shall** be capable of being   542-04-02
disconnected to enable the resistance of the earthing
arrangements to be measured.

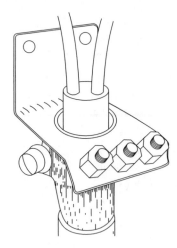

Figure 7.3 Earthing conductor

All installations shall have a main earthing terminal that connects the following to the earthing conductor:

- circuit protective conductors
- main bonding conductors

- functional earthing conductors (if required)
- lightning protection system bonding conductor (if any) (542-04-01).

For TT or IT systems, the main earthing terminal shall be connected via an earthing conductor to an earth electrode.   542-01-04

All joints in earthing conductors shall be: 542-04-02

- capable of disconnection only by means of a tool
- mechanically strong
- capable of maintaining electrical continuity.

The thickness of tape (or strip) used as an earthing conductor shall be capable of withstanding mechanical damage and corrosion (see BS 7430).   542-03-01

Buried earthing conductors shall have a cross-sectional area not less than the values shown in Table 54A (p. 113) of the IEE Wiring Regulations.   542-03-01

The connection of earthing conductors to the earth electrode shall be: 542-03-03

- soundly made
- electrically and mechanically satisfactory
- suitably protected against corrosion
- labelled in accordance with Figure 7.4.

Safety Electrical Connection – Do Not Remove

Figure 7.4 Earthing and bonding notice

Note: This label shall be permanently fixed at or near the connection point of every earthing conductor to an earth electrode.

If electrical earth monitoring is used, the operating coil shall be connected in the pilot conductor and **not** in the protective earthing conductor.   543-03-05

Protective multiple earthing

Where Protective Multiple Earthing (PME) exists, the cross-sectional area of the main equipotential bonding conductor) shall be in accordance with the values shown in Table 54H (p. 120) of the IEE Wiring Regulations.

Note: Local distributors' network conditions may require a larger conductor.

7.2.12 PEN conductors

A separate metal cable enclosure shall **not** be used as a PEN conductor.   543-02-10

PEN conductors shall **not** be isolated or switched.   460-01-04

Automatic disconnection using an RCD shall **not** be applied to a circuit incorporating a PEN conductor.   471-08-07

The following conductors may serve as PEN conductors provided that the part of the installation concerned is not supplied through an RCD: 546-02-02

- conductor of a cable not subject to flexing and with a cross-sectional area not less than 10 mm² (for copper) or 16 mm² for aluminium
- outer conductor of a concentric cable with a cross-sectional area not less than 4 mm².

PEN conductors may only be used if: 546-02-01

- authorisation for use of a PEN conductor has been obtained by the distributor or
- the installation is supplied by a privately owned transformer or convertor and there is no metallic connection (other than the earthing connection) to the distributor's network, or
- the installation is supplied from a private generating plant.

The outer conductor of a concentric cable shall not be common to more than one circuit.   546-02-03

The conductance of the outer conductor of a concentric cable shall not be less than: 546-02-04

- the internal conductor for a single-core cable
- one internal conductor for a multicore cable in a multiphase or multipole circuit
- the internal conductors for a multicore cable serving a number of points contained within one final circuit (or where the internal conductors are connected in parallel).

The continuity of all joints in the outer conductor of a concentric cable (and at a termination of those joints) shall be supplemented by an additional conductor – additional, that is, to any other means used for sealing and clamping the outer conductor.	546-02-05
Isolation or switching devices shall not be included in the outer conductor of a concentric cable.	546-02-06
Separate terminals (or bars) shall be provided for the protective and neutral conductors at the point of separation.	546-02-08
PEN conductors shall be connected to the protective earthing conductor and the neutral conductor's terminals.	546-02-08

 Note: If neutral and protective functions are provided by separate conductors, those conductors shall not then be re-connected together beyond that point.

The PEN conductors of all cables shall be insulated or have an insulating covering that is suitable for the highest voltage to which it may be subjected.	546-02-07
A PEN conductor (when insulated) shall either be: • green-and-yellow throughout its length with blue markings at the terminations, or • blue throughout its length with green-and-yellow markings at the terminations.	514-04-03
In locations where there is a risk of fire owing to the nature of processed or stored materials, a PEN conductor shall not be used.	482-02-07
Isolators and switches shall not break a PEN conductor.	460-01-03
Terminations and joints in a live conductor or a PEN conductor shall only be made inside:	526-03-02

- an equipment enclosure or accessory
- an enclosure made out of material complying with the relevant glow-wire test requirements of BS 6458-2.1
- an enclosure formed or completed with non-combustible building material according to BS 476-4
- an enclosure that is formed by part of the building structure, with an ignitability characteristic 'P' as specified in BS 476 Part 5.

When an RCD is used in a TN-C-S system: 413-02-07

- PEN conductors shall not be used on the load side
- the protective conductor to the PEN conductor shall be on the source side of the RCD.

7.2.13 Bonding conductors

A permanent label/warning notice – with the words   514-13-01
shown in Figure 7.5 – shall be permanently fixed at or
near the bonding conductor's connection point to an
extraneous part.

Safety Electrical Connection – Do Not Remove

Figure 7.5 Earthing and bonding notice

Source supplies may supply more than one item of   413-06-05
equipment provided that:

- all exposed conductive parts of the separated circuit are connected together by an insulated and non-earthed equipotential bonding conductor
- the non-earthed equipotential bonding conductor is not connected to a protective conductor (or to an exposed conductive part of any other circuit or to any extraneous conductive part)
- all socket outlets are provided with a protective conductor contact (that is connected to the equipotential bonding conductor)
- all flexible equipment cables (other than Class II equipment) have a protective conductor for use as an equipotential bonding conductor.

Special installations and locations

The Regulations list the following specific requirement for bonding in special installations and locations (Table 7.8).

Table 7.8 Requirements for bonding in special installations and locations

Installation or location	Requirement	
Locations containing a bath or a shower	In zones A and B the surface wiring system shall not use metallic conduits or metallic trunkings, an exposed metallic cable sheath or an exposed earthing or bonding conductor.	602-06-01

7.2.14 Main equipotential bonding conductor

Main equipotential bonding conductors shall (for each installation) be connected to the main earthing terminal of that installation. These shall include the following:

- water service pipes **but also see Building Regulations requirements in Chapter 2!**
- gas installation pipes
- other service pipes and ducting
- central heating and air conditioning systems
- exposed metallic structural parts of the building
- the lightning protective system (413-02-02).

 Note: Where an installation serves more than one building the above requirement shall be applied to each building (547-02-02).

The cross-sectional area of the main equipotential bonding conductor: 547-02-01

- shall not be less than half the cross-sectional area of the installation's earthing conductor and not less than 6 mm^2
- need not exceed 25 mm^2 if copper.

Where PME conditions apply, the main equipotential bonding conductor shall be selected in accordance with the neutral conductor of the supply and the values shown in Table 54H (p. 120) of the IEE Wiring Regulations.   547-02-01

 Note: Local distributors' network conditions may require a larger conductor.

Earth-free local equipotential bonding

 This method shall **only** be used in special circumstances.   413-05-01

Local equipotential bonding conductors shall not be   413-05-03
in electrical contact with earth.

An equipotential bonding conductor shall be used to   413-05-02
connect together simultaneously accessible exposed
conductive parts and extraneous conductive parts.

Supplementary equipotential bonding

Supplementary bonding conductors connecting:

- two exposed conductive parts or   547-03-01
- an exposed conductive part to an extraneous   and
  conductive part   547-03-02

shall have a conductance (if sheathed or mechanically
protected) not less than the protective conductor
connected to an exposed conductive part.

Note: If mechanical protection is not provided, then the cross-sectional area
shall be greater than $4 \, mm^2$.

Supplementary bonding conductors connecting two   547-03-03
conductive parts shall have a cross-sectional area not less
than $2.5 \, mm^2$ (if sheathed or mechanically protected) or
greater than $4 \, mm^2$ if mechanical protection is not provided.

Supplementary bonding not attached to a fixed appliance   547-03-04
shall be provided by a supplementary conductor or a
conductive part of a permanent and reliable environment.

Fixed appliances supplied by a flexible cord contained   547-03-05
in a circuit protective conductor are considered to have
been provided with a supplementary bonding connection
to its exposed conductive parts.

The resistance of the supplementary bonding conductor   413-02-28
between simultaneously accessible exposed conductive parts
and extraneous conductive parts shall be in accordance with:

$$R < \frac{50}{I_a}$$

where I_a is the operating current of the protective device.

Except for swimming pools, construction site installations and agricultural and horticultural premises, where the formula is changed to:

$$R < \frac{25}{I_a}$$

Special installations and locations

The Regulations list the following specific requirements for supplementary equipotential bonding conductors in special installations and locations (Table 7.9).

Table 7.9 Requirements for supplementary equipotential bonding conductors in special installations and locations

Installation or location	Requirement	
Locations containing a bath or a shower	Local supplementary equipotential bonding shall be provided to connect together the terminals of the protective conductor for each circuit supplying Class I and Class II equipment in zones 1, 2 or 3 with extraneous conductive parts in those zones.	601-04-01
	Note: These extraneous conductive parts may include: • metallic pipes supplying services and metallic waste pipes (e.g. water, gas) • metallic central heating pipes • air conditioning systems • accessible metallic structural parts of the building • metallic baths and shower basins. Supplementary equipotential bonding may be provided in close proximity to the location.	
Swimming pools	Local, supplementary, equipotential bonding shall connect all extraneous conductive parts (and the protective conductors of all exposed conductive parts) within zones A, B and C.	602-03-02
	If there is a metal grid in a solid floor, then this shall be connected to the local supplementary equipotential bonding.	602-03-02
Agricultural and horticultural premises	Metallic grids laid in the floor for supplementary bonding shall be connected to the protective conductors of the installation.	605-08-03

7.2.15 Circuit protective conductors

Installations that provide protection against indirect   471-08-08
contact by automatically disconnecting the supply
shall have a circuit protective conductor run to (and
terminated at) each point in the wiring and at
each accessory.

Circuits supplying Class II equipment shall have a   471-09-02
circuit protective conductor run to, and terminated at,
each point in the wiring and at each accessory.

The metalwork of exposed Class II equipment shall be   471-09-02
mounted so that it is not in electrical contact with any
part of the installation connected to a protective
conductor.

7.2.16 Conductors in parallel

Conductors in parallel **shall** be protected against   432-05-01
overcurrent.

A single protective device may be used to protect   473-02-05
conductors in parallel against the effects of fault
current occurring.

When two or more conductors are connected in parallel   523-02-01
(or when parallel cables are non-twisted single-core cables
with cross-sectional areas greater than 50 mm^2 in copper
or 70 mm^2 in aluminium and are connected in parallel), in the
same phase, the load current shall be shared equally
between them.

Fault current protective devices shall be provided: 473-02-05

- at the supply end of each parallel conductor where
  two conductors are in parallel
- at the supply and load ends of each parallel conductor
  where more than two conductors are in parallel.

7.2.17 Insulated conductors

The Regulations list the following specific requirements for insulated conductors in special installations and locations (Table 7.10).

Table 7.10 Requirements for insulated conductors in special installations and locations

Installation or location	Requirement	
Caravans and motor caravans	Stranded insulated conductors and flexible single-core insulated conductors in non-metallic conduit may be used in electrical installations in caravans and motor caravans.	608-06-01
	The cross-sectional area of conductors shall be greater than 1.5 mm².	608-06-02

7.2.18 Overhead conductors

It is only permissible to dispense with protective measures against indirect contact if:

- the overhead line insulator brackets (and any metal parts connected to them) are not within arm's reach
- the steel reinforcement of steel-reinforced concrete poles is not accessible
- exposed conductive parts (including small isolated metal parts such as bolts, rivets, nameplates not exceeding 50 mm × 50 mm and cable clips) cannot be gripped or contacted by a major surface of the human body
- there is no risk of fixing screws used for non-metallic accessories coming into contact with live parts
- all metal enclosures are mechanically protected
- unearthed street furniture supplied from an overhead line is inaccessible whilst in normal use.

Flexible and non-flexible cables, flexible cords and any conductors that are used as an overhead line that is operating at low voltage shall comply with the relevant British or European harmonised Standard.	521-01-01 521-01-03
Non-flexible cables (and flexible cords not forming part of a portable appliance or luminaire), which are sheathed with lead, PVC or an elastomeric material, may include a catenary wire (or hard-drawn copper conductor) for aerial use or when suspended.	521-01-01
Suspended cables which have insulated conductors with earthed metallic coverings are considered to be an 'underground cable'.	443-02-02
• Installations that are supplied by low-voltage networks that contain overhead lines or • installations that contain overhead lines	443-02-01

and whose location is subject to less than 25 thunderstorm   443-02-02
days per year, do not require any additional protection
against overvoltages of atmospheric origin, provided that
they meet the required minimum equipment impulse
withstand voltages shown in Table 44A (p. 61) of the
IEE Wiring Regulations.

 Note: Suspended cables that have insulated conductors with earthed metallic coverings are considered to be an 'underground cable'.

Special installations and locations

The Regulations list the following specific requirements for overhead conductors in special installations and locations (Table 7.11).

Table 7.11 Requirements for overhead conductors in special installations and locations

Installation or location	Requirement	
Caravan parks	All overhead conductors shall be: • insulated • at least 2 m away from the boundary of any caravan pitch • not less than 6 m in vehicle movement areas and 3.5 m in all other areas. **Note:** Poles and other overhead wiring supports shall be protected against any reasonably foreseeable vehicle movement.	608-12-03

7.3 Conduits, cable ducting, cable trunking, busbar or busbar trunking

Conduit and trunking **shall** comply with the   527-01-05
resistance to flame propagation requirements of BS EN
50085 or BS EN 50086.
Buried cables, conduits and ducts shall be: 522-06-03

• suitably identified
• at a sufficient depth to avoid being damaged by any
  reasonably foreseeable disturbance.

In locations with combustible constructional materials, conduits and trunking systems shall be in accordance with (and meet the fire resistance tests of) BS EN 50086-1 and BS EN 50085-1 respectively.	482-03-04
Non-sheathed cables for fixed wiring shall be enclosed in conduit, ducting or trunking.	521-07-03
The metallic sheaths and/or non-magnetic armour of single-core cables in the same circuit shall either be bonded together:	523-05-01

- at both ends of their run (solid bonding) or
- at one point in their run (single point bonding).

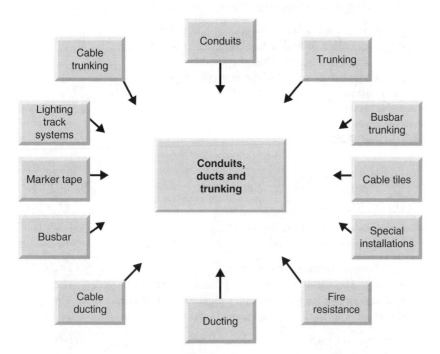

Figure 7.6 Conduits and conduit fittings

 Note: Provided that, at full load, voltages from sheaths and/or armour to earth:

- do not exceed 25 volts, and
- do not cause corrosion, and
- do not cause danger or damage to property.

The sheath (trunking, ducting and/or metal conduit)   543-02-05
may be used as a protective conductor for an associated
circuit provided that:

- it is protected against mechanical, chemical or
  electrochemical deterioration
- its cross-sectional area is in accordance with
  BS EN 60439-1.

When a connection is made inside an enclosure, the   526-03-03
cores of sheathed cables (from which the sheath has
been removed) and non-sheathed cables at the termination
of conduit, ducting or trunking shall be enclosed.

Where the wiring system is buried in a structure, a conduit 522-08-02
or cable ducting system, each circuit shall be completely
erected before the cable is drawn in.

Wiring systems (such as conduit, cable ducting, cable   527-02-02
trunking, busbar or busbar trunking) that penetrate a
building construction that has specified fire resistance,
shall be internally sealed to maintain the degree of
fire resistance.

7.3.1 Conduits

Flexible or pliable conduit shall **not** be selected as a   543-02-01
protective conductor.

Conduits and conduit fittings shall comply with the   521-04-01
appropriate British Standard.

If a protective conductor is formed by conduit, the   543-02-07
earthing terminal of each accessory shall be connected
by a separate protective conductor to an earthing terminal
that is incorporated in the associated box or enclosure.

Conduits and conduit fittings shall comply with the   521-04-01
appropriate British Standard shown in Table 7.12.

Table 7.12 Conduits and conduit fittings

Steel conduit and fittings	BS 31, BS EN 60423, BS EN 50086-1
Flexible steel conduit	BS 731-1, BS EN 60423, BS EN 50086-1
Steel conduit fittings with metric threads	BS 4568, BS EN 60423, BS EN 50086-1
Non-metallic conduits and fittings	BS 4607, BS EN 60423, BS EN 50086-2-1

Orange coloured conduits shall be used to distinguish an electrical conduit from other services or other pipelines.	514-02-01
Screw or mechanical clamps may be used to ensure that joints in metallic conduits are mechanically and electrically continuous.	543-03-06

7.3.2 Trunking, ducting and fittings

Where applicable, trunking, ducting and their fittings shall comply with BS 4678 or BS EN 50085-1. Where BS 4678 does not apply, non-metallic trunking, ducting and their fittings shall be of insulating material complying with the ignitability characteristic 'P' of BS 476 Part 5.	521-05-01
The sheath of a mineral insulated cable (trunking, ducting and metal conduit) may be used as a protective conductor for the associated circuit provided that:	543-02-05
• it is protected against mechanical, chemical or electrochemical deterioration • its cross-sectional area is in accordance with BS EN 60439-1.	
If a protective conductor is formed by trunking, ducting or the metal sheath and/or armour of a cable, the earthing terminal of each accessory shall be connected by a separate protective conductor to an earthing terminal incorporated in the associated box or enclosure.	543-02-07

7.3.3 Busbar trunking systems

Busbar trunking systems shall comply with BS EN 60439-2.	521-01-02

7.3.4 Lighting track systems

Lighting track systems shall comply with BS EN 60570.	521-06-01

7.3.5 Special installations and locations

The Regulations only list the following specific requirements for conduits, cable enclosures and trunking special installations and locations (Table 7.13).

Table 7.13 Requirements for conduits, cable enclosures and trunking in special installations and locations

Installation or location	Requirement	
Locations containing a bath or a shower	In zones A and B the surface wiring system shall not use metallic conduits or metallic trunkings, an exposed metallic cable sheath or an exposed earthing or bonding conductor.	602-06-01
Highway power supplies	Ducting, marker tape and/or cable tiles used with highway power supply cables shall be suitably colour coded or marked so as to identify them from other services.	611-04-03

7.3.6 Other associated requirements

Flexible or pliable conduits shall **not** be used as a protective conductor.	543-02-01
A Band I circuit shall not be contained in the same wiring system as Band II voltage circuits unless the cables:	528-01-02

- are insulated for their respective voltages
- are installed in a separate compartment of a cable ducting or trunking system
- are installed on a tray or ladder separated by a partition
- use a separate conduit, trunking or ducting system.

Cables that are concealed in a wall or partition less than 50 mm from the surface may be enclosed in an earthed conduit, trunking or ducting.	522-06-06
Conductors entering a ferrous enclosure shall be arranged so that they are not individually surrounded by a ferrous material unless some other provision has been made to prevent eddy (induced) currents.	521-02-01
If a protective conductor is formed by conduit, trunking, ducting or the metal sheath and/or armour of a cable, the earthing terminal of each accessory shall be connected	543-02-07

by a separate protective conductor to an earthing
terminal incorporated in the associated box or enclosure.

Phase, neutral and/or protective conductors of each circuit   521-02-01
installed in ferromagnetic enclosures shall be contained
in the same enclosure.

7.3.7 Fire resistance

Conduit, cable ducting, cable trunking, busbar or busbar   527-02-02
trunking used in wiring systems and which penetrate a
building construction that has a specified fire resistance
shall be internally sealed to maintain the degree of
fire resistance.

Temporary sealing arrangements shall be provided during   527-03-01
the erection of a wiring system.

8

Special installations and locations

Whilst the Regulations apply to all electrical installations in buildings, there are also some indoor and out-of-doors special installations and locations that are subject to special requirements due to the extra dangers they pose. This chapter considers the requirements for these special locations and installations.

 These particular Regulations are additional to all of the other requirements, and not alternatives to them.

Figure 8.1 Special electrical installations and locations

In addition to the normal safety protection methods against direct and indirect contact listed in other parts of the Regulations, special installations and locations such as:

- agricultural and horticultural premises
- caravans and motor caravans
- construction sites
- electrical installations in caravan sites
- locations containing a bath or shower
- street furniture and street located equipment
- hot air saunas
- swimming pools

must also comply with the requirements for safety protection in respect of:

- electric shock
- thermal effects
- overcurrent
- undervoltage
- isolation and switching.

These requirements are described in the following sections.

8.1 General requirements

The following are intended to act as a reminder of the general requirements with regard to the design of a circuit and equipotential bonding conductors.

8.1.1 Type of demand

The number and type of circuits required for lighting,   131-03-01
heating, power, control, signalling, communication
and information technology, etc. shall depend on:

- the location and points of power demand
- the loads to be expected on the various circuits
- the daily and yearly variation of demand
- any special conditions
- requirements for control, signalling, communication
  and information technology, etc.

8.1.2 Equipotential bonding conductors

Main equipotential bonding conductors shall (for each   413-02-02
installation) be connected to the main earthing terminal of
that installation. These shall include the following:

- water service pipes
- gas installation pipes
- other service pipes and ducting
- central heating and air conditioning systems
- exposed metallic structural parts of the building
- the lightning protective system.

 Note: Where an installation serves more than one building the above requirement shall be applied to each building.

8.1.3 Heating appliances

Figure 8.2 Heating appliances

In the Regulations, the general requirements for heating appliances (e.g. water heaters, boilers, heating units, heating conductors and cables, surface and underfloor heating systems etc.) are very important to special installations and locations and the following, whilst perhaps not being a complete list, represents the most important requirements.

Accessibility

Connections and joints of heating appliances shall be accessible for inspection, testing and maintenance, unless:

- they are in a compound-filled or encapsulated joint
- the connection is between a cold tail and a heating element
- the joint is made by welding, soldering, brazing or a compression tool.

526-04-01

Electrical heating units

The equipment, system design, installation and testing of an electric surface heating system **shall** meet the requirements of BS 6351.

554-07-01

Heating appliances **shall** be fixed so as to minimise the risks of burns to livestock and of fire from combustible material.

605-10-02

Electric heating units that are embedded in the floor (and intended for heating the location) may be installed below any zone in a bathroom, provided that they are covered by an earthed metallic grid or sheath that is connected to local supplementary equipotential bonding.

601-09-04

If an electric heating unit is embedded in the floor in zone B or C of a swimming pool it shall either:

- be connected to the local supplementary equipotential bonding by a metallic sheath or
- be covered by an earthed metallic grid connected to the equipotential bonding.

602-08-04

Radiant heaters shall be fixed not less than 0.5 m from livestock and from combustible material.

605-10-02

Forced air heating systems

Forced air heating systems **shall** have two, independent, temperature-limiting devices.	424-01-01

Electric heating elements of forced air heating systems (other than those of central-storage heaters) shall: • not be capable of being activated until the prescribed airflow has been established • de-activate when the airflow is reduced or stopped.	424-01-01
Frames and enclosures of electric heating elements shall be of non-ignitable material.	424-01-02

Heating conductors and cables

Heating cables passing through (or in close proximity to) a fire hazard: • shall be enclosed in material with an ignitability characteristic 'P' as specified in BS 476 Part 5 • shall be protected from any mechanical damage.	554-06-01
Heating cables that are going to be laid (directly) in soil, concrete, cement screed, or other material used for road and building construction shall be: • capable of withstanding mechanical damage • resistant to damage from dampness or corrosion.	554-06-02
Heating cables that are going to be laid (directly) in soil, a roadway, or the structure of a building shall be installed so that they are: • completely embedded in the substance it is intended to heat • not damaged by movement (either by it, or the substance in which it is embedded) • compliant (in all respects) with the maker's instructions and recommendations.	554-06-03
The maximum loading of floor-warming cable under operating conditions is shown in Table 8.1.	554-06-04

Table 8.1 Maximum conductor operating temperatures for a floor-warming cable

Type of cable	Maximum conductor operating temperature (°C)
General-purpose PVC over conductor	70
Enamelled conductor, polychlorophene over enamel, PVC overall	70
Enamelled conductor PVC overall	70
Enamelled conductor, PVC over enamel, lead-alloy 'E' sheath overall	70
Heat-resisting PVC over conductor	85
Nylon over conductor, heat-resisting PVC overall	85
Synthetic rubber or equivalent elastomeric insulation over conductor	85
Mineral insulation over conductor, copper sheath overall	Temperature dependent on type of seal employed, outer covering etc.
Silicone-treated woven-glass sleeve over conductor	180

Hot water and/or steam appliances

Electric appliances producing hot water or steam shall be protected against overheating. 424-02-01

Locations with risks of fire due to the nature of processed or stored materials

Heating appliances **shall** be fixed. 482-02-16

Equipment enclosures (such as heaters and resistors) shall not attain surface temperatures higher than: 482-02-18

- 90°C under normal conditions, and
- 115°C under fault conditions.

Heat storage appliances shall not ignite combustible dust and/or fibres. 482-02-17

Heating appliances mounted close to combustible materials shall be protected by barriers. 482-02-16

Where heating and ventilation systems containing heating   482-02-15
elements are installed:

- the dust or fibre content and the temperature of the air
  shall not present a fire hazard
- temperature-limiting devices shall have manual reset.

Water heaters and boilers

Heaters that are intended for liquid or other substances   554-04-01
shall incorporate (or be provided with) an automatic
device to prevent a dangerous rise in temperature.

Metal parts (other than the current-carrying parts of   554-05-02
single-phase water heaters and boilers) that are in
contact with the water shall be solidly and metallically
connected to the metal water pipe supplying that
heater/boiler.

The metal water pipe should be connected to the main earthing terminal by a
circuit-protective conductor that is independent of the heater/boiler.

The heater/boiler shall be permanently connected to the   554-05-03
electricity supply via a double-pole linked switch that
is either:

- separate from and within easy reach of the heater/boiler
- or part of the boiler/heater (provided that the wiring
  from the heater or boiler is directly connected to the
  switch without use of a plug and socket outlet).

8.2 Overhead wiring systems

Whilst the only economic method of transmitting power from a grid station is
by means of lines suspended from pylons, at lower voltages there is a choice
between running them overhead or underground. The supply to most domestic
buildings (particularly in towns) is predominantly underground but for electrical installations such as agricultural buildings, the most cost-effective way is
via overhead cables. The downside of this, of course, is the potential for the
overhead cable to become a safety hazard, and protective methods must be use
to guard against this possibility.

Figure 8.3 Overhead wiring systems

8.2.1 Protection against indirect contact

Protective measures against indirect contact may only be dispensed with if:	471-13-04

- overhead line insulator brackets (and metal parts connected to them) are not within arm's reach
- the steel reinforcement of concrete poles is not accessible
- exposed conductive parts (including small isolated metal parts such as bolts, rivets, nameplates and cable clips) cannot be gripped or cannot be contacted by a major surface of the human body
- there is no risk of fixing screws used for non-metallic accessories coming into contact with live parts
- inaccessible lengths of metal conduit do not exceed 150 mm
- metal enclosures mechanically protecting equipment comply with requirements for Class II protection
- unearthed street furniture that is supplied from an overhead line is inaccessible whilst in normal use.

8.2.2 Protection against direct contact

Bare live parts (other than overhead lines) shall **not** be within arm's reach.	412-05-02

Bare and/or insulated overhead lines being used for distribution between buildings and structures shall be installed in accordance with the Electricity Safety, Quality and Continuity Regulations 2002.	412-05-01
Conductors used as an overhead line, operating at low voltage, shall comply with the relevant British and/or harmonised Standard.	521-01-01 521-01-03

 Note: If access to live equipment (from a normally occupied position) is restricted by an obstacle (such as a handrail, mesh or screen) with a degree of protection less than 1P2X or IPXXB, the extent of arm's reach shall be measured from that obstacle (412-05-03).

When protection against direct contact is required for   611-02-01
highway power supplies:

- protection by obstacles shall not be used
- protection by placing out of reach shall only apply to
  low-voltage overhead lines constructed in accordance
  with the Electricity Safety, Quality and Continuity
  Regulations 2002
- except when the maintenance of equipment is to be
  restricted to skilled persons specially trained, items
  of street furniture (or street-located equipment) that
  are within 1.5 m of a low-voltage overhead line,
  protection against direct contact shall be provided
  by some means other than placing out of reach.

8.2.3 Protection against overvoltages

No additional protection against overvoltages of
atmospheric origin is necessary for:

- installations that are supplied by low-voltage   443-02-01
  systems which do not contain overhead lines
- installations that are supplied by low-voltage networks 443-02-02
  networks which contain overhead lines and their
  location is subject to less than 25 thunderstorm
  days per year
- installations that contain overhead lines and their   443-02-02
  location is subject to less than 25 thunderstorm days
  per year

provided that they meet the required minimum equipment
impulse withstand voltages shown in Table 44A (p. 61) of
the IEE Wiring Regulations.

 Note: Suspended cables having insulated conductors with earthed metallic coverings are considered to be 'underground cables'.

Installations that are supplied by (or include) low-voltage   443-02-03
overhead lines shall incorporate protection against
overvoltages of atmospheric origin if the location is
subject to more than 25 thunderstorm days per year.

This protection shall be provided either by: 443-02-05

- a surge-protective device with a protection level not exceeding Category II or
- other means providing at least an equivalent attenuation of overvoltages.

8.2.4 Other requirements for overhead power lines and cables

Mains-operated electric fence controllers shall:

- take account of the effects of induction when in the   605-14-01
  vicinity of overhead power lines
- not be fixed to any supporting pole of an overhead   605-14-03
  power or telecommunication line.

All overhead conductors in caravan sites shall be: 608-12-03

- protected by insulation of live parts
- at least 2 m away from the boundary of any caravan pitch
- not less than 6 m in vehicle movement areas and 3.5 m in all other areas.

 Note: Poles and other overhead wiring supports shall be protected against any reasonably foreseeable vehicle movement.

8.3 Agricultural and horticultural premises

In contrast to normal domestic installations, an agricultural installation is usually prone to damp conditions and so contact with earth will be better and people and animals are more liable to electric shock. For animals (whose body resistance is much lower than humans) this situation is worsened as their contact with earth will be greater and so even a small voltage could prove lethal to them. Animals can also cause a lot of physical damage to electrical installations and animal effluents present a greater risk of corrosion. Horticultural installations are also subject to the same wet/high earth contact conditions and for these reasons, special requirements have been introduced for all agricultural or horticultural installations.

 If these premises include dwellings that are intended solely for human habitation, then the dwellings are excluded from the scope of these particular Regulations.

The Regulations list the following requirements for all fixed agricultural and horticultural installations (outdoors and indoors) and for locations where live-stock is kept (e.g. cow sheds, stables, chicken houses, piggeries etc.) plus food processing stations and storage areas for hay, straw and fertilisers).

8.3.1 Protection against direct contact

Circuits (less SELV) supplying a socket outlet **shall**   605-03-01
be protected by an RCD.

As the possibility of animals unintentionally coming in direct contact with a live installation is greater than for humans (e.g. they cannot read the warning notices!) and as livestock cannot be protected by Earthed Equipotential Bonding and Automatic Disconnection (EEBAD) (because the voltages to which they would be subjected in the event of a fault would be unsafe for them) the follow-ing protective methods have to be used.

If SELV is used, protection against direct contact shall be   605-02-02
provided by either:

- barriers or enclosures providing protection to at least   (412-03)
  1P2X or IPXXB or
- insulation that can withstand a type-test voltage of   (412-02)
  500 V a.c. RMS for 60 seconds.

8.3.2 Protection against indirect contact

In locations where livestock is intended to be kept,   413-02-01
protection against indirect contact is provided by earthed
equipotential bonding and automatic supply disconnection
which shall be applied in accordance with the type of
earthing system being used (i.e. TN, TT or IT).

Simultaneously accessible exposed conductive parts shall   413-02-03
be connected to the same earthing system individually,
in groups or collectively.

The increase in temperature and corresponding increase   413-02-05
in the resistance of circuit conductors as a result of
overcurrents shall be taken into account.

Exposed conductive parts of an installation shall be   413-02-06
connected (by a protective conductor) to the main earthing

terminal of the installation and that terminal shall be
connected to the earthed point of the supply source
depending on which type of system (i.e. TN, TT and/or IT) it is.

Installations which are part of a TN system

The maximum disconnection times to a circuit supplying   605-05-01
socket outlets and to other final circuits which supply
portable equipment intended for manual movement during
use, or hand-held Class I equipment, shall not exceed
those shown in Table 8.2.

 Note: This requirement does not apply to a final circuit supplying an item of
stationary equipment connected by means of a plug and socket outlet where
precautions are already taken to prevent the use of the socket outlet for supplying hand-held equipment, nor to the reduced low-voltage circuits (i.e. described
in section 471-15 of the Regulations).

Table 8.2 Maximum disconnection times for TN systems

Installation nominal voltage U_o (volts)	Maximum disconnection time t (seconds)
120	0.35
220 to 277	0.2
400, 480	0.05
580	0.02

 Note: If this disconnection time cannot be guaranteed, other protective measures
such as supplementary equipotential bonding may be used.

Where a fuse is used to satisfy these requirements,   413-02-10
maximum values of earth fault loop impedance (Z_s)
corresponding to a disconnection time of 0.4 s shall be in
accordance with Table 41B1 of the Regulations (see
example below) for a nominal voltage to earth (U_o) of 230 V.

**Table 8.3 Maximum earth fault loop impedance (Z_s) for fuses, for 0.2 s disconnec-
tion time with U_0 of 230 V (Reproduced courtesy of BSI)**

General purpose (gG) fuses to BS 88-2.1 and BS 88-6

Rating (amperes)	6	10	16	20	25	32	40	50
Z_s (ohms)	7.74	4.71	2.53	1.60	1.33	0.92	0.71	0.53

If a circuit breaker is used to disconnect a circuit supplying intentionally move-able installations and/or equipment either directly or through socket outlets, the maximum values of earth fault loop impedance are shown in Table 605B2 of the Regulations. As an example, the following table provides an indication of the maximum earth fault loop impedance for Type B circuit breakers (with a nominal voltage of 230 V) corresponding to a disconnection time of 0.2 s.

Table 8.4 Maximum earth fault loop impedance for circuit breakers (Reproduced courtesy of BSI)

Type B circuit breakers to BS EN 60898 and RCBOs to BS EN 61009

Rating (amperes)	6	10	20	40	60	80	100	125
Z_s (ohms)	8.00	4.80	2.40	1.20	0.76	0.60	0.48	0.38

 Note: The circuit loop impedances given in the tables above should not be exceeded when the conductors are at their normal operating temperature. If the conductors are at a different temperature when tested, then the reading should be adjusted accordingly.

The maximum disconnection time for final circuits supplying stationary equipment shall **not** exceed 5 s.   605-05-06

If protection is provided by an RCD, then: 605-05-09

$Z_s \times I_{An} \leqslant 25 \, V$

where Z_s is the earth fault loop impedance in ohms, and I_{An} is the rated residual operating current of the protective device in amperes.

Installations which are part of a TT system

If protection is provided by an RCD, then: 605-06-01

$R_A \times I_a \leqslant 25 \, V$

where R_A is the sum of the resistances of the earth electrode and the protective conductor(s) connecting it to the exposed conductive part, and I_a is the current causing the automatic operation of the protective device within 5 s.

Installations which are part of an IT system

All exposed conductive parts shall be earthed so that   605-07-01
(for each circuit):

$$R_A \times I_d \leqslant 25 \text{ V}$$

where R_A is the sum of the resistances of the earth electrode
and the protective conductor connecting it to the exposed
conductive parts, and I_d is the fault current of the first fault
of negligible impedance between a phase conductor and
an exposed conductive part.

The maximum disconnection time for a second fault in   605-07-02
an IT system shall not exceed that shown in Table 8.5.

Table 8.5 Maximum disconnection time in IT systems (2nd fault)

Installation nominal volts	Maximum disconnection time	
	Neutral not disturbed	Neutral disturbed
120/240	0.8 s	5.0 s
230/240	0.4 s	0.8 s
400/690	0.2 s	0.4 s
580/1000	0.1 s	0.2 s

8.3.3 Supplementary equipotential bonding

 In locations intended for livestock, supplementary   605-08-02
bonding **shall** connect all exposed and extraneous
conductive parts which can be touched by livestock.

Supplementary equipotential bonding must be applied to connect together all
exposed and extraneous conductive parts which are accessible to livestock and
the main protective system.

It is recommended that a metallic grid should be laid in the floor and connected
to the protective conductors of the installation.

The resistance of the supplementary bonding conductor   605-08-01
between simultaneously accessible exposed conductive

parts and extraneous conductive parts **shall** be in accordance with:

$$R < \frac{25}{I_a}$$

where I_a is the operating current of the protective device.

Metallic grids laid in the floor for supplementary bonding **shall** be connected to the protective conductors of the installation.   605-08-03

8.3.4 Protection by earthed equipotential bonding and automatic disconnection of supply

With circuits supplying fixed equipment which are outside of the earthed equipotential zone and which have exposed conductive parts that could be touched by a person who has direct contact directly with earth, the earth fault loop impedance shall ensure that disconnection occurs within the time stated in Table 8.3.   605-09-02

8.3.5 Protection against fire and harmful thermal effects

Radiant heaters **shall** be fixed not less than 0.5 m from livestock and from combustible material.   605-10-02

Fire is a particular hazard in agricultural premises as there is usually large quantities of straw and other flammable material stored in these locations.

For fire protection purposes an RCD (with a residual operating current not exceeding 0.5 A) shall be installed on equipment supplies other than that essential to the welfare of livestock.   605-10-01

Heating appliances shall be fixed so as to minimise the risks of burns to livestock and of fire from combustible material.   605-10-02

8.3.6 External influences

Installed electrical equipment shall have at least the degree of protection IP44.	605-11-01

8.3.7 Isolation and switching devices

Emergency switching and emergency stopping devices **shall**:   605-13-01

- be inaccessible to livestock
- not be impeded by livestock in the event of them panicking.

8.3.8 Electric fence controllers

Electric fences (and their associated controllers) shall **not** be liable to come into contact with any other equipment and/or conductor.   605-14-06

Electric fencing systems have been developed to stop the free movement of animals across pasture. They are semi-permanent solutions that can be extended, altered, or removed to allow grazing to be divided up and/or protected from livestock access.

The system consists of plastic or wooden posts, insulators, conductive wire/rope/tape and an electrical energiser unit. This unit can be mains, battery or solar powered and it sends short electrical impulses along a conductive wire (tape, rope, etc.) so that when the conductive wire or fence is touched by an animal the current passes through it to the ground and causes the animal to feel a shock. The shock is sufficient to alarm an animal but not to harm it and in time the animal will learn to stay away from it.

Only one controller **shall** be connected to each electric fence.   605-14-05

Mains-operated electric fence controllers shall: 605-14-01

- comply with BS EN 61011 and BS EN 6101 I -1
- take account of the effects of induction when in the vicinity of overhead power lines

- be so installed so that they are free from risk of mechanical damage or unauthorised interference   605-14-02
- not be fixed to any supporting pole of an overhead power or telecommunication line.   605-14-03

Earth electrodes that are connected to the earthing terminal of an electric fence controller shall be: 605-14-04

- separate from the earthing system of any other circuit
- situated outside the resistance area of any electrode used for protective earthing.

8.4 Caravans and motor caravans

Caravans and motor caravans are designed as leisure accommodation vehicles which are either towed (e.g. by a car) or self-propelled to a caravan site. They will often contain a bath or a shower and special requirements for such installations

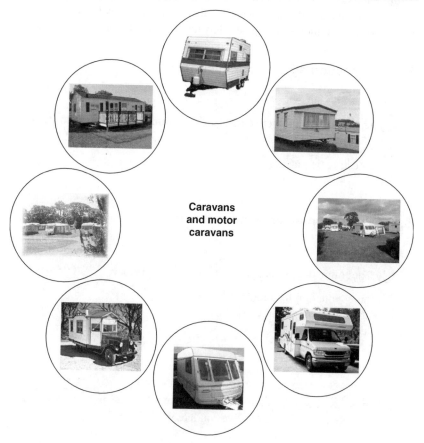

Figure 8.4 Caravans and motor caravans

will apply. In addition to the normal dangers associated with fixed electrical installations, there is the potential hazard of totally unskilled people moving the caravan/motor caravan, connecting and disconnecting the mains supply, and ensuring that it is correctly earthed.

 Caravans that are used as mobile workshops will **also** be subject to the requirements of the Electricity at Work Regulations 1989 and locations containing baths or showers for medical treatment, or for disabled persons, may have special requirements.

The Regulations include a number of special requirements for caravans and mobile homes (as listed below) but these do not deal with:

- electrical circuits and equipment covered by the Road Vehicles Lighting Regulations 1989
- installations covered by BS EN 1648 for 12 V d.c. extra-low-voltage installations in leisure accommodation vehicles and BS EN 1648-2.

 Note: A mobile home is defined as a *'transportable leisure accommodation vehicle that does not meet the requirements for use as a road vehicle'* and is usually a permanent fixture on a caravan park. They normally have recognised power supplies and earthing and the internal electrical installation is outside the scope of the Regulations. Nevertheless, potential safety hazards still have to be considered.

8.4.1 Protection against direct contact

 The following protective measures shall **not** be used: 608-02-01

- protection by obstacles
- protection by placing out of reach.

8.4.2 Protection against indirect contact

 The following protective methods shall **not** be used: 608-03-01

- non-conducting location
- earth-free equipotential bonding
- electrical separation.

When protection by automatic disconnection of supply is   608-03-02
used, an RCD that breaks all live conductors shall be

provided and the wiring system shall include a circuit
protective conductor that is connected to the: 412-06-02

- inlet protective contact, and
- electrical equipment's exposed conductive parts, and
- socket outlet's protective contacts.

If the cable does not include a protective conductor (and   608-03-03
the cable is not enclosed in conduit or trunking) then it shall:

- have a minimum cross-sectional area of 4 mm²
- be insulated.

All extraneous conductive parts of a caravan and/or motor   608-03-04
caravan that are likely to become live in the event of a fault
shall be bonded to the circuit protective conductor by a
conductor with a minimum cross-sectional area of 4 mm².

 Metal sheets forming part of the structure of the caravan or motor caravan are
not considered to be extraneous conductive parts.

8.4.3 Protection against overcurrent – final circuits

Live conductors of all final circuits **shall** be   608-04-01
protected by an overcurrent protective device.

8.4.4 Selection and erection of equipment

When there is more than one electrically independent   608-05-01
installation, each independent system shall be supplied
by a separate connecting device unless all cables are
insulated for the highest voltage present or one of the
following methods is adopted:

- each cable is enclosed within an earthed metallic screen
- the cables are insulated for their respective system voltages
- the cables are installed in a separate compartment of a
  cable ducting or cable trunking system
- the cables have an earthed metallic covering
- each conductor in a multicore cable is insulated for the
  highest voltage present in the cable.

8.4.5 Switchgear and control gear

 All installations **shall** have a main isolating switch   608-07-04
that disconnects all live conductors.

The electrical inlet to caravans and/or motor caravans   608-07-01
shall be a two-pole, earthing contact (key position 6h)
type complying with BS EN 60309-2.

Note: Electrical inlets are not limited to 16 A single-phase where the caravan
or motor caravan demands more.

Electrical inlets shall be installed: 608-07-02

- on the outside of the caravan
- in a readily accessible position
- no more than 1.8 m above ground level
- in an enclosure with a suitable cover.

A notice shall be fixed on or near the electrical inlet recess   608-07-03
containing the following information concerning the
caravan and/or motor caravan installation:

- nominal (design) voltage and frequency
- current rating.

The isolating switch for installations consisting of only   608-07-04
one final circuit may be the overcurrent protection device
in accordance with Regulation 608-04.

A notice (worded as shown in Figure 8.5) shall be   608-07-05
permanently fixed near the main isolating switch inside
the caravan or motor caravan.

8.4.6 Wiring systems

 All protective conductors, regardless of cross-   608-06-03
sectional area, **shall** be insulated.

Electrical equipment shall **not** be installed in compartments   608-06-06
where gas cylinders are stored.

Flame propagating wiring systems shall **not** be used.   608-06-01

1.1 INSTRUCTIONS FOR ELECTRICITY SUPPLY

TO CONNECT

1. Before connecting the caravan installation to the mains supply, check that:
   (a) the supply available at the caravan pitch supply point is suitable for the caravan electrical installation and appliances, and
   (b) the caravan main switch is in the OFF position.
2. Open the cover to the appliance inlet provided at the caravan supply point and insert the connector of the supply flexible cable.
3. Raise the cover of the electricity outlet provided on the pitch supply point and insert the plug of the supply cable.

THE CARAVAN SUPPLY FLEXIBLE CABLE MUST BE FULLY UNCOILED TO AVOID DAMAGE BY OVERHEATING.

4. Switch on at the caravan main switch.
5. Check the operation of residual current devices, if any, fitted in the caravan by depressing the test buttons.

IN CASE OF DOUBT OR, IF AFTER CARRYING OUT THE ABOVE PROCEDURE THE SUPPLY DOES NOT BECOME AVAILABLE, OR IF THE SUPPLY FAILS, CONSULT THE CARAVAN PARK OPERATOR OR THE OPERATOR'S AGENT OR A QUALIFIED ELECTRICIAN.

TO DISCONNECT

6. Switch off at the caravan main isolating switch, unplug both ends of the cable.

1.2 PERIODIC INSPECTION

Preferably not less than once every three years and more frequently if the vehicle is used more than normal average mileage for such vehicles, the caravan electrical installation and supply cable should be inspected and tested and a report on their condition obtained as prescribed in BS 7671 Requirements for Electrical Installations published by the Institution of Electrical Engineers.

Figure 8.5 Electrical safety notice

The following wiring systems shall be used: 608-06-01

- flexible single-core insulated conductors in non-metallic conduits
- stranded insulated conductors (with a minimum of seven strands) in non-metallic conduits
- sheathed flexible cables.

The cross-sectional area of conductors shall be greater than $1.5\,mm^2$.   608-06-02

Low-voltage system cables shall be run separately from extra-low-voltage systems cables.   608-06-04

All cables (unless enclosed in rigid conduit) shall be supported at intervals not exceeding 0.4 m for vertical runs and 0.25 m for horizontal runs.	608-06-05
All flexible conduits shall be supported at intervals not exceeding 0.4 m for vertical runs and 0.25 m for horizontal runs.	608-06-05
All wiring shall be protected against mechanical damage.	608-06-07
Wiring passing through metalwork shall be securely fixed and protected by bushes or grommets.	608-06-07

8.4.7 Accessories

The conductive parts of all accessories shall not be easily accessible.	608-08-01
Low-voltage socket outlets shall include a protective contact (unless supplied by an individual winding of an isolating transformer).	608-08-02
Plugs designed for extra-low-voltage socket outlets shall be incompatible with low-voltage socket outlets.	608-08-03
Extra-low-voltage socket outlets shall: • clearly show their voltage • prevent the insertion of a low-voltage plug.	608-08-03
Accessories exposed to moisture shall be protected to at least 1P55.	608-08-04
Appliances not connected to the supply by a plug and socket outlet shall be controlled by a switch.	608-08-05
Luminaries shall (preferably) be fixed directly to the structure or lining of the caravan or motor caravan.	608-08-06
Pendant luminaries shall be securely installed to prevent damage when the caravan or motor caravan is moved.	608-08-06
Dual voltage luminaries shall: • be fitted with separate lampholders for each voltage • clearly indicate the lamp wattage • not be damaged if both lamps are lit at the same time • provide adequate separation between LV and ELV circuits • have lamps that cannot be fitted into lampholders intended for lamps of other voltages.	608-08-07

The connector used to attach the supply to a caravan or   608-08-08
motor caravan shall comprise:

- a plug complying with BS EN 60309-2
- a flexible cord or cable, 25 m ($\pm$ 2 m) in length
  (HO7RN-F, HO5VV-F or equivalent) that includes a
  protective conductor whose cross-section meets the
  requirements of Table 8.6
- a connector meeting the requirements of   608-07-01
  BS EN 60309-2.

**Table 8.6 Cross-sectional areas of flexible cords and
cables for caravan connectors**

Rated current (A)	Cross-sectional area (mm^2)
16	2.5
25	4
32	6
63	16
100	35

8.4.8 Extra-low-voltage installation

Any part of a caravan installation that is operating at   608-08-09
extra-low voltage shall meet the requirements of protection
by SELV.

The following extra-low-voltage d.c. power sources are   608-08-09
permissible: 12 V, 24 V and 48 V.

When a.c. extra-low voltage is required, the following   608-08-09
voltages (RMS) are permissible: 12 V, 24 V, 42 V and 48 V.

8.5 Caravan parks

The requirements of the Electricity Supply Regulations do not allow the supply neutral to be connected to any metalwork in a caravan (which means that
only TT or TN-S systems may be used) and in general, the supply of electrical
energy to caravan and tent sites must ensure that:

- wherever possible the supply is via underground cables
- overhead supplies use insulated as opposed to bare cables
- cables are installed outside the area of the caravan pitch and must be at least
  3.5 m above ground level (increased to 6 m where vehicle movements are
  possible)

- each socket must have its own individual overcurrent protection in the form of a fuse or circuit breaker
- cables that are run below caravan pitches must be provided with additional protection as shown in Figure 8.6
- all sockets must be protected by an RCD complying with BS 4293, BS EN 61008-1 or BS EN 61009-1 with a 30 mA rating, either individually or in groups not exceeding three sockets.

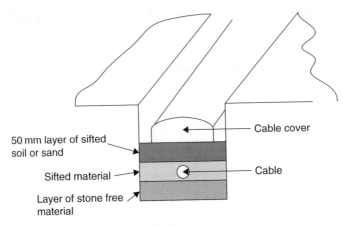

50 mm layer of sifted soil or sand

Cable cover

Sifted material

Cable

Layer of stone free material

Figure 8.6 Cable covers

The actual requirements (which only apply to installations supplying electricity to leisure accommodation vehicles in caravan parks) are as follows.

8.5.1 Protection against direct contact

 The following protection methods shall **not** be used: 608-10-01

- protection by obstacles
- protection by placing out of reach.

8.5.2 Protection against indirect contact

 The following protection methods shall **not** be used: 608-11 01

- non-conducting location
- earth-free local equipotential bonding
- electrical separation.

8.5.3 Wiring systems

 Unless mechanically protected, all underground   608-12-02
cables **shall** be installed outside of the caravan pitch
and areas where tent pegs or ground anchors may be driven.

Caravan pitch supply equipment to caravans shall   608-12-01
(preferably) be connected by underground cable.

All overhead conductors shall be: 608-12-03

- insulated and protected against direct contact
- at least 2 m away from the boundary of any caravan pitch
- not less than 6 m in vehicle movement areas and 3.5 m
  in all other areas.

 Note: Poles and other overhead wiring supports shall be protected against any
reasonably foreseeable vehicle movement.

8.5.4 Switchgear and control gear

 Socket outlets must **not** be bonded to the PME terminal.   608-13-05

Caravan pitch supply equipment shall be located adjacent   608-13-01
to the pitch and not more than 20 m from any point on the
pitch which it is intended to serve.

Socket outlets and enclosures forming part of the caravan
pitch supply equipment shall:

- comply with BS EN 60309-2
- be protected to at least IPX4
- be placed between 0.80 m and 1.50 m above the ground
  (to the lowest part of the socket outlet)
- have a current rating of not less than 16 A.

Note: At least one socket outlet shall be provided for each pitch (608-13-02).

Socket outlets shall be protected (individually) by an   608-13-04
overcurrent device.

All exposed conductive parts which are protected by a single protective device shall be connected to a common earth electrode of the following types of protective device shall be used:	413-02-18
• a residual current device (preferred method) • an overcurrent protective device	413-02-19
and each circuit shall meet the following requirement:	413-02-20

$$R_A \times I_a \leqslant 50\,\text{V}$$

where R_A is the sum of the resistances of the earth electrode and the protective conductor(s) connecting it to the exposed conductive part, and I_a is the current causing the automatic operation of the protective device within 5 s.

Socket outlets shall be protected (individually or in groups of not more than three) by an RCD.	608-13-05
Although RCDs reduce the risk of electric shock, they may **not** be used as the sole means for protecting socket outlets against direct contact.	608-13-05 412-06-01 412-06-02
Grouped socket outlets shall be on the same phase.	608-13-06

8.6 Construction sites

Electrical installations at construction sites are there primarily to provide lighting and power to enable work to proceed. As workmen will probably be working ankle deep in wet, muddy conditions and using a selection of portable tools such as drills and grinders, they will be particularly susceptible to electric shock.

There are six levels of voltage normally associated with construction sites. These are:

25 V single-phase SELV	for portable hand lamps in damp and confined situations
50 V single-phase centre-point earthed	for hand lamps in damp and confined situations
400 V three-phase	for use with fixed or transportable equipment with a load of more than 3750 watts
230 V single-phase	for site buildings and fixed lighting
110 V three-phase	for transportable equipment with a load up to 3750 watts
110 V single-phase	for transportable tools and equipment, such as floodlighting

Equipment used must be suitable for the particular supply to which it is connected, and for the application it will meet on site. Where more than one voltage is in use, plugs and sockets must be non-interchangeable to prevent misconnection.

Supplies will normally be obtained from the electrical supply company but remote sites could need an IT supply (such as a generator) and care must be taken in complying with the safety requirements for this particular source.

The following requirements apply to installations providing an electricity supply for:

- new building construction
- repair, alteration, extension or demolition of existing buildings
- engineering construction
- earthworks

and are applicable to:

- the main switchgear and protective devices
- installations of mobile and transportable electrical equipment
- the interface between the supply system and the construction site installations.

The following requirements do **not** apply to:

- construction site offices, cloakrooms, meeting rooms, canteens, restaurants, dormitories and toilets
- installations covered by BS 6907.

8.6.1 Protection against indirect contact

An IT system shall **not** be used if an alternative   604-03-01
system is available.

If an IT system is used, permanent earth fault monitoring   604-03-01
shall be provided.

If a portable generating set is used, earth fault monitoring   604-03-01
may be omitted.

TN system

Except for reduced low-voltage systems the maximum   604-04-01
disconnection times for circuits supplying movable
installations and equipment (either directly or through
socket outlets) shall be as indicated in Table 8.7.

Table 8.7 Maximum disconnection times for TN systems

Installation nominal voltage U_o (volts)	Maximum disconnection time t (seconds)
120	0.35
220 to 277	0.2
400, 480	0.05
580	0.02

 Note: If this disconnection time cannot be guaranteed other protective measures, such as supplementary equipotential bonding may be used.

Where a fuse is used to satisfy this requirement, maximum   604-04-03
values of earth fault loop impedance (Z_s) corresponding
to a disconnection time of 0.2 s shall be in accordance
with Table 604B1 of the Regulations (see example below)
for a nominal voltage to earth (U_0) of 230 V.

Table 8.8 Maximum earth fault loop impedance (Z_s) for fuses, for 0.2 s disconnection time with U_0 of 230 V (Reproduced courtesy of BSI).

General purpose (gG) fuses to BS 88-2.1 and BS 88-6

Rating (amperes)	6	10	16	20	25	32	40	50
Z_s (ohms)	7.74	4.71	2.53	1.60	1.33	0.92	0.71	0.53

 Note: The circuit loop impedances given in the table above should not be exceeded when the conductors are at their normal operating temperature. If the conductors are at a different temperature when tested, then the reading should be adjusted accordingly.

Where a circuit breaker is used to satisfy this requirement,   604-04-04
maximum values of earth fault loop impedance (Z_s)
corresponding to a disconnection time of 0.2 s shall be in
accordance with Table 604B2 of the Regulations (see
example below) for a nominal voltage to earth (U_0) of 230 V.

Table 8.9 Maximum earth fault loop impedance (Z_s) for fuses, for 0.2 s disconnection time with U_0 of 230 V (Reproduced courtesy of BSI).

Type B circuit breakers to BS EN 60898 and RCBOs to BS EN 61009

Rating (amperes)	6	10	16	20	32	50	80	125
Z_s (ohms)	8.00	4.80	3.00	2.40	1.50	0.96	0.60	0.38

The maximum disconnection time for fixed installations and reduced low-voltage systems is 5 s.

604-04-06

If these conditions cannot be fulfilled by using overcurrent protective devices, then protection shall be provided by an RCD and the following condition shall be fulfilled:

604-04-07 and 604-04-08

$$Z_s \times I_{An} \leqslant 25\,V$$

where Z_s is the earth fault loop impedance in ohms, and I_{An} is the rated residual operating current of the protective device in amperes.

TT system

Each circuit shall meet the following requirement:

604-05-01

$$R_A \times I_a \leqslant 25\,V$$

where R_A is the sum of the resistances of the earth electrode and the protective conductor(s) connecting it to the exposed conductive part, and I_a is the current causing the automatic operation of the protective device within 5 s.

IT system

All exposed conductive parts shall be earthed so that (for each circuit):

604-06-01

$$R_A \times I_d \leqslant 25\,V$$

where R_A is the sum of the resistances of the earth electrode and the protective conductor connecting it to the exposed conductive parts, and I_d is the fault current of the first fault of negligible impedance between a phase conductor and an exposed conductive part.

The maximum disconnection times for IT systems (2nd fault) shall be no greater than that shown in Table 8.10.

604-06-02

Table 8.10 Maximum disconnection time in IT systems (2nd fault)

Installation	Maximum disconnection time	
	Neutral not disturbed	Neutral disturbed
120/240	0.4	1
220/380 to 277/480	0.2	0.5
400/690	0.06	0.2
580/1000	0.02	0.08

8.6.2 Supplementary equipotential bonding

The resistance of the supplementary bonding conductor   604-07-01
between simultaneously accessible exposed conductive
parts and extraneous conductive parts shall be in
accordance with:

$$R < \frac{25}{I_a}$$

where I_a is the operating current of the protective device.

8.6.3 Selection and erection of equipment

All assemblies used for distributing electricity on   604-09-01
construction and demolition sites shall comply with the
requirements of BS 4363 and BS EN 60439-4.

 Note: All other equipment shall be protected appropriate to the external influence.

8.6.4 Supplies

Except for control or signalling circuits (and inputs   604-02-01
from stand-by supplies) equipment shall:

- be compatible with the particular supply from which it
  is energised
- only contain components connected to the same installation.

The nominal voltages shown in Table 8.11 shall not be   604-02-02
exceeded.

 The only exception is if large equipment requires a high-voltage supply for functional reasons.

Table 8.11   Nominal voltages for equipment at construction site installations

SELV	Portable hand lamps in confined or damp locations
110 V, 1-phase, centre point earthed	Reduced low-voltage system Portable hand lamps for general use Portable hand-held tools and local lighting up to 2 kW
110 V, 3-phase, star point earthed	Reduced low-voltage system Portable hand-held tools and local lighting up to 2 kW Small mobile plant, up to 3.75 kW
230 V, 1-phase	Fixed floodlighting
400 V, 3-phase	Fixed and movable equipment, above 3.75 kW

8.6.5 Wiring systems

Cables shall **not** be installed across a site road or a walkway unless they are adequately protected against mechanical damage.   604-10-02

Reduced low-voltage systems shall use low temperature 300/500 V thermoplastic (PVC) or equivalent flexible cables.   604-10-03

Applications such as fixed floodlighting, and fixed/moveable equipment above 3.75 kV exceeding the reduced low-voltage system shown in Table 8.11 shall use H07 RN-F type flexible cable that is resistant to abrasion and water.   604-10-03

8.6.6 Isolation and switching devices

Emergency switching **shall** be provided for supplies to equipment that need to have live conductors disconnected in order to remove a hazard.   604-11-03

Supply and distribution assemblies shall be capable of isolating and switching the incoming supply.   604-11-02

Circuits supplying current using equipment shall be fed from a distribution assembly that includes: 604-11-05

• overcurrent protective devices
• protection against indirect contact
• socket outlets (if required).

It shall not be possible to interconnect different safety and stand-by supplies.   604-11-06

8.6.7 Plugs and socket outlets

 Plug and socket outlets **shall** comply with   604-12-02
BS EN 60309-2.

Luminaire supporting couplers shall not be used.   604-12-03

Socket outlets shall be part of an assembly complying   604-12-01
with BS 4363 and BS EN 60439-4.

Socket outlets rated at 32 A or less supplying portable   604-08-03
equipment for use outdoors shall be provided with
supplementary protection (i.e. an RCD) except for socket
outlets supplied by a circuit protected by:

- SELV
- electrical separation
- automatic disconnection and reduced low-voltage systems.

Where socket outlets and permanently connected hand-held   604-08-04
equipment socket outlets are protected by an RCD, they
shall be supplied from a separate isolating transformer.

8.6.8 Cable couplers

Cable couplers **shall** comply with BS EN 60309-2.   604-13-01

8.7 Equipment with high protective conductor current

Protective conductor current (as the name implies) is the amount of electric current
that flows in a protective conductor under normal operating conditions and is used
as a measure for Class I devices where insulation resistance cannot be measured
and where all exposed conductive parts are connected to the protective conductor.

Specific requirements concerning protective conductor current are as follows.

No special precaution is necessary for equipment with a   607-02-01
protective conductor current not exceeding 3.5 mA.

Equipment with a protective conductor current of between   607-02-02
3.5 mA and 10 mA shall be either permanently connected

to the fixed wiring of the installation (without the use of a plug and socket outlet) or be connected by a plug and socket complying with BS EN 60309-2.

Equipment with a protective conductor current exceeding 10 mA shall be connected to the supply by either: 607-02-03

- a permanent connection to the wiring of the installation (which is the preferred option) or
- via a flexible cable with a plug and socket outlet complying with BS EN 60309-2.

The wiring of final and distribution circuits to equipment with a protective conductor current exceeding 10 mA shall have a protective connection complying with one or more of the following: 607-02-04 and 607-02-05

- a single protective conductor with a cross-sectional area greater than 10 mm^2
- a single (mechanically protected) copper protective conductor with a cross-sectional area greater than 4 mm^2
- two individual protective conductors (in which case the ends of the protective conductors shall be terminated independently of each other at tall distribution board, junction boxes and socket outlets)
- a BS 4444 earth monitoring system that will automatically disconnect the supply to the equipment in the event of a continuity fault
- connection (i.e. of the equipment) to the supply by means of a double wound transformer which has its secondary winding connected to the protective conductor of the incoming supply and the exposed conductive parts of the above.

Final circuits whose protective conductor current is likely to exceed 10 mA shall be provided with a high-integrity protective conductor connection.   607-03-01

The following types of final circuit are acceptable: 607-03-01

- a ring final circuit with a ring protective conductor
- a radial final circuit with a single protective conductor
- other circuits complying with the general requirements for protection of safety (as detailed in Regulation 607-02).   607-02

Distribution boards shall have a notice which clearly indicates circuits which have a high protective conductor current. This information shall be positioned so as to be visible to a person who is modifying or extending the circuit.   607-03-02

The cross-sectional area of protective conductors shall   607-04-01
not be less than:

$$S = \frac{vI^2t}{k}$$

where S is the nominal cross-sectional area of the
conductor in mm^2; I is the value in amperes (RMS for a.c.)
of fault current for a fault of negligible impedance, which
can flow through the associated protective device (also
see Regulation 413-02-05); t is the operating time of the
disconnecting device in seconds; and k is a factor taking
account of the resistivity, temperature coefficient and heat
capacity of the conductor material.

 Note: Values of k for protective conductors in various use or service are as
given in Tables 54B, 54C, 54D, 54E and 54F of the Regulations.

Equipment with a protective conductor current exceeding   607-05-01
3.5 mA shall be supplied from an installation forming part
of a TT system.

The product of the total protective conductor current (in   607-05-01
amperes) and twice the resistance of the installation earth
electrodes (in ohms) shall not exceed 50.

Equipment with a protective conductor current exceeding   607-06-01
3.5 mA shall not be connected directly to an IT system.

If more than one item of equipment has a protective   607-07-01
conductor current exceeding 3.5 mA and is going to be
supplied from an installation incorporating a residual
current device, the circuit used shall ensure that the residual
current (including switch-on surges), will not trip the device.

8.8 Highway power supplies

The requirements of this section of the Regulations apply to highway distribution circuits, street furniture (i.e. street and footpath lighting, traffic signs, traffic control and traffic surveillance equipment etc.) and street-located equipment
(i.e. telephone kiosks, bus shelters, advertising signs and car park ticket dispensers
etc.). They also apply to similar equipment located in other areas (e.g. private
roads, car parks etc.) used by the public but not designated as a highway or part
of a building.

These requirements do **not** apply to electric supply company's equipment (as defined by the Electricity Safety, Quality and Continuity Regulations 2002) or the overhead or underground supplies that they provide to feed the equipment. It does, however, apply to all overhead or underground supplies that connect street furniture or street-located equipment to the electric company's point of supply.

8.8.1 Protection against direct contact

Protection by obstacles shall **not** be used for highway power supplies.   611-02-01

Protection by placing out of reach shall only apply to low-voltage overhead lines constructed in accordance with the Electricity Safety, Quality and Continuity Regulations 2002.   611-02-01

 Except when the maintenance of equipment is to be restricted to skilled persons specially trained. In which case, for items of street furniture (or street-located equipment) that are within 1.5 m of a low-voltage overhead line, protection against direct contact with the overhead line shall be provided by some means other than placing out of reach.

Any door used to provide access to electrical equipment in street furniture or street-located equipment: 611-02-02

- shall not be considered as providing protection by barriers or enclosures
- shall be locked with a key or secured by use of a tool if located less than 2.50 m above ground level
- shall be protected against direct contact (i.e. when the door is open) by an intermediate barrier providing protection to at least IP2X or IPXXB which can only be removed by the use of a tool.

Access to the light source for luminaries located less than 2.80 m above ground level shall only be possible after removing a barrier or enclosure that requires the use of a tool.   611-02-02

8.8.2 Protection against indirect contact

For highway power supplies, protection against   611-02-03
indirect contact shall **not** be provided by:

- a non-conducting location
- earth-free equipotential bonding
- electrical separation.

Where protection against indirect contact is provided by   611-02-05
using earthed equipotential bonding and automatic
disconnection, metallic structures such as fences and
grids not connected to or forming part of the street furniture
or street-located equipment shall not be connected as an
extraneous conductive part to the main earthing terminal.

Where protection against indirect contact is provided by   611-02-06
Class II equipment or equivalent insulation, no protective
conductor shall be provided and conductive parts of the
lighting column, street furniture or street-located equipment
shall not be intentionally connected to the earthing
system.

8.8.3 Isolation and switching devices

Where it is intended that isolation and switching shall   611-03-01
only be carried out by instructed persons, a suitably rated
fuse carrier may be used as the isolating/switching device.

If the distributor's cut-out is used for isolating a highway   611-03-02
power supply, then the approval of the distributor shall be
obtained.

8.8.4 Fixed equipment

Circuits feeding fixed equipment that is used in highway   611-02-04
power supplies shall have a maximum disconnection time of 5 s.

8.8.5 External influences

All electrical equipment shall have a degree of protection   611-05-02
not less than 1P33.

8.8.6 Temporary supplies

Temporary supplies that have been taken from street   611-06-01
furniture and street located equipment shall:

- not reduce the safety of the permanent installation
- generally be in accordance with the requirements for
  construction site installations (see section 604 of the
  Regulations).

All temporary supply units shall have an externally   611-06-02
mounted label stating the maximum sustained current
that can be supplied from that unit.

8.8.7 Records

Detailed records **shall** be provided (together with   611-04-01
an Electrical Installation Certificate) on completion
of installations that include highway distribution circuits
and highway power supplies.

Drawings shall be prepared showing the position and   611-04-05
depth of all underground cables supplying highway power
supplies, street furniture and/or street-located equipment.

8.8.8 Identification of cables

Unless installed in a conduit or duct, cables buried in the   611-04-02
ground shall be marked by cable covers or marking tape.

Buried conduits and ducts shall be suitably identified.   611-04-02

 Note: Except where the method of cable installation does not permit.

All ducting, marker tape and/or cable tiles used with highway power supply cables shall be suitably colour-coded or marked so as to identify them from other services.

611-04-03

8.8.9 Notices

A notice concerning the requirement for periodic inspection and testing need not be applied if the installation is subject to a programmed inspection and testing procedure.

611-04-04

8.9 Hot air saunas

Saunas, similar to bathrooms and showers, are primarily used by people who are unclothed and wet and thus very vulnerable to electric shock due to their reduced body resistance (i.e. absence of shoes means less protection from shock, whilst water on their skin will tend to short-circuit its natural protection). Special measures are, therefore, needed to ensure that the possibility of direct and/or indirect contact is reduced.

The Regulations requirements for hot air saunas are based on four zones which take into account the limitations of walls, doors, fixed partitions, ceilings and floors and the electric heater itself. The four zones are shown in Figure 8.7.

8.9.1 General

All equipment shall be protected to at least 1P24.

603-06-01

Only the equipment shown in Table 8.12 may be installed in each of the four temperature zones.

603-06-02

8.9.2 Protection against both direct and indirect contact

When SELV is used, protection against direct contact **shall** be provided either by:

603-03-01

- insulation of live parts (capable of withstanding 500 V a.c. RMS for 60 seconds) or by
- barriers or enclosures (providing protection to at least IP24 or IPX4B).

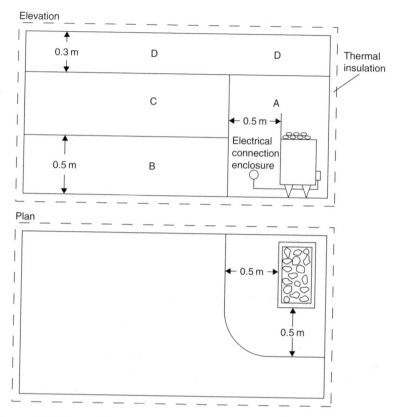

Figure 8.7 Zones of ambient temperature

Table 8.12 Permissible equipment

Temperature zone	Equipment
Zone A	Sauna heaters and equipment directly associated with it
Zone B	No special requirement (concerning heat resistance of equipment)
Zone C	Equipment suitable for an ambient temperature of 125°C
Zone D	Luminaries (and their associated wiring) suitable for an ambient temperature of 125°C Control devices for the sauna heater (and their associated wiring) suitable for an ambient temperature of 125°C

8.9.3 Protection against direct contact

The following protective measures against direct contact shall **not** be used:

603-04-01

- protection by means of obstacles
- protection by placing out of reach.

8.9.4 Protection against indirect contact

The following protective measures against indirect   603-05-01
contact shall **not** be used:

- protection by non-conducting location
- protection by means of earth-free local equipotential bonding.

8.9.5 Wiring systems

Only flexible cords with 180°C thermosetting (rubber)   603-07-01
insulation that is protected against mechanical damage
shall be used.

8.9.6 Switchgear, control gear and accessories

All switchgear (other than a thermostat or a thermal   603-08-01
cut-out) not actually built into the sauna heater shall be
installed outside the hot air sauna.

8.9.7 Accessories

Accessories shall not be installed within the hot air sauna   603-08-02
unless they comply with the requirements of Table 8.13.

8.9.8 Other fixed equipment

Luminaries shall be mounted so as to prevent overheating.   603-09-01

8.10 Locations containing a bath or shower

Special requirements exist for locations containing   601-01-01
baths and showers for medical use or for disabled persons
(see Building Regulations Part M).

When people use bathrooms and showers, most of the time they are naturally unclothed and wet and thus very vulnerable to electric shock due to their reduced body resistance (i.e. absence of shoes means less protection from shock, whilst water on their skin will tend to short-circuit its natural protection). Special measures are, therefore, needed to ensure that the possibility of direct and/or indirect contact is reduced.

The Regulations requirements for locations containing baths, showers and cabinets containing a shower and/or bath and the surrounding zones. They do not apply to emergency facilities in industrial areas and laboratories. The requirements are based on four zones which take into account the limitations of walls, doors, fixed partitions, ceilings and floors (see Figure 8.8).

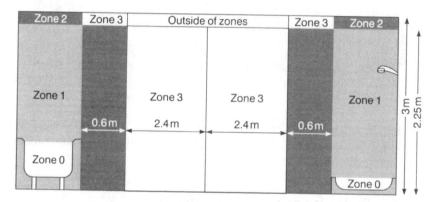

Figure 8.8 Zone dimensions

Table 8.13 Limitations of zones

Zone 0	The interior of the bath tub or shower basin which is limited by the floor and by the plane above the floor.
Zone 1	The areas directly above the bath which is limited by: • the upper plane of zone 0 • the horizontal plane 2.25 m above the floor • the vertical plane(s) circumscribing the bath tub or shower basin.
Zone 2	The vertical area directly above the bath which is limited by: • the vertical plane(s) external to zone 1 • the parallel vertical plane(s) 0.60 m external to zone 1 • the floor and the horizontal plane 2.25 m above the floor.
Zone 3	The outside zone which is limited by: • the vertical plane(s) external to zone 2 • the parallel vertical plane(s) 2.40 m external to zone 2 • the floor and the horizontal plane 2.25 m above the floor.

 These requirements also apply to installations in caravans or motor caravans.

8.10.1 Application of protective measures against electric shock

 Protection by obstacles and placing out of reach are **not** permitted.   601-05-02

Protection by non-conducting location and earth-free local equipotential bonding are **not** permitted.   601-05-03

In zone 0, only protection by SELV (at a nominal voltage not exceeding 12 V a.c. RMS or 30 V ripple-free d.c.) is permitted.   601-05-01

The safety source shall be installed outside zones 0, 1 and 2.   601-05-01

Note: The following requirements do not apply to zone 3, where a cabinet containing a shower is installed in a room other than a bathroom or shower room (e.g. in a bedroom).

8.10.2 Protection against electric shock

Where SELV or PELV is used (whatever the nominal voltage) protection against direct contact shall be provided by: 601-03-02

- barriers or enclosures providing protection to at least 1P2X or IPXXB, or
- insulation capable of withstanding 500 V a.c. RMS for 1 minute.

8.10.3 Supplementary equipotential bonding

Local supplementary equipotential bonding shall be provided to connect together the terminals of the protective conductor for each circuit supplying Class I and Class II equipment in zones 1, 2 or 3 with extraneous conductive parts in those zones.   601-04-01

Note: Extraneous conductive parts include:

- metallic pipes supplying services and metallic waste pipes (e.g. water, gas)
- metallic central heating pipes
- air conditioning systems
- accessible metallic structural parts of the building
- metallic baths and shower basins.

8.10.4 Selection and erection of equipment

Electrical equipment shall have the minimum degree of protection shown in Table 8.14.   601-06-01

Table 8.14 Equipment protection

Zone	IP code	Alternatives
Zone 0	IPX7	Or ◖ ◖ if the equipment is not IP coded
Zones 1 and 2	IPX4	Or △◖ if the equipment is not IP coded
	But where water jets are likely to be used for cleaning purposes (e.g. in communal baths and/or communal showers) IPX5	Or △◖△◖ if the equipment is not IP coded
Zone 3 Where water jets are likely to be used for cleaning purposes (e.g. in communal baths and/or communal showers)	IPX5	Or △△ if the equipment is not IP coded

8.10.5 Wiring systems

Surface wiring systems (and wiring systems embedded in the walls) at a depth not exceeding 50 mm and which are not: 601-07-01

- protected by an earthed metallic conductor (complying with BS 5467, BS 6346, BS 6724, BS 7846, BS EN 60702-1 or BS 8436) or

- of an insulated concentric construction (complying with BS 4553-1, BS 4553-2 or BS 4553-3) or
- enclosed in earthed conduit, trunking or ducting

shall be limited to the rules shown in Table 8.15.   601-07-02

Table 8.15 Zonal limitation on wiring systems

Zone	Limitation
Zone 0	Fixed electrical equipment situated in that zone
Zone 1	Fixed electrical equipment situated in zones 0 and 1
Zone 2	Fixed electrical equipment in zones 0, 1 and 2

8.10.6 Fixed current-using equipment

The requirements shown in Table 8.16 shall apply.   601-09-01, 02 and 03

Table 8.16 Zonal limitation on fixed current using equipment

Zone	Limitation
Zone 0	Only fixed current-using equipment may be installed.
Zone 1	The following fixed current-using equipment may be installed: • a water heater • a shower pump and • SELV current-using equipment • other fixed current-using equipment, provided that: – it is suitable for the conditions of that zone, and the supply circuit is additionally protected by a residual current protective device.
Zone 2	The following fixed current-using equipment may be installed: • a water heater • a shower pump • a luminaire, fan, heating appliance or unit for a whirlpool bath • SELV current-using equipment • other fixed current-using equipment, provided that it is suitable for the conditions of that zone.
Zone 3	Current-using equipment other than fixed current-using equipment protected by an RCD.

Electric heating units that are embedded in the floor   601-09-04
(and intended for heating the location) may be installed
below any zone provided that they are covered by an earthed

metallic grid or by an earthed metallic sheath connected
to the local supplementary equipotential bonding.

8.10.7 Switchgear and control gear

The requirements shown in Table 8.17 shall apply.   601-08-01

Table 8.17 Zonal limitation on Switchgear and control gear

Zone	Limitation
Zone 0	Switchgear or accessories shall not be installed.
Zone 1	Only switches of SELV circuits (supplied at a nominal voltage not exceeding 12 V a.c. RMS or 30 V ripple-free d.c.) shall be installed. **Note:** The safety source shall be installed outside zones 0, 1 and 2.
Zone 2	Only switches and socket outlets of SELV circuits and shaver supply units complying with BS EN 60742 shall be installed. **Note:** The safety source shall be installed outside zones 0, 1 and 2.
Zone 3	Only SELV socket outlets complying with Regulation 411-02 and shaver supply units complying with BS EN 60742 shall be installed. **Note:** Portable equipment can only be connected as above.

Only insulating pull cords of cord-operated switches meeting   601-08-01
the requirements of BS 3676 are permitted in zones 1 and 2.

Socket outlets (other than a SELV socket outlet or shaver   601-08-02
supply unit) located in shower cubicles that are installed
in rooms other than a bathroom or shower room, and that
is outside zones 0, 1, 2 or 3, shall be protected by an RCD.

8.10.8 Water heaters having immersed and uninsulated heating elements

The heater or boiler shall be permanently connected to   554-05-03
the electricity supply via a double-pole linked switch and
this linked switch shall either be:

• separate from and within easy reach of the heater/
  boiler or

- part of the boiler/heater (provided that the wiring from the heater or boiler is directly connected to the switch without use of a plug and socket outlet).

8.11 Restrictive conductive locations

A restrictive conductive location is one in which the surroundings consist mainly of metallic or conductive parts such as a large metal container or boiler. People employed inside these locations (e.g. a person working inside the boiler whilst using an electric drill or grinder) would have their freedom of movement physically restrained and a large proportion of their body would be in contact with the sides of that location and, therefore, prone to shock hazards.

8.11.1 Protection against direct and indirect contact

When SELV is used, protection against direct contact shall be provided by: 606-02-01

- a barrier or an enclosure with a protection degree of IP2X or IPXXB, or
- insulation capable of withstanding a 500 V a.c. rms for 60 seconds.

8.11.2 Protection against direct contact

Protection by means of obstacles and/or placing   606-03-01 out of reach is **not** permitted.

8.11.3 Protection against indirect contact

Protection against indirect contact shall be provided by: 606-04-01

- SELV or
- automatic disconnection or
- electrical separation or
- the use of Class II equipment adequately protected to an IP code.

Socket outlets and supplies to a hand lamp shall be protected by SELV.	606-04-02
If an equipment (e.g. measurement or control apparatus) requires a functional earth, then equipotential bonding shall be provided between all exposed conductive parts, all extraneous conductive parts inside the restrictive conductive location and the functional earth.	606-04-03
Socket outlets and supplies to a hand-held tool shall be protected by SELV or by electrical separation.	606-04-04
Supplies to fixed equipment shall be protected by:	606-04-05

- SELV or
- automatic disconnection or
- electrical separation or
- the use of Class II equipment adequately protected to an IP code.

All safety and isolating sources (other than those specified in Regulation 411-02-02(iii)), such as a battery or other form of electrochemical source, shall be situated outside the restrictive conductive location, unless it is part of a fixed installation within a permanent restrictive conductive location.	606-04-06

8.12 Swimming pools

Special requirements exist for swimming pools for medical use or for disabled persons (see Building Regulations Part M).	601-01-01

Swimming pools are, by design, wet areas and people using them are normally wet which will increase their vulnerability to electric shock. Special measures are, therefore, needed to ensure that all possibility of direct and/or indirect contact is reduced.

The Regulations requirements for swimming pools are based on three zones which take into account the actual basin of the pool and the surrounding area. Limitations of walls, doors, fixed partitions, ceilings and floors and the electric heater itself. The three zones are shown in Figure 8.9.

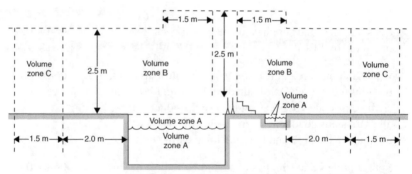

Note: The dimensions are measured taking account of walls and fixed partitions

Figure 8.9   Zone dimensions for swimming pools and paddling pools

Where:

- zone A is the interior of the basin, chute or flume and includes the portions of essential apertures in its walls and floor which are accessible to persons in the basin
- zone B is limited:
  – by the vertical plane 2 m from the rim of the basin, and
  – by the floor or surface expected to be accessible to persons, and
  – by the horizontal plane 2.5 m above that floor or surface, except where the basin is above ground, when it shall be 2.5 m above the level of the rim of the basin.

 Note: Where the building containing the swimming pool contains diving boards, spring boards, starting blocks or a chute, zone B includes also the zone limited by the vertical plane spaced 1.5 m from the periphery of diving boards, spring boards and starting blocks, and within that zone, by the horizontal plane 2.5 m above the highest surface expected to he occupied by persons, or to the ceiling or roof if they exist.

- zone C is limited by:
  – the vertical plane circumscribing zone B, and the parallel vertical plane 1.5 m external to zone B, and
  – the floor or surface expected to be occupied by persons and the horizontal plane 2.5 m above that floor or surface.

The following requirements apply to basins of swimming pools and paddling pools and their surrounding zones.

8.12.1 Protection against electric shock

> The following protective measures shall **not** be used   602-04-02
> in any zone:
>
> - protection by means of obstacles
> - protection by means of placing out of reach

- protection by means of a non-conducting location
- protection by means of earth-free local equipotential bonding.

Where SELV is used (irrespective of the nominal voltage), protection against direct contact shall be provided by: 602-03-01

- barriers or enclosures providing protection to at least 1P2X or IPXXB, or
- insulation capable of withstanding a type-test voltage of 500 V a.c. RMS for 60 seconds.

Local, supplementary, equipotential bonding shall connect all extraneous conductive parts (and the protective conductors of all exposed conductive parts) within zones A, B and C.   602-03-02

 This requirement does not apply to equipment supplied by SELV circuits.

If there is a metal grid in a solid floor it shall be connected to the local supplementary bonding.   602-03-02

In zones A and B, only SELV may be used as a means of protection against electric shock.   602-04-01

The safety source shall be installed outside zones A, B and C, except:

- if floodlights are installed, then each floodlight shall be supplied from its own transformer (or an individual secondary winding of a multi-secondary transformer) having an open circuit voltage not exceeding 18 V
- automatic disconnection of supply by means of a residual current device may be used to protect socket outlets.

8.12.2 Degree of protection of enclosures

Equipment shall have the minimum degrees of protection as shown in Table 8.18.   602-05-01

Table 8.18 Equipment protection in swimming pools

Zone	Degree of protection
Zone A	IPX8
Zone B	IPX5
	IPX4 for swimming pools (where water jets are not likely to be used for cleaning)
Zone C	IPX2 for indoor pools
	IPX4 for outdoor pools
	IPX5 for swimming pools (where water jets are likely to be used for cleaning)

8.12.3 Wiring systems

In zones A and B the surface wiring system shall not use metallic conduits or metallic trunkings, an exposed metallic cable sheath or an exposed earthing or bonding conductor. 602-06-01

Zones A and B shall contain only wiring necessary to supply equipment situated in those zones. 602-06-02

Accessible metal junction boxes shall not be installed in zones A and/or B. 602-06-03

8.12.4 Switchgear, control gear and accessories

In zones A and B, switchgear, control gear and accessories shall not be installed, unless socket outlets comply with BS EN 60309-2 and they are: 602-07-01

- installed more than 1.25 m outside of zone A
- installed at least 0.3 m above the floor
- protected by either:
  – a residual current device, or
  – electrical separation with the safety isolating transformer placed outside zones A, B and C.

In zone C, socket outlets, switches or accessories are only permitted if they are: 602-07-02

- protected individually by electrical separation, or
- protected by SELV, or
- protected by an RCD, or
- a shaver socket complying with BS 3535.

This requirement does not apply to the insulating cords of cord-operated switches complying with BS 3676.

8.12.5 Other equipment

Socket outlets shall comply with BS EN 60309-2.	602-08-01
In zones A and B, only current-using equipment specifically intended for use in swimming pools shall be installed.	602-08-02
In zone C, equipment shall be protected by:	602-08-03

- electrical separation
- SELV or
- an RCD.

 This requirement does not apply to instantaneous water heaters.

If an electric heating unit is embedded in the floor in zone B or C, it shall either:	602-08-04

- be connected to the local supplementary equipotential bonding by a metallic sheath or
- be covered by an earthed metallic grid connected to the equipotential bonding.

9

Inspection and testing

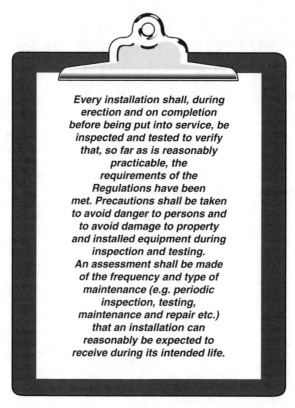

Every installation shall, during erection and on completion before being put into service, be inspected and tested to verify that, so far as is reasonably practicable, the requirements of the Regulations have been met. Precautions shall be taken to avoid danger to persons and to avoid damage to property and installed equipment during inspection and testing.
An assessment shall be made of the frequency and type of maintenance (e.g. periodic inspection, testing, maintenance and repair etc.) that an installation can reasonably be expected to receive during its intended life.

Figure 9.1 Mandatory requirements for installation, maintenance and repair

To meet these requirements it is essential for any electrician engaged in inspection, testing and certification of electrical installations to have a **full** working knowledge of the IEE Wiring Regulations.

The electrician must also have above-average experience and knowledge of the type of installation under test in order to carry out **any** inspection and testing. Without this prerequisite, it could be quite dangerous – particularly concerning installations such as that shown in Figure 9.2!

Figure 9.2 Example of a non-conforming electrical installation!

9.1 Basic safety requirements

The fundamental safety requirements of the IEE Wiring Regulations are as follows.

9.1.1 Mandatory requirements

- protective safety measures shall be applied in every installation, part installation and/or equipment
- installations shall comply with the requirements for safety protection in respect of:
  - electric shock
  - thermal effects
  - overcurrent
  - fault current
  - undervoltage
  - isolation and switching
- there shall be no detrimental influence between various protective measures used in the same installation, part installation or equipment.

9.1.2 Protection from electric shock

- protection against electric shock shall be provided
- protection against both direct contact and indirect contact shall be provided
- persons and livestock shall be protected against dangers that may arise from contact with live parts of the installation
- persons and livestock shall be protected against dangers that may arise from contact with exposed conductive parts during a fault
- live parts shall be completely covered with insulation which:
  - can only be removed by destruction
  - is capable of durably withstanding electrical, mechanical, thermal and chemical stresses normally encountered during service

428 Wiring Regulations in Brief

- live parts shall be inside enclosures (or behind barriers) protected to at least IP2X or IPXXB
- bare (or insulated) overhead lines being used for distribution between buildings and structures shall be installed in accordance with the Electricity Safety, Quality and Continuity Regulations 2002
- bare live parts (other than overhead lines) shall not be within arm's reach
- bare live parts (other than an overhead line) shall not be within 2.5 m of:
  - an exposed conductive part
  - an extraneous conductive part
  - a bare live part of any other circuit
- simultaneously accessible exposed conductive parts shall be connected to the same earthing system either individually, in groups or collectively
- exposed parts of electrical equipment shall be located (or guarded) so as to prevent accidental contact and/or injury to persons or livestock.

1 mA – 2 mA	Barely perceptible, no harmful effects
5 mA – 10 mA	Throw off, painful sensation
10 mA – 15 mA	Muscular contraction, can't let go!
20 mA – 30 mA	Impaired breathing
50 mA and above	Ventricular fibrillation and death

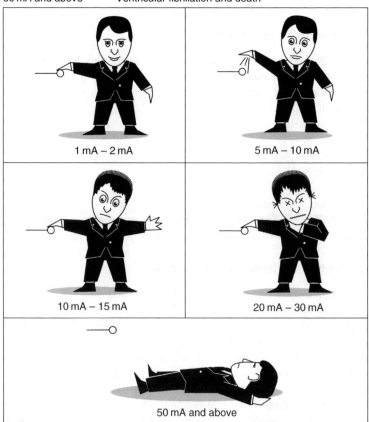

Figure 9.3 The effects of electric shock

9.2 Fundamental safety requirements

The following are précised details of the most important elements of the IEE Wiring Regulations that meet these fundamental design requirements. For your assistance, the relevant section of the Regulations has also been included to enable you to confirm the requirement if you have a need to.

9.2.1 Design

Electrical installations shall be designed for:

- the protection of persons, livestock and property (131-01-01)
- the proper functioning of the electrical installation (131-01-01)
- protection against mechanical and thermal damage (Building Regulations Approved Document P0.1a), and
- protection of people from an electric shock or fire hazard (Building Regulations Approved Document P0.1a).

9.2.2 Characteristics of available supply or supplies

Detailed design characteristics shall be available for all supplies. These shall include:

- nature of current (a.c. and/or d.c.)
- purpose and number of conductors

For a.c.	For d.c.
• phase conductor(s)	• outer conductor
• neutral conductor	• middle conductor
• protective conductor	• earthed conductor
• PEN conductor	• live conductor
	• protective conductor
	• PEN conductor

- values and tolerances of:
  - earth fault loop impedance
  - nominal voltage and voltage tolerances
  - nominal frequency and frequency tolerances
  - maximum current allowable
  - particular requirements of the distributor
  - prospective short-circuit current
  - protective measures inherent in the supply (e.g. earth, neutral or midwire) (131-02-01).

9.2.3 Electricity distributor's responsibilities

The electricity distributor is responsible for:

- evaluating and agreeing proposals for new installations or significant alterations to existing ones (Building Regulations Approved Document P1.4)

- ensuring that their equipment on consumers' premises:
  - is suitable for its purpose
  - is safe in its particular environment
  - clearly shows the polarity of the conductors (Building Regulations Approved Document P3.9)
- installing the cut-out and meter in a safe location (Building Regulations Approved Document P1.3)
- ensuring that the cut-out and meter is mechanically protected and can be safely maintained (Building Regulations Approved Document P1.3)
- providing an earthing facility for all new connections (Building Regulations Approved Document P3.8)
- maintaining the supply within defined tolerance limits (Building Regulations Approved Document P3.8)
- providing certain technical and safety information to the consumer to enable them to design their installations (Building Regulations Approved Document P3.8).

9.2.4 Installation and erection

- All electrical joints and connections shall meet stipulated requirements concerning conductance, insulation, mechanical strength and protection (133-01-04).
- Conductors shall be identified by colour, lettering and/or numbering (133-01-03).
- Connections and joints shall be accessible for inspection, testing and maintenance, unless:
  - they are in a compound-filled or encapsulated joint
  - the connection is between a cold tail and a heating element
  - the joint is made by welding, soldering, brazing or compression tool (526-04-01).
- Design temperatures shall not be exceeded by the installation of electrical equipment (133-01-05).
- Electrical equipment shall be arranged so that it is fully accessible (i.e. for operation, inspection, testing, maintenance and repair) and that there is sufficient space for later replacement (131-12-01).
- Equipment used for the supply of safety services shall be arranged to allow easy access for periodic inspection, testing and maintenance (561-01-04).
- Exposed parts of electrical equipment shall be located (or guarded) so as to prevent accidental contact and/or injury to persons or livestock (133-01-06).
- Good workmanship and proper materials shall be used (133-01-01).
- Installed electrical equipment shall minimise the risk of igniting flammable materials (133-01-06).
- Installed equipment must be accessible for operational, inspection and maintenance purposes (513-01-01).
- Installations shall be divided into circuits in order to:
  - avoid danger and minimise inconvenience in the event of a fault
  - facilitate safe operation, inspection, testing and maintenance (314-01-01).

- The process of erection shall not impair the characteristics of electrical equipment (133-01-02).

9.2.5 Identification and notices

Wiring shall be marked and/or arranged so that it can be quickly identified for inspection, testing, repair or alteration of the installation (514-01-02).

9.2.6 Inspection and testing

- Every electrical installation must be inspected and tested during erection and on completion **before** being put into service.
- Details of the general design characteristics of the electrical installation must be made available. These shall include the result of the assessment of general characteristics (711-01-02).
- Information (e.g. diagrams, charts, tables and/or schedules) must be made available to the person carrying out the inspection and testing and these (as a minimum) shall indicate:
  - the type and composition of each circuit (points of utilisation served, number and size of conductors, type of wiring), and
  - the method used for compliance with Regulation 413-01-01
  - the data (where appropriate) required by Regulation 413-02-04
  - the identification (and location) of all protection, isolation and switching devices
  - circuits or equipments that are susceptible to a particular test (711-01-02).
- If the inspection and tests are satisfactory, a signed Electrical Installation Certificate together with a Schedule of Inspections and a Schedule of Test Results (see Section 9.71) are to be given to the person responsible for ordering the work (711.01.01).
- Precautions shall be taken to avoid danger to persons and to avoid damage to property and installed equipment during inspection and testing (711-01-01).

9.2.7 Maintenance

An assessment shall be made of the frequency and type of maintenance (e.g. periodic inspection, testing, maintenance and repair etc.) that an installation can reasonably be expected to receive during its intended life (341-01-01).

9.2.8 Building Regulations requirements

- All proposals to carry out electrical installation work **must** be notified to the local authority's Building Control Body before work begins, **unless** the proposed installation work is undertaken by a person who is a competent person registered with an electrical self-certification scheme and does not include the provision of a new circuit.

- Any work that involves adding a new circuit to a dwelling needs to be either notified to the Building Control Body (who will then inspect the work) or needs to be carried out by a competent person who is registered under a Government Approved Part P Self Certification Scheme (see Fig 2.3 on page 24).

 Note: Where a person who is **not** registered to self-certify intends to carry out the electrical installation, then a Building Regulation (i.e. a Building Notice or Full Plans) application will need to be submitted together with the appropriate fee, based on the estimated cost of the electrical installation. The Building Control Body will then arrange to have the electrical installation inspected at first fix stage and tested upon completion.

- Reasonable provision shall be made in the design, installation, inspection and testing of electrical installations in order to protect persons from fire or injury (Building Regulations Approved Document P1).
- Sufficient information shall be provided so that persons wishing to operate, maintain or alter an electrical installation can do so with reasonable safety (Building Regulations Approved Document P2).
- Work involving any of the following will also have to be notified to the Building Control Body:
  - locations containing a bath tub or shower basin
  - swimming pools or paddling pools
  - hot air saunas
  - electric floor or ceiling heating systems
  - garden lighting or power installations
  - solar photovoltaic (p.v.) power supply systems
  - small-scale generators such as microCHP units
  - extra-low-voltage lighting installations, other than pre-assembled, CE-marked lighting sets.

 Note: Whilst Part P of the Building Regulations makes requirements for the safety of fixed electrical installations, this does not cover system functionality (such as electrically powered fire alarm systems, fans and pumps), which are covered in other parts of the Building Regulations and other legislation.

Conservation of fuel and power

- Energy efficiency measures shall be provided which:
  - provide lighting systems that utilise energy-efficient lamps with manual switching controls or, in the case of external lighting fixed to the building, automatic switching, or both manual and automatic switching controls as appropriate, such that the lighting systems can be operated effectively as regards the conservation of fuel and power
  - provide information, in a suitably concise and understandable form (including results of performance tests carried out during the works) that shows building occupiers how the heating and hot water services can be operated and maintained (Building Regulations Approved Document L1).

Responsibility for achieving compliance with the requirements of Part L rests with the person carrying out the work. That 'person' may be, for example, a developer, a main (or sub) contractor, or a specialist firm directly engaged by a private client.

 The person responsible for achieving compliance should either themselves provide a certificate, or obtain a certificate from the subcontractor, that commissioning has been successfully carried out. The certificate should be made available to the client and the building control body.

Access and facilities for disabled people

- In addition to the requirements of the Disability Discrimination Act 1995, precautions need to be taken to ensure that:
  - new non-domestic buildings and/or dwellings (e.g. houses and flats used for student living accommodation etc.)
  - extensions to existing non-domestic buildings
  - non-domestic buildings that have been subject to a material change of use (e.g. so that they become a hotel, boarding house, institution, public building or shop)
- are capable of allowing people, regardless of their disability, age or gender to:
  - gain access to buildings
  - gain access within buildings
  - be able to use the facilities of the buildings (both as visitors and as people who live or work in them) (Building Regulations Approved Document M).

Extensions, material alterations and material changes of use

Where any electrical installation work is classified as an extension, a material alteration or a material change of use, the work must consider and include:

- confirmation that the mains supply equipment is suitable and can carry the additional loads envisaged (Building Regulations Approved Document P2.1b–P2.2)
- the amount of additions and alterations that will be required to the existing fixed electrical installation in the building (Building Regulations Approved Document P2.1a)
- the earthing and bonding systems are satisfactory and meet the requirements (Building Regulations Approved Document P2.1a–P2.2c)
- the necessary additions and alterations to the circuits which feed them (Building Regulations Approved Document P2.1a)
- the protective measures required to meet the requirements (Building Regulations Approved Document P2.1a–P2.2b)

- the rating and the condition of existing equipment (belonging to both the consumer and the electricity distributor) is sufficient (Building Regulations Approved Document P2.2a).

 Note: Appendix C to Part P of the Building Regulations offers guidance on some of the older types of installations that might be encountered during alteration work and Appendix D provides guidance on the application of the now harmonised European cable identification system.

9.3   What inspections and tests have to be completed and recorded?

As shown at the beginning of this chapter:

- every installation must be:
  - inspected and tested during erection and on completion before being put into service
  - inspected and tested to verify that, so far as is reasonably practicable, the requirements of the Regulations have been met (711-01-01)
- precautions must be taken to avoid danger to persons and to avoid damage to property and installed equipment during inspection and testing (711-01-01)
- an assessment must be made of the frequency and type of maintenance (e.g. periodic inspection, testing, maintenance and repair etc.) that an installation can reasonably be expected to receive during its intended life (341-01-01).

Details of the inspections and tests required to confirm compliance with the Regulations are contained in Sections 9.4 and 9.5 below.

9.4   Inspections

An inspection is, most generally, an official examination or formal evaluation exercise involving measurements, tests, and gauges applied to certain characteristics with regard to an object or activity. The results of an inspection are usually compared to specified requirements (e.g. a British Standard) in order to determine whether an item or activity is in line with the relevant standard(s) and achieves certain criteria and characteristics. Inspections are usually non-destructive.

9.4.1 General

Inspections are normally completed with the installation disconnected from the supply and shall precede testing (712-01-01).

Inspections shall be made to verify that the installed electrical equipment:

- complies with the requirements of the applicable British Standard, or harmonised Standard appropriate to the intended use of the equipment (511-01-01)

Note: Equipment complying with a foreign national standard may be used **only** if it provides the same degree of safety afforded by a British or Harmonised Standard.

- is correctly selected and erected in accordance with the Regulations
- is not visibly damaged or defective so as to impair safety (712-01-02).

9.4.2 Inspection checklist

In accordance with Sections 712 and 713 of the Regulations (and for compliance with the Building Regulations) the inspection shall include the following items:

- access to switchgear and equipment
- cable routing
- choice and setting of protective and monitoring devices
- connection of accessories and equipment
- connection of conductors
- connection of single-pole devices for protection or switching in phase conductors
- continuity all protective conductors
- continuity of all ring final circuit conductors
- earth electrode resistance
- earth fault loop impedance
- erection methods
- functional testing
- identification of conductors
- insulation of non-conducting floors and walls
- insulation resistance
- labelling of protective devices, switches and terminals
- polarity
- presence of danger notices and other warning signs
- presence of diagrams, instructions and similar information
- presence of fire barriers, suitable seals and protection against thermal effects
- prevention of mutual (i.e. detrimental) influence
- presence of undervoltage protective devices
- prospective fault current
- protection against electric shock:
  - capability of equipment to withstand mechanical, chemical, electrical and thermal influences and stresses normally encountered during service
  - exposed conductive parts
  - insulating enclosures

- insulation of operational electrical equipment
- verification of the quality of the insulation
- protection against electric shock by direct and/or indirect contact:
  - SELV
  - limitation of discharge of energy
- protection against direct current:
  - barriers or an enclosure
  - insulation of live parts
  - obstacles
  - PELV
  - placing out of reach
- protection against external influences
- protection against indirect contact:
  - automatic disconnection of supply
  - earth-free local equipotential bonding
  - earthed equipotential bonding
  - earthing and protective conductors
  - earthing arrangements for combined protective and functional purposes
  - non-conducting location (absence of protective conductors)
  - electrical separation
  - main equipotential bonding conductors
  - use of Class II equipment or equivalent insulation
  - supplementary equipotential bonding conductors
- selection of conductors for current-carrying capacity and voltage drop
- selection of equipment appropriate to external influences
- site applied insulation:
  - protection against direct contact
  - protection against indirect contact
  - supplementary insulation.

 Note: The Building Regulations specifically state that inspections shall include *the design, construction, inspection and testing of any new electrical installation or new work associated with an alteration or addition to an existing installation.*

In addition to the above list of mandatory inspections for compliance with the IEE Wiring and the Building Regulations, the following are some of the additional inspections that electricians usually complete during initial and periodic inspections and tests of electrical installations:

- cables and conductors (current-carrying capacity, insulation and/or sheath)
- correct connection of accessories and equipment
- electrical joints and connections (to ensure that they meet stipulated requirements concerning conductance, insulation, mechanical strength and protection)
- emergency switching
- insulation

- insulation-monitoring devices (design, installation and security)
- inspection of any other electrical installations
- isolation and switching devices (and their correct location)
- locations with risks of fire due to the nature of processed and/or stored materials
- plug and socket outlets
- protection against electric shock – special installations or locations
- protection against earth insulation faults
- protection against mechanical damage
- protection against overcurrent
- protection by extra-low-voltage systems (other than SELV)
- protection by non-conducting location
- protection by residual current devices
- protection by separation of circuits
- supplies
- supplies for safety services
- wiring systems (selection and erection, temperature variations).

 Note: Details concerning tests to confirm compliance with the Regulations are contained in Section 9.5.

9.5 Testing

Testing any electrical installation (even the most simple ones) can be very dangerous unless it is carried out safely – dangerous not just to the tester himself but also to bystanders and other people.

As a minimum, the electrician must:

- have an above-average experience and knowledge of the type of installation under test
- have a thorough understanding of the correct application and use of the relevant test instruments (and their associated leads, probes and accessories)
- ensure that the test equipment being used has recently been inspected, correctly maintained and (where necessary) calibrated either against a workshop standard or a national standard
- observe the safety measures and procedures set out in HSE Guidance Note GS38 concerning the safe use of instruments and their accessories.

The following are précised details of the most important elements of the IEE Wiring Regulations that an electrician must test for in order to confirm that the electrical installation meets the fundamental design requirements of the IEE Wiring Regulations and is installed in conformance with the requirements of that British Standard.

 For your assistance the relevant section of the Regulations has also been included to enable you to 'officially' confirm the requirement if you have a need to.

9.5.1 Sequence of tests

Initial inspection and tests

In accordance with both the Wiring Regulations and the Building Regulations, all new installations (plus additions and/or alterations to existing circuits) need an initial verification to:

- ensure equipment and accessories meet the requirements of the relevant standard
- comply with the requirements of BS 7671
- comply with the requirements of the Building Regulations
- ensure that the installation is not damaged so as to impair safety.

 The following tests **shall** be carried out (and in the following order) before the installation is energised (713-01-01):

- a continuity test of all protective conductors (including main and supplementary equipotential bonding) (713-02)
- a continuity test of all ring final circuit conductors (713-03)
- a measurement of the insulation resistance between live conductors and between each live conductor and earth (713-04-01)
- a measurement of the insulation resistance of the main switchboard and each distribution circuit (713-04-02)
- confirmation that insulation for protection against direct and/or indirect contact meets requirements (713-05)
- verification that the separation of circuits is protected by SELV, PELV and/or electrical separation meets requirements (713-06)
- a measurement of the insulation resistance of live parts from those of other circuits and those of other circuits and from earth (713-06)
- confirmation that functional extra-low-voltage circuits meet all the test requirements for low-voltage circuits (713-06-05)
- a test to ensure that the amount of protection against direct contact that is provided by a barrier or an enclosure (provided during erection) meets requirements (713-07)
- verification (by measurement) that the amount of protection against indirect contact provided by a non-conducting location meets requirements (713-08)
- a polarity test to verify that, fuses, single-pole control and protective devices, lampholders and wiring meet requirements (713-09).

 Other tests

The following tests shall be carried out when the installation is energised:

- a measurement of the electrode resistance to earth for earthing systems incorporating an earth electrode (713-10)
- a measurement of earth loop impedance (713-11)
- a measurement of prospective short-circuit and earth fault (713-12)

- functional tests to verify the effectiveness of residual current devices and test assemblies (e.g. switchgear, control gear, drives, controls and interlocks) to show that they are properly mounted, adjusted and installed in accordance with the Regulations (713-13).

9.5.2 Protective measures

The Regulations stipulate (713-02-01) that a continuity test of all protective conductors (including main and supplementary equipotential bonding) shall be made to ensure that the correct degree of protection is being provided by the following protective measures by confirming, checking and testing:

Protection against overload current	That the protective device is capable of breaking any overload current flowing in the circuit conductors before the current can damage the insulation of the conductors (433-01-01).
Protection by earth-free local equipotential bonding	That earth-free local equipotential bonding prevents the appearance of a dangerous voltage between simultaneously accessible parts in the event of failure of the basic insulation (471-11-01).
Protection by electrical separation	That equipment used as a fixed source of supply has been manufactured so that the output is separated from the input and from the enclosure by insulation for protection against indirect contact (413-06-02, 471-12-01 and 471-14-05).

 Note: This form of protection is intended for an individual circuit and is aimed at preventing shock current through contact with exposed conductive parts, which might be energised by a fault in the basic insulation of that circuit.

Protection by extra-low-voltage systems (other than SELV)	That if an extra-low-voltage system complies with the requirements for SELV, it is not connected to a live part or a protective conductor forming part of another system and not connected to:

- earth
- an exposed conductive part of another system
- a protective conductor of any system, or
- an extraneous conductive part.

Protection against direct contact has been provided by either:

- insulation capable of withstanding 500 V a.c. r.m.s. for 60 seconds or
- barriers or enclosures with a degree of protection of at least IP2X or IPXXB (471-14-02).

 This form of protection against direct contact is **not** required if the equipment is within a building in which main equipotential bonding is applied and the voltage does not exceed:

- 25 V a.c. r.m.s. or 60 V ripple-free d.c. when the equipment is normally only used in dry locations and large-area contact of live parts with the human body is not to be expected
- 6 V a.c. r.m.s. or 15 V ripple-free d.c. in all other cases.

Note: When an extra-low-voltage circuit is used to supply equipment whose insulation does not comply with the minimum test voltage required for the primary circuit, then the insulation of that equipment shall be reinforced to withstanding a voltage of 1500 V a.c. r.m.s. for 60 seconds (471-14-03).

Protection by insulation of live parts	That the insulation protection has been designed to prevent contact with live parts (471-04-01).

 Whilst, generally speaking, this method is for protection against direct contact, it also provides a degree of protection against indirect contact.

Protection by non-conducting location (Regulation 413-04)	That this form of protection prevents simultaneous contact with parts which may be at different potentials through failure of the basic insulation of live parts (471-10-01).

 Whilst this protection is **not** recognised in the Regulations for general use, it may be applied in special situations provided that they are under effective supervision.

 Protection by non-conducting location shall not be used in installations and locations subject to increased risk of shock such as agricultural and horticultural premises, caravans, swimming pools etc.

Protection by residual current devices

That:

- parts of a TT system that are protected by a single residual current device have been placed at the origin of the installation unless that part between the origin and the device complies with the requirements for protection by using Class II equipment or an equivalent insulation (531-04-01)

 Where there is more than one origin this requirement applies to each origin.

- installations forming part of an IT system have been protected by a residual current device supplied by the circuit concerned or make use of an insulation-monitoring device (473-01-05).

Protection by SELV

That circuit conductors for each SELV system have been physically separated from those of any other system. Where this proves impracticable, SELV circuit conductors have been:

- insulated for the highest voltage present
- enclosed in an insulating sheath additional to their basic insulation (411-02-06).

Protection by the use of Class II equipment or equivalent insulation	That this form of protection prevents a fault in the basic insulation causing a dangerous voltage to appear on the exposed metalwork of electrical equipment.

A typical method for testing the continuity of protective conductors is illustrated in Figure 9.4 and basically this involves bridging the phase conductor to the protective conductor at the distribution board (so as to include all of the circuit) and then testing between phase and earth terminals at each point of the circuit.

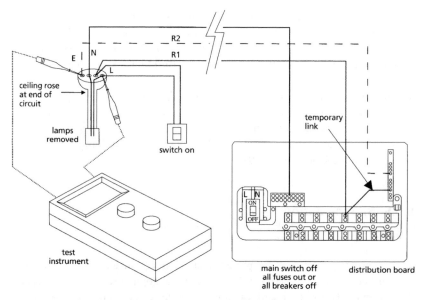

Figure 9.4 Connections for testing continuity of protective conductors (courtesy of IEE).

9.5.3 Design requirements

Check to confirm that the number and type of circuits required for lighting, heating, power, control, signalling, communication and information technology, etc. have taken consideration of:

- the location and points of power demand
- the loads to be expected on the various circuits
- the daily and yearly variation of demand
- any special conditions
- requirements for control, signalling, communication and information technology, etc. (131-03-01).

9.5.4 Electricity distributor

Confirm that the electricity distributor has:

- evaluated and agreed proposals for new installations or significant alterations to existing ones (Building Regulations Approved Document P1.4)
- maintained the supply within defined tolerance limits (Building Regulations Approved Document P3.8)
- provided an earthing facility for all new connections (Building Regulations Approved Document P3.8)
- installed the cut-out and meter in a safe location (Building Regulations Approved Document P1.3)

- ensured that the cut-out and meter is mechanically protected and can be safely maintained (Building Regulations Approved Document P1.3)
- provided certain technical and safety information to the consumer to enable them to design their installations (Building Regulations Approved Document P3.8)
- ensured that their equipment on consumers' premises:
  - is suitable for its purpose
  - is safe in its particular environment
  - clearly shows the polarity of the conductors (Building Regulations Approved Document P3.9).

9.5.5 Site insulation

Confirm that:

- insulation applied on site to protect against direct contact is capable of withstanding, without breakdown or flashover, an applied test voltage as specified in the British Standard for similar type-tested equipment (713-05-01)
- supplementary insulation applied to equipment during erection to protect against indirect contact is tested to ensure that the insulating enclosure:
  - protects to at least 1P2X or IPXXB, and
  - is capable of withstanding, without breakdown or flashover, an applied test voltage as specified in the British Standard for similar type-tested equipment (713-05-02).

9.5.6 Insulation resistance

The Regulation requires that:

- the insulation resistance (measured with all its final circuits connected but with current-using equipment disconnected) between live conductors and between each live conductor and Earth, is not be less than that shown in Table 9.1 (713-04-01 and 713-04-02)
- the separation of live parts from those of other circuits and from earth is in accordance with the values shown in Table 9.1, by measuring the insulation resistance (713-06-02, 03 and 04).

This test (more usually referred to as *meggering*) is, therefore, aimed at ensuring that the insulation of conductors, accessories and equipment is still capable of preventing dangerous leakage current between conductors and between conductors and earth.

To determine the insulation resistance between live conductors, test between the live (phase and neutral) conductors at the distribution board (see Figure 9.5).

The resistance readings obtained should be greater than the minimum values shown in Table 9.1.

Table 9.1 Minimum values of insulation resistance

System	Test voltage	Minimum insulation resistance
SELV and PELV	250 V d.c.	0.25 MΩ
LV up to 500 V	500 V d.c.	0.5 MΩ
Over 500 V	1000 V d.c.	1.0 MΩ

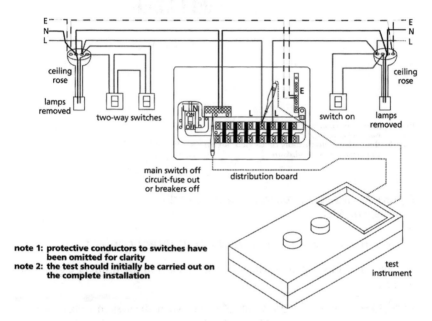

Figure 9.5 Insulation resistance tests between live conductors of a circuit (courtesy of IEE).

For checking the insulation resistance to earth:

- single phase – test between the live conductors (phase and neutral) and the circuit protective conductors at the distribution board as shown in Figure 9.6
- three phase – test to earth all live conductors (including the neutral) connected together. Resistance figures should be greater than the minimum figures shown in Table 9.1.

Note:

(1) Measurements shall be carried out with direct current (713-04-03).
(2) When the circuit includes electronic devices, only a measurement to protective earth shall be made with the phase and neutral connected together (713-04-04).
(3) Precautions may be necessary to avoid damage to electronic devices.

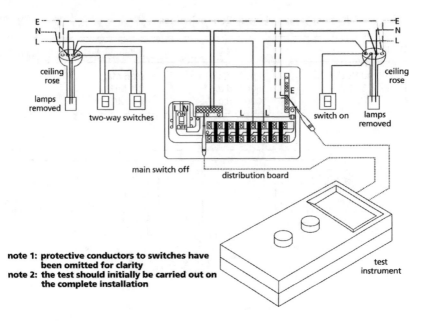

note 1: **protective conductors to switches have been omitted for clarity**
note 2: **the test should initially be carried out on the complete installation**

Figure 9.6 Insulation resistance tests to earth (courtesy of IEE).

9.5.7 Protection against direct and indirect contact

The two methods for protecting against shock from both direct and indirect contact are:

• SELV (separated extra-low voltage) – i.e. where the system voltage does not exceed extra low (e.g. 50 V a.c., 120 V ripple-free d.c.) and associated wiring etc. is separated from all other circuits of higher voltage
• limitation of discharge of energy – i.e. where equipment is arranged so that the current that can flow through the body (or livestock) is limited to a safe level (e.g. electric fences).

SELV

The system is not deemed to be SELV if any exposed conductive part of an extra-low voltage system is capable of coming into contact with an exposed conductive part of any other system. In addition, a system which does **not** use a device such as an autotransformer, potentiometer, semiconductor device etc., to provide electrical separation, is also not deemed to be a SELV system.

Confirm that for protection by SELV:

• the nominal circuit voltage does **not** exceed extra-low voltage
• the supply is from one of the following:
  – a safety isolating transformer complying with BS 3535
  – a motor generator with windings providing electrical separation equivalent to that of the safety isolating transformer specified above

- a battery or other form of electrochemical source
- a source independent of a higher voltage circuit (e.g. an engine-driven generator)
- electronic devices which (even in the case of an internal fault) restrict the voltage at the output terminals so that they do not exceed extra-low voltage (411-02-02).

Confirm and test that:

- a mobile source for SELV has been selected and erected in accordance with the requirements for protection by the use of Class II equipment or by equivalent insulation (411-02-04)
- all live parts of a SELV system are:
  - electrically separated from that of any other higher-voltage system

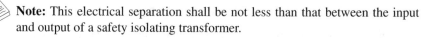 **Note:** This electrical separation shall be not less than that between the input and output of a safety isolating transformer.

  - not connected to earth
  - not connected to a live part or a protective conductor forming part of another system (411-02-05)
- circuit conductors for each SELV system are physically separated from those of any other system and where this proves impracticable, SELV circuit conductors are:
  - insulated for the highest voltage present and (where this proves impracticable)
  - enclosed in an insulating sheath additional to their basic insulation (411-02-06)
- conductors of systems with a higher voltage than SELV are separated from the SELV conductors by an earthed metallic screen or an earthed metallic sheath (411-02-06)
- SELV circuit conductors that are contained in a multicore cable with other circuits having different voltages are insulated, individually or collectively, for the highest voltage present in the cable or grouping (411-02-06)
- electrical separation between live parts of a SELV system (including relays, contactors and auxiliary switches) and any other system are maintained (411-02-06)
- exposed conductive parts of a SELV system are **not** connected to:
  - earth
  - an exposed conductive part of another system
  - a protective conductor of any system
  - an extraneous conductive part (411-02-07)

Note: Except where that electrical equipment is mainly required to be connected to an extraneous conductive part (in which case, measures shall be incorporated so that the parts cannot attain a voltage exceeding extra-low voltage).

- if the nominal voltage of a SELV system exceeds 25 V a.c. r.m.s. or 60 V ripple-free d.c., protection against direct contact is provided by one or more of the following:
  - a barrier (or an enclosure) capable of providing protection to at least 1P2X or IPXXB
  - insulation capable of withstanding a type-test voltage of 500 V a.c. r.m.s. for 60 seconds (411-02-09)
- the socket outlet of a SELV system is:
  - compatible with the plugs used for other systems in use in the same premises
  - does not have a protective conductor contact (411-02-10)
- luminaire-supporting couplers which have a protective conductor contact are **not** installed in a SELV system (411-02-11).

Protection – by limitation of discharge of energy

Protection against both direct and indirect contact shall be deemed to be provided when the equipment incorporates a means of limiting the amount of current which can pass through the body of a person or livestock to a value lower than that likely to cause danger (411-04-01).

9.5.8 Protection against direct contact

Electric shock caused by direct contact (i.e. when a body part directly touches live parts of equipment or systems that are intended to be live) is particularly dangerous, as the full voltage of the supply can be developed across the body (as shown in Figure 9.3). On the whole, however, if an electrical installation has been designed and installed correctly, then there shouldn't be too much risk from direct contact – but carelessness (such as changing an electric light bulb without switching the mains off first) or overconfidence such as working on a circuit with the power on are the prime causes of injuries and death from electric shock.

The main protective methods against direct contact causing an electric shock are:

- barriers or an enclosure
- insulation of live parts
- by obstacles and placing out of reach
- PELV and FELV
- placing out of reach.

The use of a residual current device (RCD) cannot prevent direct contact, but may be used to supplement other protective means that are used.

Protection by a barrier or an enclosure provided during erection

Test to ensure that the degree of protection against direct contact provided by a barrier or an enclosure (provided during erection) is not less than IP2X or IPXXB or IP4X, as appropriate (713-07-01).

Protection by insulation of live parts

Complete a functional test to verify that protection by insulation of live parts does prevent contact with a live part (471-04-01).

 Note: Whilst, generally speaking, this basic form of insulation protection is for protection against direct contact, it also provides a degree of protection against indirect contact.

Protection by obstacles

Complete a functional test to verify that protection by obstacles prevents unintentional contact with a live part, but **not** intentional contact by deliberate circumvention of the obstacle (471-06-01).

 The application shall be limited to protection against direct contact and in an area accessible only to skilled persons.

 For some installations and locations where an increased risk of shock exists, this protective measure shall not be used (see Part 6).

PELV and FELV systems

Protective extra-low voltage (PELV) and functional extra-low voltage (FELV) systems shall provide protection against electric shock and meet the following requirements (471-14-01):

- if an extra-low-voltage system complies with the requirements for SELV except for the requirements of Regulations 411-02-05(ii) which stipulates that

ALL live parts of a SELV system **shall**:

- be electrically separated from that of any other higher-voltage system

 Note: This electrical separation shall be not less than that between the input and output of a safety isolating transformer.

- not be connected to earth
- not be connected to a live part or a protective conductor forming part of another system

- and 411-02-07 which stipulates that:

exposed conductive parts of a SELV system shall **not** be connected to:

- earth
- an exposed conductive part of another system
- a protective conductor of any system
- an extraneous conductive part

- then protection against direct contact shall be provided by either:
  - barriers or enclosures with a degree of protection of at least 1P2X or IPXXB, or
  - insulation capable of withstanding 500 V a.c. r.m.s. for 60 seconds

 This form of protection against direct contact is not required if the equipment is within a building in which main equipotential bonding is applied and the voltage does not exceed:

  - 25 V a.c. r.m.s. or 60 V ripple-free d.c. when the equipment is normally only used in dry locations and large-area contact of live parts with the human body is not to be expected
  - 6 V a.c. r.m.s. or 15 V ripple-free d.c. in all other cases (471-14-02)
- for extra-low-voltage systems that do not comply with the requirements for SELV in some respect (other than that specified in Regulation 471-14-02 above) then protection against direct contact shall be provided by one or more of the following:
  - barriers or enclosures
  - insulation corresponding to the minimum voltage required for the primary circuit (471-14-03, 471-14-02)
- when an extra-low-voltage circuit is used to supply equipment whose insulation does not comply with the minimum test voltage required for the primary circuit, then the insulation of that equipment shall be reinforced to withstand a voltage of 1500 V a.c. r.m.s. for 60 seconds (471-14-03)
- if the primary circuit of the functional extra-low-voltage source is protected by automatic disconnection, then the exposed conductive parts of equipment in that functional extra-low-voltage system shall be connected to the protective conductor of the primary circuit (471-14-04)
- if the primary circuit of the functional extra-low-voltage source is protected by electrical separation, then the exposed conductive parts of equipment in that functional extra-low-voltage system shall be connected to the non-earthed protective conductor of the primary circuit (471-14-05)
- all socket outlets and luminaire-supporting couplers in a functional extra-low voltage system shall use a plug which is dimensionally different with those used for any other system in use in the same premises (471-14-06).

Protection by placing out of reach

Check to insure that:

- bare (or insulated) overhead lines being used for distribution between buildings and structures are installed in accordance with the Electricity Safety, Quality and Continuity Regulations 2002 (412-05-01)
- bare live parts (other than overhead lines) are not within arm's reach (412-05-02)

 Notes:
(1) If access to live equipment (from a normally occupied position) is restricted by an obstacle (i.e. such as a handrail, mesh or screen) with a degree of

protection less than 1P2X or IPXXB, the extent of arm's reach shall be measured from that obstacle (412-05-03).

(2) If a bulky or long conducting object is normally handled in these areas, the distances required by Regulations 412-05-02 and 412-05-03 shall be increased accordingly (412-05-04).

- bare live parts (other than an overhead line) are not within 2.5 m of:
    - an exposed conductive part
    - an extraneous conductive part
    - a bare live part of any other circuit (412-05-02)
- if a bulky or long conducting object is normally handled in these areas, the distances required by Regulations 412-05-02 and 412-05-03 shall be increased accordingly (412-05-04).

 No additional protection against overvoltages of atmospheric origin is necessary for:

- installations that are supplied by low-voltage systems which do not contain overhead lines
- installations that are supplied by low-voltage networks which contain overhead lines and their location is subject to less than 25 thunderstorm days per year
- installations that contain overhead lines and their location is subject to less than 25 thunderstorm days per year

provided that they meet the required minimum equipment impulse withstand voltages shown in Table 44A (Page 61) of the IEE Wiring Regulations.

 Suspended cables having insulated conductors with earthed metallic coverings are considered to be an 'underground cable' (443-02-02).

- check to ensure that installations that are supplied by (or include) low-voltage overhead lines incorporate protection against overvoltages of atmospheric origin or (if the location is subject to more than 25 thunderstorm days per year) protection against overvoltages of atmospheric origin shall be provided in the installation of the building by:
    - a surge protective device with a protection level not exceeding Category II (443-02-03) or
    - by other means providing an equivalent attenuation of overvoltages (443-02-05)
- where protective measures against indirect contact only have been dispensed with, confirm that:
    - overhead line insulator brackets (and metal parts connected to them) are not within arm's reach
    - the steel reinforcement of steel reinforced concrete poles is not accessible
    - exposed conductive parts (including small isolated metal parts such as bolts, rivets, nameplates not exceeding 50 mm × 50 mm and cable clips) cannot be gripped or cannot be contacted by a major surface of the human body

- there is no risk of fixing screws used for non-metallic accessories coming into contact with live parts
- inaccessible lengths of metal conduit do not exceed $150\,mm^2$
- metal enclosures mechanically protecting equipment comply with the relevant British Standard
- unearthed street furniture that is supplied from an overhead line is inaccessible whilst in normal use (471-13-04).

9.5.9 Protection against indirect contact

Indirect contact (i.e. touching conductive parts which are not meant to be live, but which have become live due to a fault) is the other main cause of electric shock. Again this is particularly dangerous and the main protection against indirect contact is for the electrical installation to be correctly earthed and for the circuit to be fitted with some form of overcurrent cut-out device.

The main protective methods against indirect contact causing an electric shock are:

- use of Class II equipment
- automatic disconnection of supply
- earth-free local equipotential bonding
- earthed equipotential bonding
- earthing and protective conductors
- earthing arrangements for combined protective and functional purposes
- non-conducting location (absence of protective conductors)
- electrical separation
- main equipotential bonding conductors
- use of Class II equipment or equivalent insulation
- supplementary equipotential bonding conductors.

Use of Class II equipment

Class II equipment (often referred to as double insulated equipment) is typical of modern equipment intended to be connected to fixed electrical installations (e.g. household appliances, portable tools and similar loads) and where all live parts are insulated so as to prevent a fault in the basic insulation causing a dangerous voltage to appear on the exposed metalwork of electrical equipment.

To verify compliance, confirm that:

- circuits supplying Class II equipment have a circuit protective conductor that is run to (and terminated at) each point in the wiring and at each accessory (471-09-02)

Except suspended lampholders which have no exposed conductive parts.

- the metalwork of exposed Class II equipment is mounted so that it is not in electrical contact with any part of the installation that is connected to a protective conductor (471-09-02)

- when Class II equipment is used as the sole means of protection against indirect contact, the installation or circuit concerned is under effective supervision whilst in normal use (471-09-03).

This form of protection shall **not** be used for circuits that include socket outlets or where a user can change items of equipment without authorisation.

Earthed equipotential bonding and automatic disconnection of supply

The intention of this form of protection is to prevent a dangerous voltage occurring between simultaneously accessible conductive parts. For installations and locations with increased risk of shock (e.g. those in Part 6 of the Regulations such as agricultural and horticultural buildings, saunas etc.), additional measures may be required. For example:

- automatic disconnection of supply by means of a residual current device (RCD) with a rated residual operating current (I_{sn}) not exceeding 30 mA
- supplementary equipotential bonding
- reduction of maximum fault clearance time.

Confirm and test that:

- installations which are part of a TN system meet the requirements for earth fault loop impedance and for circuit protective conductor impedance as specified in Regulations 413-02-08 and 413-02-10 to 413-02 regarding specified times for automatic disconnection of supplies (471-08-02)
- for circuits supplying fixed equipment which are outside of the earthed equipotential zone and which have exposed conductive parts that could be touched by a person who has direct contact directly with earth, that the earth fault loop impedance ensures that disconnection occurs within the time stated in Table 41A (Page 44) of the IEE Wiring Regulations
- if the installation is part of a TF system, all socket outlet circuits are protected by a residual current device (471-08-06)
- automatic disconnection using a residual current device is not be applied to a circuit incorporating a PEN conductor (471-08-07)
- installations that provide protection against indirect contact by automatically disconnecting the supply have a circuit protective conductor run to (and terminated at) each point in the wiring and at each accessory (471-08-08).

Except suspended lampholders which have no exposed conductive parts.

Earth-free local equipotential bonding

Earth-free local equipotential bonding is effectively a Faraday cage, where all metal is bonded together (but **not** to earth!) so as to prevent the appearance of a dangerous voltage occurring between simultaneously accessible parts in the event of failure of the basic insulation.

Confirm that:

- earth-free local equipotential bonding has **only** used in special situations which are earth-free
- a warning notice (warning that earth-free local equipotential bonding is being used) has been fixed in a prominent position adjacent to every point of access to the location concerned (471-11-01).

For some installations and locations with an increased shock risk (e.g. agricultural and horticultural, saunas etc.), earth-free local equipotential bonding shall be used.

Earthing conductors

A protective conductor may consist of one or more of the following:

- a single-core cable
- a conductor in a cable
- an insulated or bare conductor in a common enclosure with insulated live conductors
- a fixed bare or insulated conductor
- a metal covering (for example, the sheath, screen or armouring of a cable)
- a metal conduit or other enclosure or electrically continuous support system for conductors
- an extraneous conductive part.

Verify and test that:

- the thickness of tape or strip conductors is capable of withstanding mechanical damage and corrosion (see BS 7430, 542-03-01)
- the connection of earthing conductors to the earth electrode are:
  - soundly made
  - electrically and mechanically satisfactory
  - labelled in accordance with the Regulations
  - suitably protected against corrosion (542-03-03)
- all installations have a main earthing terminal to connect the following to the earthing conductor:
  - the circuit protective conductors
  - the main bonding conductors
  - functional earthing conductors (if required)
  - lightning protection system bonding conductor (if any) (542-04-01)
- earthing conductors are capable of being disconnected to enable the resistance of the earthing arrangement to be measured (542-04-02)
- any joint is:
  - capable of disconnection only by means of a tool
  - mechanically strong
  - ensures the maintenance of electrical continuity

- confirm that unless a protective conductor forms part of a multicore cable (or cable trunking or a conduit is used as a protective conductor) that the cross-sectional area, up to and including $6\,mm^2$, has been protected throughout, by a covering of at least equivalent in insulation to that of a single-core non-sheathed cable having a voltage rating of at least 450/750 V (543-03-02).

Where PME (protective multiple earthing) conditions apply, verify and test that:

- the main equipotential bonding conductor has been selected in accordance with the neutral conductor of the supply and the dimensions shown in Table 54H (p. 120) of the IEE Wiring Regulations

Local distributor's network conditions may require a larger conductor.

- buried earthing conductors have a cross-sectional area not less than that stated in Table 54A (p.113) of the IEE Wiring Regulations.

Protective conductors

All protective conductors – particularly main equipotential and supplementary bonding conductors – must be tested for continuity using a low resistance ohmmeter.

- Verify and test that the cross-sectional area of all protective conductors (less equipotential bonding conductors) is not less than

$$S = \frac{\sqrt{I^2 t}}{k}$$

or as per Table 54F in the IEE Wiring Regulations (543-01-01).
- If the protective conductor:
  - is not an integral part of a cable, or
  - is not formed by conduit, ducting or trunking, or
  - is not contained in an enclosure formed by a wiring system
  then the cross-sectional area shall be not less than:
  - $2.5\,mm^2$ copper equivalent if protection against mechanical damage is provided or
  - $4\,mm^2$ copper equivalent if mechanical protection is not provided; (543-01-01).
- Protective conductors buried in the ground shall have a cross-sectional area not less than that stated in Table 54H of the IEE Wiring Regulations (p. 120) (543-01-01).

Earthing arrangements for combined protective and functional purposes

The following conductors may serve as a PEN conductor provided that the part of the installation concerned is not supplied through a Residual Current Device:

- a conductor of a cable not subject to flexing and with a cross-sectional area not less than $10\,mm^2$ (for copper) or $16\,mm^2$ for aluminium (this applies to a fixed installation by the way)

- the outer conductor of a concentric cable where that conductor has a cross-sectional area not less than $4\,\text{mm}^2$ (546-02-02).

Verify and test (where necessary) that combined protective and neutral (PEN) conductors have only been used if:

- authorisation to use a PEN conductor has been obtained by the distributor or
- the installation is supplied by a privately owned transformer or convertor and there is no metallic connection (except for the earthing connection) with the distributor's network, or
- the installation is supplied from a private generating plant (546-02-01)
- the outer conductor of a concentric cable is not common to more than one circuit (546-02-03 and 546-02-08)
- the conductance of the outer conductor of a concentric cable (and the terminal link or bar):
  - for a single-core cable, is not less than the internal conductor
  - for a multicore cable in a multiphase or multipole circuit, is not less than that of one internal conductor
  - for a multicore cable serving a number of points contained within one final circuit (or where the internal conductors are connected in parallel) is not less than that of the internal conductors (546-02-04)
- the continuity of all joints in the outer conductor of a concentric cable (and at a termination of that joint) is supplemented by an additional conductor, additional (that is) to any means used for sealing and clamping the outer conductor (546-02-05)
- isolation devices or switching have **not** been inserted in the outer conductor of a concentric cable (546-02-06)
- PEN conductors of all cables have been insulated or have an insulating covering suitable for the highest voltage to which it may be subjected (546-02-07)
- if neutral and protective functions are provided by separate conductors, those conductors are not then re-connected together beyond that point (546-02-08)
- separate terminals (or bars) have been provided for the protective and neutral conductors at the point of separation (546-02-08)
- PEN conductors have been connected to the terminals or bar intended for the protective earthing conductor and the neutral conductor (546-02-08).

 Note: Where earthing is required for protective as well as functional purposes, then the requirements for protective measures shall take precedence (546-01-01).

Non-conducting floors and walls

This form of protection (as the name implies) consists of an area in which the floor, walls and ceiling are all insulated and within which protective conductors and socket outlet do not have an earthing connection.

Test that the insulation of extraneous conductive parts:

- is not less than 0.5 MΩ (when tested at 500 V d.c)
- is able to withstand a test voltage of at least 2 kV a.c. r.m.s.
- does not pass a leakage current exceeding 1 mA in normal use (713-08-02).

Check that:

- the degree of protection against indirect contact provided by a non-conducting location is verified by measuring the resistance of the location's floors and walls to the installation's main protective conductor at not less than three points on each relevant surface (713-08-01)

Note: One of these measurements shall be not less than 1 m and not more than 1.2 m from any extraneous conductive part in the location. The other two measurements shall be made at greater distances.

- the insulation of extraneous conductive parts (to satisfy the requirements for protection be non-conducting location) are:
  - not less than 0.5 MΩ when tested at 500 V d.c. and
  - are able to withstand a test voltage of at least 2 kV a.c. r.m.s. and
  - do not pass a leakage current exceeding 1 mA in normal use (713-08-02).

Electrical separation

This form of protection is intended for an individual circuit and is aimed at preventing shock current through contact with exposed conductive parts which might be energised by a fault in the basic insulation of that circuit.
    Verify that:

- protection by electrical separation has been applied to the supply of individual items of equipment by means of a transformer complying with BS 3535 (the secondary of which is not earthed) or a source affording equivalent safety (471-12-01)
- protection by electrical separation has been used to supply several items of equipment from a single separated source (but **only** for special situations) (471-12-01)
- equipment used as a fixed source of supply is either:
  - selected and/or installed with Class II or equivalent protection (413-06-02) or
  - manufactured so that the output is separated from the input and from the enclosure by insulation satisfying the conditions for Class II (413-06-02)
- the supply source to the circuit is either:
  - an isolating transformer complying with BS 3535 or
  - a motor generator

- mobile supply sources (fed from a fixed installation) are selected and/or installed with Class II or equivalent protection (413-06-02)
- source supplies are only supplying more than one item of equipment provided that:
  - all exposed conductive parts of the separated circuit are connected together by an insulated and non-earthed equipotential bonding conductor
  - the non-earthed equipotential bonding conductor is not connected to a protective conductor, or to exposed conductive part of any other circuit or to any extraneous conductive part
  - all socket outlets are provided with a protective conductor contact (that is connected to the equipotential bonding conductor)
  - all flexible equipment cables (other than Class H equipment) have a protective conductor for use as an equipotential bonding conductor
  - exposed conductive parts which are fed by conductors of different polarity (and which are liable to a double fault occurring) are fitted with an associated protective device
  - any exposed conductive part of a separated circuit cannot come in contact with an exposed conductive part of the source (413-06-05)
- live parts of a separated circuit are not connected (at any point) to another circuit or to earth (413-06-03)
- live parts of a separate circuit are electrically separated from all other circuits (see BS 3535, 413-06-03)

 Note: Live parts of relays, contactors etc. included in a separated circuit (and between a separated circuit and other live parts of other circuits) shall be similarly electrically separated.

- separated circuits, preferably, use a separate wiring system

 If this is not feasible, multicore cables (without a metallic sheath) or insulated conductors (in an insulating conduit) may be used (413-06-03).

- the voltage of an electrically separated circuit does not exceed 500 V (413-06-02)
- all parts of a flexible cable (or cord) that is liable to mechanical damage is visible throughout its length (413-06-03)
- for circuits supplying a single piece of equipment, no exposed conductive part of the separated circuit is connected:
  - to the protective conductor of the source
  - to any exposed conductive part of any other circuit (413-06-04)
- a warning notice (warning that protection by electrical separation is being used) is fixed in a prominent position adjacent to every point of access to the location concerned (471-12-01).

Main equipotential bonding conductors

All main equipotential (and supplementary) bonding conductors must be tested for continuity.

The normal approach is to just connect the leads from a low-resistance ohmmeter to the ends of the bonding conductor as shown in Figure 9.7 – making sure that one end is disconnected from its bonding clamp, otherwise the measurement may include the resistance of parallel paths of other earthed metalwork.

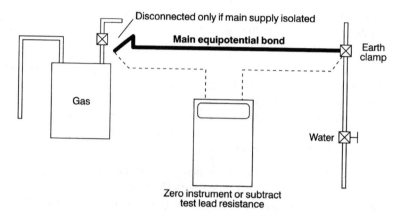

Figure 9.7 Continuity testing main and equipotential bonding conductors

Confirm that main equipotential bonding conductors have (for each installation) been connected to the main earthing terminal of that installation. These can include the following:

- water service pipes
- gas installation pipes
- other service pipes and ducting
- central heating and air conditioning systems
- exposed metallic structural parts of the building
- the lightning protective system (413-02-02).

 Note: Where an installation serves more than one building, the above requirement shall be applied to each building.

Supplementary equipotential bonding

Using a low-resistance ohmmeter (similar to that employed for testing main equipotential bonding conductors as described above) test all supplementary equipotential bonding conductors for continuity, particularly in locations intended for livestock to confirm that:

- supplementary bonding connects all exposed and extraneous conductive parts which can be touched by livestock (605-08-02)
- metallic grids laid in the floor for supplementary bonding are connected to the protective conductors of the installation (605-08-03).

Locations with increased risk of shock

For installations and locations with increased risk of shock (e.g. saunas, bathrooms and agricultural/horticultural premises etc.) certain additional measures may be required, such as:

- automatic disconnection of supply by means of a residual current device with a rated residual operating current (I_{sn}) not exceeding 30 mA
- supplementary equipotential bonding
- reduction of maximum fault clearance time.

In these cases, test to confirm that circuits supplying fixed equipment which are outside of the earthed equipotential zone (and which have exposed conductive parts that could be touched by a person who has direct contact directly with earth) that the earth fault loop impedance ensures that disconnection occurs within the time stated in Table 41A (p. 44) of the IEE Wiring Regulations.

Test, measure and confirm that:

- if the installation is part of a TF system, all socket outlet circuits have been protected by a residual current device (471-08-06)
- automatic disconnection using a residual current device has **not** been applied to a circuit incorporating a PEN conductor (471-08-07)
- installations that provide protection against indirect contact by automatically disconnecting the supply have a circuit protective conductor run to (and terminated at) each point in the wiring and at each accessory (471-08-08)

 Except suspended lampholders which have no exposed conductive parts.

- one or more of the following types of protective device have been used:
  - an overcurrent protective device
  - a residual current device (413-02-07)
- where a residual current device is used in a TN-C-S system, a PEN conductor has not been used on the load side (413-02-07)
- the protective conductor to the PEN conductor is on the source side of the residual current device (413-02-07)
- the maximum disconnection times to a circuit supplying socket outlets and to other final circuits which supply portable equipment intended for manual movement during use, or hand-held Class I equipment does not exceed those shown in Table 9.2 (413-02-09)

Note: This requirement does not apply to a final circuit supplying an item of stationary equipment connected by means of a plug and socket outlet where precautions are already taken to prevent the use of the socket outlet for supplying hand-held equipment, nor to reduced low-voltage circuits.

- where a fuse is used to satisfy this disconnection requirement, maximum values of earth fault loop impedance (Z_s) corresponding to a disconnection time of 0.4 s are as stated in Table 9.2 for a nominal voltage to earth (U_o) of 230 V (413-02-10)
- for a distribution circuit and a final circuit supplying only stationary equipment, the maximum disconnection time of 5 s is not exceeded (413-02-13).

Table 9.2 British Standards for fuse links

	Standard	Current rating	Voltage rating	Breaking capacity	Notes
1	BS 2950	Range 0.05–25 A	Range 1000 V (0.05 A) to 32 V (25 A) a.c. and d.c.	Two or three times current rating	Cartridge fuse links for telecommunication and light electrical apparatus. Very low breaking capacity
2	BS 646	1, 2, 3 and 5 A	Up to 250 V a.c. and d.c.	1000 A	Cartridge fuse intended for fused plugs and adapters to BS 546: 'round-pin' plugs
3	BS 1362 cartridge	1, 2, 3, 5, 7, 10 and 13 A	Up to 250 V a.c.	6000 A	Cartridge fuse primarily intended for BS 1363: 'flat pin' plugs
4	BS 1361 HRC cut-out fuses	5, 15, 20, 30, 45 and 60 A	Up to 250 V a.c.	16,500 A–33,000 A	Cartridge fuse intended for use in domestic consumer units. The dimensions prevent interchangeability of fuse links which are not of the same current rating
5	BS 88 motors	Four ranges, 2–1200 A	Up to 660 V, but normally 250 or 415 V a.c. and 250 or 500 V d.c.	Ranges from 10,000 to 80,000 A in four AC and three DC categories	Part 1 of Standard gives performance and dimensions of cartridge fuse links, whilst Part 2 gives performance and requirements of fuse carriers and fuse bases designed to accommodate fuse links complying with Part 1
6	BS 2692	Main range from 5 to 200 A; 0.5 to 3 A for voltage transformer protective fuses	Range from 2.2 to 132 kV	Ranges from 25 to 750 MVA (main range) 50 to 2500 MVA (VT fuses)	Fuses for a.c. power circuits above 660 V
7	BS 3036 rewirable	5, 15, 20, 30, 45, 60, 100, 150 and 200 A	Up to 250 V to earth	Ranges from 1000 to 12,000 A	Semi-enclosed fuses (the element is a replacement wire) for a.c. and d.c. circuits
8	BS 4265	500 mA to 6.3 A 32 mA to 2 A	Up to 250 V a.c.	1500 A (high breaking capacity) 35 A (low breaking capacity)	Miniature fuse links for protection of appliances of up to 250 V (metric standard)

Note:
(1) The circuit loop impedances given in the table above should not be exceeded when the conductors are at their normal operating temperature. If the conductors are at a different temperature when tested, then the reading should be adjusted accordingly.
(2) See appropriate British Standard for types and rated currents of fuses other than those mentioned in Table 9.2 above.

Protection by separation of circuits

Ensure that:

- the separation of circuits shall be verified for protection by:
  - SELV
  - PELV
  - electrical separation (713-06-01)
- the separation of live parts from those of other circuits and those of other circuits from earth is verified by measuring that the insulation resistance in accordance with values shown in Table 9.3 (713-06-02)

Table 9.3 Minimum values of insulation resistance

System	Test voltage	Minimum insulation resistance
SELV and PELV	250 V d.c.	0.25 MΩ
LV up to 500 V	500 V d.c.	0.5 MΩ
Over 500 V	1000 V d.c.	1.0 MΩ

- functional extra-low-voltage circuits meet all the test requirements for low-voltage circuits (713-06-05).

Polarity

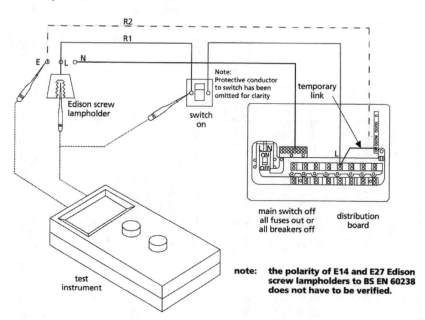

Figure 9.8 Polarity test on a lighting circuit (courtesy of IEE).

Complete a polarity test to verify that:

- fuses and single-pole control and protective devices are only connected in the phase conductor
- circuits (other than BS EN 60238 E14 and E27 lampholders) which have an earthed neutral conductor centre contact bayonet (and Edison screw lampholders) have their outer or screwed contacts connected to the neutral conductor
- wiring has been correctly connected to socket outlets and similar accessories (713-09-01).

9.5.10 Additional tests with the supply connected

Other than insulation tests (see above) the following tests are to be completed with the supply connected:

- re-check of polarity
- earth electrode resistance
- earth fault loop impedance
- prospective fault current.

Polarity

Repeat the polarity test shown above to verify that:

- fuses and single-pole control and protective devices are only connected in the phase conductor
- circuits (other than BS EN 60238 E14 and E27 lampholders) which have an earthed neutral conductor centre contact bayonet (and Edison screw lampholders) have their outer or screwed contacts connected to the neutral conductor
- wiring has been correctly connected to socket outlets and similar accessories (713-09-01).

Earth electrode resistance

If the earthing system incorporates an earth electrode as part of the installation, measure the electrode resistance to earth (using test equipment similar to that shown in Appendix 9.1 (p. 532) (713-10-01).

If the electrode under test is being used in conjunction with an RCD protecting an installation, test (prior to energising the remainder of the installation) between the phase conductor at the origin of the installation and the earth electrode with the test link open.

 Note: The resulting impedance reading (i.e. the electrode resistance) should then be added to the resistance of the protective conductor for the protected circuits.

Earth fault loop impedance

This is an extremely important test to ensure that under earth fault conditions, overcurrent devices disconnect fast enough to reduce the risk of electric shock.

The Regulations stipulate that *'if the protective measures employed require a knowledge of earth fault loop impedance, then the relevant impedances must be measured'* (713-11-01).

Where a fuse is used, maximum values of earth fault loop impedance (Z_s) corresponding to a disconnection time of 0.4 s are stated in Table 9.4 for a nominal voltage to earth (U_o) of 230 V (413-02-10).

 Note: See appropriate British Standard for types and rated currents of fuses other than those mentioned in Table 9.4.

Using test equipment (similar to that listed in Appendix 9.1) complete the following tests:

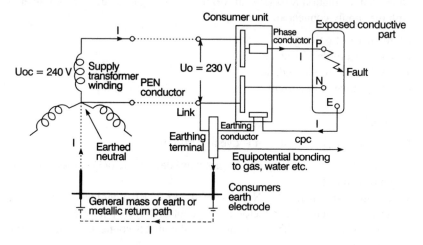

Figure 9.9 Testing earth fault loop impedance

Ensure that all main equipotential bonding is in place, connect the test equipment to the phase, neutral and earth terminals at the remote end of the circuit under test. Press to test and record.

 Note: The circuit loop impedances given in Table 9.4 should not be exceeded when the conductors are at their normal operating temperature. If the conductors are at a different temperature when tested, then the reading should be adjusted accordingly.

Prospective fault current

Verify, test and ensure that:

- prospective fault currents (under both short-circuit and earth fault conditions):
  - have been assessed for each supply source (551-02-02)
  - are calculated at every relevant point of the complete installation either by enquiry or by measurement (434-02-01)

Table 9.4 British Standards for fuse links

	Standard	Current rating	Voltage rating	Breaking capacity	Notes
1	BS 2950	Range 0.05–25 A	Range 1000 V (0.05 A) to 32 V (25 A) a.c. and d.c.	Two or three times current rating	Cartridge fuse links for telecommunication and light electrical apparatus. Very low breaking capacity
2	BS 646	1, 2, 3 and 5 A	Up to 250 V a.c. and d.c.	1000 A	Cartridge fuse intended for fused plugs and adapters to BS 546: 'round-pin' plugs
3	BS 1362 cartridge	1, 2, 3, 5, 7, 10 and 13 A	Up to 250 V a.c.	6000 A	Cartridge fuse primarily intended for BS 1363: 'flat pin' plugs
4	BS 1361 HRC cut-out fuses	5, 15, 20, 30, 45 and 60 A	Up to 250 V a.c.	16,500 A–33,000 A	Cartridge fuse intended for use in domestic consumer units. The dimensions prevent interchangeability of fuse links which are not of the same current rating
5	BS 88 motors	Four ranges, 2–1200 A	Up to 660 V, but normally 250 or 415 V a.c. and 250 or 500 V d.c.	Ranges from 10,000 to 80,000 A in four AC and three DC categories	Part 1 of Standard gives performance and dimensions of cartridge fuse links, whilst Part 2 gives performance and requirements of fuse carriers and fuse bases designed to accommodate fuse links complying with Part 1
6	BS 2692	Main range from 5 to 200 A; 0.5 to 3 A for voltage transformer protective fuses	Range from 2.2 to 132 kV	Ranges from 25 to 750 MVA (main range) 50 to 2500 MVA (VT fuses)	Fuses for a.c. power circuits above 660 V
7	BS 3036 rewirable	5, 15, 20, 30, 45, 60, 100, 150 and 200 A	Up to 250 V to earth	Ranges from 1000 to 12,000 A	Semi-enclosed fuses (the element is a replacement wire) for a.c. and d.c. circuits
8	BS 4265	500 mA to 6.3 A 32 mA to 2 A	Up to 250 V a.c.	1500 A (high breaking capacity) 35 A (low breaking capacity)	Miniature fuse links for protection of appliances of up to 250 V (metric standard)

- are measured at the origin and at other relevant points in the installation (713-12-01)
- protection of wiring systems against overcurrent takes into account minimum and maximum fault current conditions (533-03-01)
- fault current protective devices are provided:
  - at the supply end of each parallel conductor where two conductors are in parallel
  - at the supply and load ends of each parallel conductor where more than two conductors are in parallel (473-02-05)
- fault current protective devices are:
  - less than 3 m in length between the point where the value of current-carrying capacity is reduced and the position of the protective device
  - installed so as to minimise the risk of fault current
  - installed so as to minimise the risk of fire or danger to persons (473-02-02)
- fault current protective devices are placed (on the load side) at the point where the current-carrying capacity of the installation's conductors is likely to be lessened owing to:
  - the method of installation
  - the cross-sectional area
  - the type of cable or conductor used
  - inherent environmental conditions (473-02-01)
- conductors are capable of carrying fault current without overheating (130-05-01).

 Note: A single protective device may be used to protect conductors in parallel against the effects of fault current occurring (473-02-05).

Ensure that:

- fault current protection devices are capable of breaking (432-04-01)
- the breaking capacity rating of each device is not less than the prospective short-circuit current or earth fault current at the point at which the device is installed (434-03-01 and 434-03-02)

 Note: A lower breaking capacity is permitted if another protective device is installed on the supply side.

- the characteristics of each device used for overload current and/or for fault current protection has been co-ordinated so that the energy let through (i.e. by the fault current protective device) does not exceed the overload current protective device's limiting values (435-01-01)
- devices providing protection against both overload current and fault current are capable of breaking (432-02-01)

 Note: Overload current protection devices may have a breaking capacity below the value of the prospective fault current at the point where the device is installed (432-03-01).

- circuit breakers used as fault current protection devices:
  - are capable of making any fault current up to and including the prospective fault current (432-04-01)
  - break any fault current flowing before that current causes danger due to thermal or mechanical effects produced in circuit conductors or associated connections (434-01-01)
  - break and make up any overcurrent up to and including the prospective fault current at the point where the device is installed (432-02-0)
- safety services with sources that are incapable of operating in parallel, are protected against electric shock and fault current (565-01-02).

9.5.11 Insulation tests

When an electrical installation fails an insulation test, the installation must be corrected and the test made again. If the failure influences any previous tests that were made, then those tests must also be repeated.

Locations with a risk of fire due to the nature of processed and/or stored materials

Test that wiring systems (less those using mineral insulated cables and busbar trunking arrangements) have been protected against earth insulation faults as follows:

- in TN and TT systems, by residual current devices having a rated residual operating current (I_{An}) not exceeding 300 mA
- in IT systems, by insulation-monitoring devices with audible and visible signals (482-02-06).

 Note:

(1) Adequate supervision is required to facilitate manual disconnection as soon as appropriate.
(2) The disconnection time of the overcurrent protective device, in the event of a second fault, shall not exceed 5 s.

Protection against electric shock – insulation tests

Confirm, measure and test that:

- circuit conductors for each SELV system are physically separated from those of any other system or (i.e. where this proves impracticable) confirm that SELV circuit conductors are:
  - insulated for the highest voltage present
  - enclosed in an insulating sheath additional to their basic insulation (411-02-06)
- equipment is capable of withstanding all mechanical, chemical, electrical and thermal influences stresses normally encountered during service (412-02-01)

Note: Paint, varnish, lacquer or similar products are not generally considered to provide adequate insulation for protection against direct contact in normal service.

- exposed conductive parts that might attain different potentials through failure of the basic insulation of live parts have been arranged so that a person will not come into simultaneous contact with two exposed conductive parts, or an exposed conductive part and any extraneous conductive part (413-04-02)

This may be achieved if the location has an insulating floor and insulating walls and one or more of the following arrangements applies:

- the distance between any separated exposed conductive parts (and between exposed conductive parts and extraneous conductive parts) is not less than 2.5 m (1.25 m for parts out of arm's reach)
- if protective obstacles (that are not connected to earth or to exposed conductive parts and which are made out of insulating material) are used between exposed conductive parts and extraneous conductive parts
- the insulation is of acceptable electrical and mechanical strength (413-04-07)
- if the nominal voltage of a SELV system exceeds 25 V a.c. r.m.s. or 60 V ripple-free d.c., then protection against direct contact has been provided by one (or more) of the following:
  - insulation capable of withstanding a type-test voltage of 500 V a.c. r.m.s. for 60 seconds or
  - a barrier (or an enclosure – see Regulation 412-03) capable of providing protection to at least 1P2X or IPXXB (411-02-09)
- in IT systems, an insulation-monitoring device has been provided so as to indicate the occurrence of a first fault from a live part to an exposed conductive part or to earth (413-02-24)
- insulating enclosures:
  - are not pierced by conductive parts (other than circuit conductors) likely to transmit a potential (413-03-07)
  - do not contain any screws of insulating material – the future replacement of which by metallic screws could impair the insulation provided by the enclosure (413-03-07)
  - do not adversely affect the operation of the equipment protected (413-03-03)

Note: Where the insulating enclosure has to be pierced by conductive parts (e.g. for operating handles of built-in equipment and for screws) protection against indirect contact shall not be impaired.

- live parts are completely covered with insulation which:
  - can only be removed by destruction
  - is capable of durably withstanding electrical, mechanical, thermal and chemical stresses normally encountered during service (412-02-01)

- the basic insulation of operational electrical equipment is at least the degree of protection 1P2X or IPXXB (413-03-04)
- where insulation has been applied during the erection of the installation, the quality of the insulation has been verified (412-02-01).

 Where the risk of electric shock is increased by a reduction in body resistance and/or by contact with earth potential, confirm that protection has been provided by insulation of live parts, protection by obstacles, protection by barriers or enclosures or SELV (471-15-05).

Protection against electric shock – special installations or locations – verification tests

Where SELV or PELV is used (whatever the nominal voltage) in locations containing a bath, shower, hot air sauna and/or in a restrictive conductive location, confirm that protection against direct contact has been provided by:

- insulation capable of withstanding a type-test voltage of 500 V a.c. r.m.s. for 1 minute or
- barriers and/or enclosures providing protection to at least 1P2X or IPXXB (601-03-01, 602-03-01 and 606-02-01).

The above requirements do not apply to a location in which freedom of movement is not physically constrained.

RCDs and RCBOs – verification tests

Test (on the load side of the RCD and as near as practicable to its point of installation and between the phase conductor of the protected circuit and their associated circuit protective conductor) that:

- general purpose RCDs:
  - do not open with a leakage current flowing equivalency to 50% of the rated tripping current
  - open in less that 200 ms with a leakage current flowing equivalency to 100% of the rated tripping current of the RCD
- general purpose RCCDs to BS EN 61008 or RCBOs to BS EN 61009
  - do not open with a leakage current flowing equivalency to 50% of the rated tripping current of the RCD
  - open in less that 300 ms with a leakage current flowing equivalency to 100% of the rated tripping current of the RCD (unless it is a Type S (or selective) device that incorporates an intentional time delay, in which case it should trip between 130 ms and 500 ms)
- RCD-protected socket outlets to BS 7288:
  - do not open with a leakage current flowing equivalency to 50% of the rated tripping current of the RCD
  - open in less that 200 ms with a leakage current flowing equivalency to 100% of the rated tripping current of the RCD.

Selection and erection of wiring systems

Test to confirm that:

- all electrical joints and connections to ensure that they meet stipulated requirements concerning conductance, insulation, mechanical strength and protection (133-01-04)
- cables that are run in thermally insulated spaces are not covered by the thermal insulation (523-04-01)
- the current-carrying capacity of cables that are installed in thermally insulated walls or above a thermally insulated ceiling conforms with Appendix 4 to the Regulations (523-04-01)
- the current-carrying capacity of cables that are totally surrounded by thermal insulation for less than 0.5 m has been reduced according to the size of cable, length and thermal properties of the insulation (523-04-01)
- the insulation and/or sheath of cables connected to a bare conductor or busbar is capable of withstanding the maximum operating temperature of the bare conductor or busbar (523-03-01)
- wiring systems are capable of withstanding the highest and lowest local ambient temperature likely to be encountered – or are provided with additional insulation suitable for those temperatures (522-01-01 and 522-02-02)
- wiring systems have been selected and erected so as to minimise (i.e. during installation, use and maintenance) damage to the sheath and insulation of cables and insulated conductors and their terminations (522-08-01).

Site-applied insulation

Test to confirm that:

- insulation applied on site to protect against direct contact is capable of withstanding, without breakdown or flashover, an applied test voltage as specified in the British Standard for similar type-tested equipment (713-05-01)
- supplementary insulation applied to equipment during erection (i.e. to protect against indirect contact):
  - protects to at least 1P2X or IPXXB, and
  - is capable of withstanding, without breakdown or flashover, an applied test voltage as specified in the British Standard for similar type-tested equipment (713-05-02).

Supplies for safety services (IT systems)

In an IT system, confirm that continuous insulation monitoring has been provided to give audible and visible indications of a first fault (561-01-03).

9.5.12 Periodic inspections and tests

Periodic inspection and testing of all electrical installations shall be carried out to confirm that the installation is in a satisfactory condition for continued service

and this inspection shall consist of careful scrutiny of the installation (dismantled or otherwise) using appropriate tests (731-01-02 and 03).

The aim of periodic inspection and testing is to:

- confirm that the safety of the installation has not deteriorated or has been damaged
- ensure the continued safety of persons and livestock against the effects of electric shock and burns
- identify installation defects and non-compliance with the requirements of the Regulations, which may give rise to danger
- protect property being damaged by fire and heat caused by a defective installation (731-01-04).

Precautions shall be taken to ensure that inspection and testing does not cause:

- danger to persons or livestock
- damage to property and equipment (even if the circuit is defective) (731-01-05).

The frequency of periodic inspection and testing of installations will depend on:

- the type of installation, its use and operation
- the frequency and quality of maintenance and
- the external influences to which it is subjected (133-03-01 and 732-01-01).

 Periodic inspection and testing of supervised installations may take the form of continuous monitoring and maintenance by skilled persons. Appropriate records shall be kept (732-01-02).

9.5.13 Verification tests

All completed installations (including additions and/or alteration to existing installations) shall be inspected and tested for conformance to BS 7621 (133-02-01).

Accessibility of connections

Confirm that all connections and joints are accessible for inspection, testing
and maintenance, unless:

- they are in a compound-filled or encapsulated joint
- the connection is between a cold tail and a heating element
- the joint is made by welding, soldering, brazing or a compression tool (526-04-01).

Appliances producing hot water or steam

Confirm that electric appliances producing hot water or steam have been protected against overheating (424-02-01).

Cables and conductors for low voltage

Confirm that flexible and non-flexible cables, flexible cords (and conductors used as an overhead line) operating at low voltage shall comply with the relevant British or harmonised Standard (521-01-01 and 521-01-03).

Electric surface heating systems

Confirm that the equipment, system design, installation and testing of all electric surface heating systems meet the requirements of BS 6351 (554-07-01).

Emergency switching – verification tests

For any exterior installation, confirm that the switch is placed outside the building, adjacent to the equipment (where this is not possible, a notice showing the position of the switch shall be placed adjacent to the equipment and a notice fixed near the switch shall indicate its use) (476-03-07).

Forced air heating systems

Confirm (by inspection and test) that electric heating elements of forced air heating systems (other than those of central-storage heaters):

- are incapable of being activated until the prescribed airflow has been established (424-01-01)
- deactivate when the airflow is reduced or stopped (424-01-01)
- do not have two, independent, temperature-limiting devices (424-01-01)
- have frames and enclosures which are constructed out of non-ignitable material (424-01-02).

Heating cables

Check that:

- heating conductors and cables that pass through (or are in close proximity to) a fire hazard:
  - are enclosed in material with an ignitability characteristic 'P' as specified in BS 476 Part 5
  - are protected from any mechanical damage (554-06-01)
- heating cables that have been laid (directly) in soil, concrete, cement screed, or other material used for road and building construction are:
  - capable of withstanding mechanical damage
  - constructed of material that will be resistant to damage from dampness and/or corrosion (554-06-02)
- heating cables that have been laid (directly) in soil, a road, or the structure of a building are installed so that it:
  - is completely embedded in the substance it is intended to heat
  - is not damaged by movement (by it or the substance in which it is embedded.

- complies in all respects with the maker's instructions and recommendations (554-06-03)
- the maximum loading of floor-warming cable under operating conditions is no greater than the temperatures shown in Table 9.5 (554-06-04).

Table 9.5 Maximum conductor operating temperatures for a floor-warming cable

Type of cable	Maximum conductor operating temperature (°C)
General-purpose PVC over conductor	70
Enamelled conductor, polychlorophene over enamel, PVC overall	70
Enamelled conductor PVC overall	70
Enamelled conductor, PVC over enamel, lead-alloy 'E' sheath overall	70
Heat-resisting PVC over conductor	85
Nylon over conductor, heat-resisting PVC overall	85
Synthetic rubber or equivalent elastomeric insulation over conductor	85
Mineral insulation over conductor, copper sheath overall	Temperature dependent on type of seal employed, outer covering etc.
Silicone-treated woven-glass sleeve over conductor	180

Heating and ventilation systems

In locations where heating and ventilation systems containing heating elements are installed and where there is a risk of fire due to the nature of processed or stored materials, check to ensure that:

- the dust or fibre content and the temperature of the air does not present a fire hazard (482-02-15)
- temperature-limiting devices have a manual reset (482-02-15)
- heating appliances are fixed (482-02-16)
- heating appliances mounted close to combustible materials are protected by barriers to prevent the ignition of such materials (482-02-16)
- heat storage appliances are incapable of igniting combustible dust and/or fibres (482-02-17)
- enclosures of equipment such as heaters and resistors do not attain higher surface temperatures than:
  - 90°C under normal conditions, and
  - 115°C under fault conditions (482-02-18).

Identification of conductors by letters and/or numbers

Test to confirm that all individual conductors and groups of conductors:

- have been identified by a label containing either letters or numbers that are clearly legible

- have numerals that contrast, strongly, with the colour of the insulation
- have had numerals 6 and 9 underlined (514-03).

Plugs and socket outlets

Inspect and test to ensure that any plug and socket outlet used in single-phase a.c. or two-wire d.c. circuits that does **not** comply with BS 1363, BS 546, BS 196 or BS EN 60309-2 has either been designed specially for that purpose or:

- the plug and socket outlet used for an electric clock has been specially designed for that purpose and the plug has a fuse not exceeding 3 amperes which complies with BS 646 or BS 1362
- plug and socket outlet used for an electric shaver is either part of the shaver supply unit that complies with BS 3535 or in a room (other than a bathroom) that complies with BS 4573 (553-01-05).

Water heaters

Confirm that water heaters (or boilers) having immersed and uninsulated heating elements are permanently connected to the electricity supply via a double-pole linked switch, which is either:

- separate from and within easy reach of the heater/boiler or
- part of the boiler/heater (provided that the wiring from the heater or boiler is directly connected to the switch without use of a plug and socket outlet) (554-05-03).

Functional testing

The following are amongst the most important functional tests that should be completed:

- verify the effectiveness of residual current devices providing protection against indirect contact (or supplementary protection against direct contact) by a test simulating a typical fault condition (713-13-01)
- functionally test assemblies (such as switchgear and control gear assemblies, drives, controls and interlocks) to show that they are properly mounted, adjusted and installed in accordance with the Regulations (713-13-02).

9.5.14 Electrical connections

Of main concern are:

- connections between conductors and between a conductor and equipment
- main earthing terminals or bars
- final and distribution circuits.

Test and verify that:

- connections between conductors (and between a conductor and equipment) provide durable electrical continuity and adequate mechanical strength (526-01-01)

- the earthing conductor of main earthing terminals (or bars) is capable of being disconnected to enable the resistance of the earthing arrangements to be measured

 For convenience (and if required) this may be combined with the main earthing terminal or bar.

- all joints:
  - are capable of disconnection only by means of a tool
  - are mechanically strong
  - ensure the maintenance of electrical continuity (542-04-02)
- the wiring of final and distribution circuits to equipment with a protective conductor current exceeding 10 mA have a protective connection complying with one or more of the following:
  - a single protective conductor with a cross-section greater than $10 \, \text{mm}^2$
  - a single (mechanically protected) copper protective conductor with a cross-section greater than $4 \, \text{mm}^2$
  - two individual protective conductors
  - a BS 4444 earth-monitoring system that will automatically disconnect the supply to the equipment in the event of a continuity fault
- connection (i.e. of the equipment) to the supply by means of a double wound transformer, which has its secondary winding connected to the protective conductor of the incoming supply and the exposed conductive parts of the above (607-02-04).

9.5.15 Tests for compliance with the Building Regulations

As shown in Table 9.6, there are four types of installation that have to be inspected and tested for compliance with the Building Regulations.

Table 9.6 Types of installation

Type of inspection	When is it used?	What should it contain?	Remarks
Minor Electrical Installation Works Certificate	For new work associated with an alteration or addition to an existing installation	Relevant provisions of Part 7 of BS 7671	
Full Electrical Installation	For the design, construction, inspection and testing of an installation	A schedule of inspections and test results as required by Part 7 (of BS 7671) A certificate, including guidance for recipients (standard form from Appendix 6 of BS 7671)	For safety reasons, the electrical installation will need to be inspected at appropriate intervals by a competent person

(continued)

Table 9.6 (*continued*)

Type of inspection	When is it used?	What should it contain?	Remarks
Electrical Installation Certificate (short form)	For use when one person is responsible for the design, construction, inspection and testing of an installation	A schedule of inspections and a schedule of test results as required by Part 7 (of BS 7671)	For safety reasons, the electrical installation will need to be inspected at appropriate intervals by a competent person
Periodic Inspection Report	For the inspection of an existing electrical installation	A schedule of inspections and a schedule of test results as required by Part 7 (of BS 7671)	For safety reasons, the electrical installation will need to be inspected at appropriate intervals by a competent person

Figure 9.10 indicated how to choose what type of inspection is required.

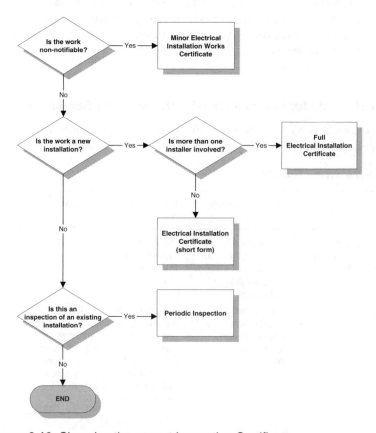

Figure 9.10 Choosing the correct Inspection Certificate

9.5.16 Additional tests required for special installations and locations

Part 6 of the Regulations contains additional requirements in respect of installations where the risk of electric shock is increased by a reduction in body resistance or by contact with earth potential (e.g. locations containing a bath or shower, swimming pools, hot air sauna, construction installations, agricultural and horticultural premises, caravans, motor caravans and highway power supplies etc.).

For inspection purposes these have been listed below in the form of check sheets.

9.5.17 Agricultural and horticultural premises

All fixed agricultural and horticultural installations (outdoors and indoors) and locations where livestock is kept (such as stables, chicken houses, piggeries, feed-processing locations, lofts and storage areas for hay, straw and fertilisers) shall be inspected to confirm that they comply with Part 605 of the Regulations.

 Note: If these premises include dwellings that are intended solely for human habitation, then the dwellings are excluded from the scope of these particular Regulations.

Protection against direct contact

- if SELV has been used, check that protection against direct contact has been provided by either:
  - barriers or enclosures supplying protection to at least 1P2X or IPXXB or
  - insulation that can withstanding a type-test voltage of 500 V a.c. r.m.s. for 60 seconds (605-02-02)
- check that circuits (less SELV) supplying a socket outlet are protected by a residual current device (605-03-01).

Protection against indirect contact

- check that all livestock locations have been protected against indirect contact using either:
  - earthed equipotential bonding or
  - automatic supply disconnection (605-04-01).

TN systems

For TN systems, check that:

- the maximum disconnection times of protective devices used for automatic disconnection (605-05-01)

- the maximum disconnection times to a circuit supplying socket outlets and to other final circuits which supply portable equipment intended for manual movement during use, or hand-held Class I equipment (605-05-02)
- meet the requirements of Table 9.7.

Table 9.7 Maximum disconnection times for TN systems

Installation nominal voltage U_o (volts)	Maximum disconnection time t (seconds)
120	0.35
220 to 277	0.2
400, 480	0.05
580	0.02

Note: If the disconnection times for installations, which are part of a TN system, cannot be met, then an overcurrent protective device shall be used (605-09-01).

When a fuse is used to satisfy this requirement check that the maximum values of earth fault loop impedance corresponding to a disconnection time of 0.2 s shall comply with the requirements of Table 9.8 for a nominal voltage to earth of 230 V (605-05-03).

Note: See appropriate British Standard for types and rated currents of fuses other than those mentioned in Table 9.8 below.

Table 9.8 Maximum earth fault loop impedance for fuses, for 0.2 s disconnection time (reproduced courtesy of the BSI).

General purpose (gG) fuses to BS 88-2.1 and BS 88-6							
Rating (amperes)	6	10	16	20	25	32	40	50
Z_s (ohms)	7.74	4.71	2.53	1.60	1.33	0.92	0.71	0.53

Note: The circuit loop impedances given in the table above should not be exceeded when the conductors are at their normal operating temperature. If the conductors are at a different temperature when tested, then the reading should be adjusted accordingly.

Where a circuit breaker has been used as a protective device, check that the maximum value of earth fault loop impedance meets the requirements of Table 9.8, or the values specified in Table 605B2 on p. 154 of the Regulations (605-05-04).

For a distribution circuit and a final circuit supplying only stationary equipment check that the disconnection time does not exceed 5 s (605-05-06).

Note: Where the disconnection time for a final circuit exceeds that required by Table 9.8 and another final circuit requiring a disconnection time according to Table 9.8 is connected to the same distribution board or distribution circuit, then:

- either the impedance of the protective conductor between the distribution board and the point at which the protective conductor is connected to the main equipotential bonding shall not exceed $25 Z_s/U_o$ ohms
- or there shall be equipotential bonding at the distribution board.

If protection is provided by a residual current device then see that the earth fault loop impedance $Z_s \times I_{\Delta n}$ shall not exceed 25 V (605-05-09).

Installations which are part of a TT system

If an installation is part of a TT system, check that the sum of the resistances of the earth electrode and the protective conductor(s) connecting it to the exposed conductive part does not exceed 25 V (605-06-01).

Installations which are part of an IT system

If an installation is part of a IT system, check that:

- the sum of the resistances of the earth electrode and the protective conductor(s) connecting it to the exposed conductive part does not exceed 25 V (605-07-01)
- the maximum disconnect time is in accordance with Table 9.9 (605-07-02).

Table 9.9 Maximum disconnection time in IT systems (2nd fault)

Installation	Maximum disconnection time	
	Neutral not disturbed	Neutral disturbed
120/240	0.4 s	1.0 s
220/280 to 277/480	0.2 s	0.5 s
400/690	0.06 s	0.2 s
580/1000	0.02 s	0.08 s

Supplementary equipotential bonding

In locations intended for livestock see that:

- supplementary bonding connects all exposed and extraneous conductive parts which can be touched by livestock (605-08-02)
- metallic grids laid in the floor for supplementary bonding are connected to the protective conductors of the installation (605-08-03).

Protection against fire and harmful thermal effects

For fire protection purposes etc., check that:

- a residual current device (with a residual operating current not exceeding 0.5 A) has been installed on all equipment supplies not essential to the welfare of livestock (605-10-01)
- heating appliances have been fixed so as to minimise the risks of burns to livestock and of fire from combustible material (605-10-02)
- radiant heaters have been fixed not less than 0.5 m from livestock and from combustible material (605-10-02).

Selection of equipment

Check that all installed electrical equipment has at least the degree of protection IP44 (605-11-01).

Devices for isolation and switching

Ensure that emergency switching and emergency stopping devices:

- are inaccessible to livestock
- cannot be impeded by livestock in the event of them panicking (605-13-01).

Electric fence controllers

Ensure that persons and livestock are protected by checking that mains operated electric fence controllers:

- comply with BS EN 61011 and BS EN 6101 I -1
- take account of the effects of induction when in the vicinity of overhead power lines (605-14-01)
- are installed so that they are free from risk of mechanical damage or unauthorised interference (605-14-02)
- are not fixed to any supporting pole of an overhead power or telecommunication line (605-14-03)
- earth electrodes that are connected to the earthing terminal of an electric fence controller are:
  - separate from the earthing system of any other circuit
  - situated outside the resistance area of any electrode used for protective earthing (605-14-04)
- only have one controller connected to each electric fence (605-14-05)
- electric fences (and their associated controllers) are not liable to come into contact with any other equipment and/or conductor (605-14-06).

9.5.18 Locations containing a bath or shower

Locations containing baths or showers for medical treatment, or for disabled persons, may have special requirements.

Protection against electric shock

Where SELV or PELV is used (whatever the nominal voltage) check to see that protection against direct contact is provided by:

- barriers or enclosures providing protection to at least 1P2X or IPXXB, or
- insulation capable of withstanding a type-test voltage of 500 V a.c. r.m.s. for 1 minute (601-03-02).

Inspect to see that the following protective measures against electric shock have been met:

- in zone 0, only protection by SELV (at a nominal voltage not exceeding 12 V a.c. r.m.s. or 30 V ripple-free d.c.) is permitted (601-05-01)
- the safety source is installed outside zones 0, 1 and 2 (601-05-01)
- protection by obstacles and placing out of reach have not been used (601-05-02)
- protection by non-conducting location and earth-free local equipotential bonding have not been used (601-05-03).

Supplementary equipotential bonding

Ensure that local supplementary equipotential bonding is provided to connect together the terminals of the protective conductor for each circuit supplying Class I and Class II equipment in zones 1, 2 or 3 (unless a cabinet containing a shower is installed in a room other than a bathroom or shower room, e.g. in a bedroom) with extraneous conductive parts in those zones (601-04-01).

 Note: 'extraneous conductive parts' include:

- metallic pipes supplying services and metallic waste pipes (e.g. water, gas)
- metallic central heating pipes
- air conditioning systems
- accessible metallic structural parts of the building
- metallic baths and shower basins.

 Supplementary equipotential bonding may be provided in close proximity to the location.

Ensure that earth-free local equipotential bonding has **only** been used in special situations which are earth-free and that an appropriate notice has been fixed in a prominent position adjacent to every point of access to the location concerned (471-11-01).

Equipment

Check to see that electrical equipment have the minimum degree of protection as shown in Table 9.10.

Table 9.10 Equipment protection

Zone	IP code	Alternatives
Zone 0	IPX7	Or 💧 💧 if the equipment is not IP coded
Zones 1 and 2	IPX4	Or ⚠ if the equipment is not IP coded
	But where water jets are likely to be used for cleaning purposes (e.g. in communal baths and/or communal showers) IPX5	Or ⚠⚠ if the equipment is not IP coded
Zone 3 Where water jets are likely to be used for cleaning purposes (e.g. in communal baths and/or communal showers)	IPX5	Or ΔΔ if the equipment is not IP coded

Wiring systems

Inspect surface wiring systems (and wiring systems embedded in the walls at a depth not exceeding $50\,mm^2$) shall either be:

- protected by an earthed metallic conductor complying with BS 5467, BS 6346, BS 6724, BS 7846, BS EN 60702-1 or BS 8436
- of an insulated concentric construction complying with BS 4553-1, BS 4553-2 or BS 4553-3
- enclosed in earthed conduit, trunking or ducting
- mechanically protected against penetration of the cable by nails, screws etc.
- installed within 150 mm from the top of the wall or partition
- within 150 mm of an angle formed by two adjoining walls or partitions

or are limited to the rules shown in Table 9.11.

Table 9.11 Zonal limitation on wiring systems

Zone	Limitation	
Zone 0	Fixed electrical equipment situated in that zone	601-07-01
Zone 1	Fixed electrical equipment situated in zones 0 and 1	601-07-02
Zone 2	Fixed electrical equipment in zones 0, 1 and 2	(522-06-06(i), (ii) or (iii))

Switchgear and control gear

Check to see that the following restrictions have been met (Table 9.12).

Table 9.12 Zonal limitation on switchgear and control gear

Zone	Limitation	
Zone 0	Switchgear or accessories have not been installed.	601-08-01
Zone 1	Only switches of SELV circuits (supplied at a nominal voltage not exceeding 12 V a.c. r.m.s. or 30 V ripple-free d.c.) have been installed. **Note:**The safety source shall be installed outside zones 0, 1 and 2.	
Zone 2	Only switches and socket outlets of SELV circuits and shaver supply units complying with BS EN 60742 have been installed. **Note:** The safety source shall be installed outside zones 0, 1 and 2.	
Zone 3	Only SELV socket outlets complying with Regulation 411-02 and shaver supply units complying with BS EN 60742 have been installed. **Note:** Portable equipment can only be connected as above.	

Confirm that:

- socket outlets (other than SELV socket outlets and/or shaver supply units complying with BS EN 60742) have not been installed outside zones 0, 1, 2 and 3 (601-08-01)
- only insulating pull cords of cord operated switches meeting the requirements of BS 3676 have been permitted in zones 1 and 2 (601-08-01)
- socket outlets (other than a SELV socket outlet or shaver supply unit) that are located in shower cubicles that have been installed in rooms other than a bathroom or shower room (and which are outside zones 0, 1, 2 or 3) are protected by a residual current device (601-08-02).

Fixed current-using equipment

In accordance with sections 601-09-01 to 04, confirm that fixed current-using equipment has been installed in the correct zones as shown in Table 9.13.

Water heaters having immersed and uninsulated heating elements

The heater or boiler shall be permanently connected to the electricity supply via a double-pole linked switch

This linked switch shall either be:

- separate from and within easy reach of the heater/boiler or
- part of the boiler/heater (provided that the wiring from the heater or boiler is directly connected to the switch without use of a plug and socket outlet) (554-05-03).

Table 9.13 Zonal limitations for fixed current equipment

Zone	Type of equipment
0	only fixed current-using equipment which can reasonably be located there
1	a water heater a shower pump and SELV current-using equipment other fixed current-using equipment, provided that: • it is suitable for the conditions of that zone, • and the supply circuit is additionally protected by a residual current protective device
2	a water heater a shower pump a luminaire, fan, heating appliance or unit for a whirlpool bath (complying with the relevant standards) SELV current-using equipment other fixed current-using equipment, provided that it is suitable for the conditions of that zone
3	current-using equipment (other than fixed current-using equipment) that is protected by a residual current device

9.5.19 Electrical installations in caravans and motor caravans

It should be noted that the requirements of this section do **not** apply to:

• electrical circuits and equipment covered by the Road Vehicles Lighting Regulations 1989
• installations covered by BS EN 1648-1 and BS EN 1648-2
• internal electrical installations of mobile homes, fixed recreational vehicles, transportable sheds and the like, temporary premises or structures.

Protection against direct contact

Check that the following protective measures have not been used:

• protection by obstacles
• protection by placing out of reach (608-02-01).

Protection against indirect contact

Confirm that the following protective methods have not been used:

• non-conducting location
• earth-free equipotential bonding
• electrical separation (608-03-01).

Confirm that when protection by automatic disconnection of supply is used, it is done by a residual current device complying with BS 4293, BS EN 61008-1 or BS EN 61009-1 that breaks all live conductors and that the wiring system includes a circuit protective conductor that is connected to the:

- inlet protective contact, and
- electrical equipment's exposed conductive parts, and
- socket outlet's protective contacts (608-03-02).

If the cable does not include a protective conductor (and the cable is not enclosed in conduit or trunking), confirm that it:

- has a minimum cross-sectional area of $4\,mm^2$
- is insulated (608-03-03).

Confirm that all extraneous conductive parts of a caravan or motor caravan that are likely to become live in the event of a fault are bonded to the circuit protective conductor by a conductor with a minimum cross-sectional area of $4\,mm^2$ (608-03-04).

Confirm that all metal sheets forming part of the structure of the caravan or motor caravan are not considered to be 'extraneous conductive parts' (608-03-04).

Protection against overcurrent

Confirm that:

- all live conductors of all final circuits are protected by an overcurrent protective device (608-04-01)
- persons and livestock are protected from being injured by excessive temperatures and/or electromagnetic stresses caused by overcurrents occurring in live conductors (130-04-01)
- property has been protected from being damaged by excessive temperatures and/or electromagnetic stresses caused by overcurrents occurring in live conductors (130-04-01).

Selection and erection of equipment

If there is more than one electrically independent installation, confirm that each independent system is:

- supplied by a separate connecting device
- separated in accordance with Regulation 528-01-02 (608-05-01).

Wiring systems

Ensure that only the following wiring systems have been used:

- flexible single-core insulated conductors in non-metallic conduits
- stranded insulated conductors (with a minimum of seven strands) in non-metallic conduits sheathed flexible cables (608-06-01).

Flame propagating wiring systems shall **not** be used.

Confirm by inspection that:

- the cross-sectional area of conductors shall be greater than $1.5\,mm^2$ (608-06-02)
- all protective conductors (regardless of cross-sectional area) have been insulated (608-06-03)
- low-voltage-system cables have been run separately from extra-low-voltage systems cables (608-06-04)
- all cables (unless enclosed in rigid conduit) have been supported at intervals not exceeding 0.4 m for vertical runs and 0.25 m for horizontal runs (608-06-05)
- all flexible conduits have been supported at intervals not exceeding 0.4 m for vertical runs and 0.25 m for horizontal runs (608-06-05)
- that no electrical equipment has been installed in compartments where gas cylinders are stored (608-06-06)
- all wiring is protected against mechanical damage (608-06-07)
- wiring that passes through metalwork has been securely fixed and protected by bushes or grommets (608-06-07).

Electrical inlets

Confirm by inspection that:

- the electrical inlet to a caravan and/or motor caravan is a two-pole, earthing contact (key position 6h type) complying with BS EN 60309-2 (608-07-01)

Note: Electrical inlets are not limited to 16 A single-phase where the caravan or motor caravan demands more.

- electrical inlets have been installed:
  - on the outside of the caravan
  - in a readily accessible position
  - no more than 1.8 m above ground level
  - in an enclosure with a suitable cover (608-07-02)
- a notice has been fixed on or near the electrical inlet recess containing the following information concerning the caravan and/or motor caravan installation:
  - nominal (design) voltage and frequency
  - current rating (608-07-03)
- all installations have a main isolating switch that disconnects all live conductors (608-07-04)

Note: The isolating switch is for an installations consisting of only one final circuit then it may be the overcurrent protection device.

- a notice (worded as shown in Figure 9.11) has been permanently fixed near the main isolating switch inside the caravan or motor caravan (608-07-05).

1.1   INSTRUCTIONS FOR ELECTRICITY SUPPLY

TO CONNECT

1. Before connecting the caravan installation to the mains supply, check that:
    (a)   the supply available at the caravan pitch supply point is suitable for the caravan electrical installation and appliances, and
    (b)   the caravan main switch is in the OFF position.
2. Open the cover to the appliance inlet provided at the caravan supply point and insert the connector of the supply flexible cable.
3. Raise the cover of the electricity outlet provided on the pitch supply point and insert the plug of the supply cable.

THE CARAVAN SUPPLY FLEXIBLE CABLE MUST BE FULLY UNCOILED TO AVOID DAMAGE BY OVERHEATING.

4. Switch on at the caravan main switch.
5. Check the operation of residual current devices, if any, fitted in the caravan by depressing the test buttons.

IN CASE OF DOUBT OR, IF AFTER CARRYING OUT THE ABOVE PROCEDURE THE SUPPLY DOES NOT BECOME AVAILABLE, OR IF THE SUPPLY FAILS, CONSULT THE CARAVAN PARK OPERATOR OR THE OPERATOR'S AGENT OR A QUALIFIED ELECTRICIAN.

TO DISCONNECT

6. Switch off at the caravan main isolating switch, unplug both ends of the cable.

1.2 PERIODIC INSPECTION

Preferably not less than once every three years and more frequently if the vehicle is used more than normal average mileage for such vehicles, the caravan electrical installation and supply cable should be inspected and tested and a report on their condition obtained as prescribed in BS 7671 Requirements for Electrical Installations published by the Institution of Electrical Engineers.

Figure 9.11 Electrical safety notice

Accessories

Accessories used in caravans and/or motor caravans shall be inspected to confirm that:

- these accessories do not have easily reachable conductive parts (608-08-01)
- low-voltage socket outlets include a protective contact (unless supplied by an individual winding of an isolating transformer) (608-08-02)
- plugs designed for extra-low-voltage socket outlets are incompatible with low-voltage socket outlets (608-08-03)

- extra-low-voltage socket outlets:
  - clearly show their voltage
  - prevent the insertion of a low-voltage plug (608-08-03)
- accessories exposed to moisture are protected to at least 1P55 (608-08-04)
- appliances not connected to the supply by a plug and socket outlet are controlled by a switch (608-08-05)
- luminaries are (preferably) fixed directly to the structure or lining of the caravan or motor caravan (608-08-06)
- pendant luminaries are securely installed so as to prevent damage when the caravan or motor caravan is moved (608-08-06)
- dual-voltage luminaries:
  - are fitted with separate lampholders for each voltage
  - clearly indicate the lamp wattage
  - cannot be damaged if both lamps are lit at the same time
  - provide adequate separation between LV and ELV circuits, and
  - have lamps that cannot be fitted into lampholders intended for lamps of other voltages (608-08-07)
- connectors used to attach the supply to a caravan or motor caravan comprise:
  - a plug complying with BS EN 60309-2
  - a connector meeting the requirements of BS EN 60309-2
  - a flexible cord or cable, 25 m ($\pm$2 m) in length (HO7RN-F, HO5VV-F or equivalent) that includes a protective conductor whose cross-section meets the requirements of Table 9.14 (608-08-08).

Table 9.14 Cross-sectional areas of flexible cords and cables for caravan connectors

Rated current (A)	Cross-sectional area (mm^2)
16	2.5
25	4
32	6
63	16
100	35

Extra-low-voltage installation

- any part of a caravan installation that is operating at extra-low voltage shall comply with the requirements for SELV (608-08-09)
- the following extra-low-voltage d.c. power sources are permissible:
  - 12 V, 24 V and 48 V (608-08-09)
- when a.c. extra-low voltage is required, the following voltages (rms) are permissible:
  - 12 V, 24 V, 42 V and 48 V (608-08-09).

9.5.20 Electrical installations in caravan parks

The following requirements apply to installations supplying electricity to leisure accommodation vehicles in caravan parks.

Protection against direct contact

Check that these protection methods have not been used:

- protection by obstacles
- protection by placing out of reach (608-10-01).

Protection against indirect contact

Check that the following protection methods have not been used:

- non-conducting location
- earth-free local equipotential bonding
- electrical separation (608-11-01).

Selection and erection of equipment

Check that the following requirements for wiring systems have been implemented:

- the caravan pitch supply equipment to caravans is (preferably) connected by underground cable (608-12-01)
- all underground cables (unless mechanically protected) are to be installed outside of any caravan pitch and areas where tent pegs or ground anchors may be driven (608-12-02)
- all overhead conductors have been:
  - protected by insulation of live parts (in accordance with Regulation 412-02)
  - at least 2 m away from the boundary of any caravan pitch
  - not less than 6 m in vehicle movement areas and 3.5 m in all other areas (608-12-03).

 Note: Poles and other overhead wiring supports shall be protected against any reasonably foreseeable vehicle movement (608-12-03).

Switchgear and control gear

The following shall be inspected to ensure that:

- all caravan pitch supply equipment are:
  - located adjacent to the pitch
  - within 20 m from any point on the pitch which it is intended to serve (608-13-01)

- socket outlets and enclosures forming part of the caravan pitch supply equipment:
  - are in compliance with BS EN 60309-2
  - are protected to at least IPX4
  - are placed between 0.80 m and 1.50 m above the ground (to the lowest part of the socket outlet)
  - have a current rating of not less than 16 A
  - have at least one socket outlet provided for each pitch (608-13-02)
- socket outlets are protected (individually) by an overcurrent device (608-13-04)
- socket outlets are protected individually (or in groups of not more than three) by a residual current device complying with BS 4293, BS EN 61008-1 or BS EN 61009-1 (608-13-05)

Note: Although residual current devices reduce the risk of electric shock, they may not be used as the sole means of protection against direct contact (412-06-01 and 412-06-02).

- socket outlets are not bonded to the PME terminal (608-13-05)
- grouped socket outlets are on the same phase (608-13-06)
- any supply from a distributor's multiple earthed network has the protective conductor of each socket outlet circuit connected to an earth electrode and that:
  - all exposed conductive parts which are protected by a single protective device are be connected, via the main earthing terminal, to a common earth electrode
  - either a residual current device (preferred method) or an overcurrent protective device has been used
  - each circuit meets the following requirement:

$$R_A \times I_a \leqslant 50\,V \text{ (608-13-05)}.$$

9.5.21 Construction site installations

The following requirements apply to installations providing an electricity supply for:

- new building construction
- repairs, alterations, extensions or demolition of existing buildings
- engineering construction
- earthworks

and are applicable to:

- the main switchgear and protective devices
- installations of mobile and transportable electrical equipment
- the interface between the supply system and the construction site installations.

The following requirements do **not** apply to:

- construction site offices, cloakrooms, meeting rooms, canteens, restaurants, dormitories and toilets
- installations covered by BS 6907.

Supplies

Except for control and/or signalling circuits (and inputs from stand-by supplies), check that:

- equipment is compatible with the particular supply from which it is energised (604-02-01)
- equipment only contains components connected to the same installation (604-02-01)
- the nominal voltages shown in Table 9.15 are not exceeded (604-02-02).

Table 9.15 Nominal voltages for equipment at construction site installations

SELV	Portable hand lamps in confined or damp locations
110 V, 1-phase, centre point earthed	Reduced low-voltage system Portable hand lamps for general use Portable hand-held tools and local lighting up to 2 kW
110 V, 3-phase, star point earthed	Reduced low-voltage system Portable hand-held tools and local lighting up to 2 kW Small mobile plant, up to 3.75 kW
230 V, 1-phase	Fixed floodlighting
400 V, 3-phase	Fixed and movable equipment, above 3.75 kW

 The only exception is if a large equipment requires a high-voltage supply for functional reasons.

Protection against indirect contact

TN systems

For TN systems, check that:

- the maximum disconnection times of protective devices used for automatic disconnection, and
- the maximum disconnection times to a circuit supplying socket outlets and to other final circuits which supply portable equipment intended for manual movement during use, or hand-held Class I equipment

meet the requirements of Table 9.16 (604-04-01).

Table 9.16 Maximum disconnection times for TN systems

Installation nominal voltage U_o (volts)	Maximum disconnection time t (seconds)
120	0.35
220 to 277	0.2
400, 480	0.05
580	0.02

Note: If the disconnection times for installations which are part of a TN system cannot be met, then an overcurrent protective device shall be used (604-04-01, 604-04-07).

Except for reduced low-voltage systems, check that the maximum disconnection times for circuits supplying movable installations and equipment (either directly or through socket outlets) complies with the requirements of Table 9.22 for a nominal voltage to earth of 230 V (604-04-02).

Note: See appropriate British Standard for types and rated currents of fuses other than those mentioned in Table 9.17 (604-04-03).

Table 9.17 Maximum earth fault loop impedance for fuses, for 0.2 s disconnection time (reproduced courtesy of the BSI).

General purpose (gG) fuses to BS 88-2.1 and BS 88-6

Rating (amperes)	6	10	16	20	25	32	40	50
Z_s (ohms)	7.74	4.71	2.53	1.60	1.33	0.92	0.71	0.53

Note: The circuit loop impedances given in the table above should not be exceeded when the conductors are at their normal operating temperature. If the conductors are at a different temperature when tested, then the reading should be adjusted accordingly.

Where a circuit breaker has been used as a protective device, confirm (by calculation) that the maximum value of earth fault loop impedance is as specified in Table 604B2 on p. 149 of the Regulations (604-04-04).

Check that:

- if protection is provided by a residual current device then the earth fault loop impedance $Z_s \times I_{An}$ does not exceed 25 V (604-04-08)
- the maximum disconnection time for fixed installations, reduced low-voltage systems and overcurrent protective devices is 5 s (604-04-06).

If this disconnection time cannot be achieved, then protection shall be provided by a residual current device (604-04-07).

TT systems

If an installation is part of a TT system, check that the sum of the resistances of the earth electrode and the protective conductor(s) connecting it to the exposed conductive part does not exceed 25 V (604-05-01).

IT systems

If an installation is part of an IT system, check to confirm that:

- an IT system has not been used if an alternative system is available (604-03-01)
- where an IT system has been used, permanent earth fault monitoring has been provided (604-03-01)
- if protection against indirect contact has been provided by earthed equipotential bonding and automatic disconnection of supply (604-03-02)

 Note: If a portable generating set is used, earth fault monitoring may be omitted (604-03-01).

- the sum of the resistances of the earth electrode and the protective conductor(s) connecting it to the exposed conductive part does not exceed 25 V (604-06-01)
- the maximum disconnect time is in accordance with Table 9.18 (604-06-02).

Table 9.18 Maximum disconnection time in IT systems (2nd fault)

Installation	Maximum disconnection time	
	Neutral not disturbed	Neutral disturbed
120/240	0.4 s	1.0 s
220/280 to 277/480	0.2 s	0.5 s
400/690	0.06 s	0.2 s
580/1000	0.02 s	0.08 s

Supplementary equipotential bonding

Check that the resistance of the supplementary bonding conductor between simultaneously accessible exposed conductive parts and extraneous conductive parts is in accordance with:

$$R < \frac{25}{I_a}$$

where I_a is the operating current of the protective device (604-07-01).

Application of protective measures

Check that all socket outlets and permanently connected hand-held equipment (with a current rating of up to and including 32 A) are protected by obstacles (604-08-03).

Notes:

(1) This intention of this form of protection is to prevent unintentional contact with a live part, but **not** intentional contact by deliberate circumvention of the obstacle.

(2) Except where otherwise stated, it is assumed for the purpose of this Regulation that normal dry conditions exist and a person is not in direct contact with earth potential and has normal body resistance.

(3) Where socket outlets and permanently connected hand-held equipment socket outlets are protected by automatic disconnection and reduced low-voltage systems, check that they are being supplied from a separate isolating transformer (604-08-04).

Selection and erection of equipment

Check that all assemblies used for distributing electricity on construction and demolition sites comply with the requirements of BS 4363 and BS EN 60439-4 (604-09-01).

Note: All other equipment shall be protected '*appropriate to the external influence*' (604-09-02).

Wiring systems

Confirm that:

- cables installed across a site road or a walkway are protected against mechanical damage (604-10-02)
- reduced low-voltage systems use low-temperature 300/500 V thermoplastic (PVC) or equivalent flexible cables (604-10-03)
- fixed floodlighting and fixed and movable equipment (above 3.75 kW) that exceed the reduced low-voltage system use H07 RN-F type flexible cable that is resistant to abrasion and water (604-10-03).

Isolation and switching devices

Confirm that:

- supply and distribution assemblies are capable of isolating and switching the incoming supply (604-11-02)
- emergency switching has been provided for supplies that need to have live conductors disconnected in order to remove a hazard (604-11-03)
- circuits supplying current using equipment are being fed from a distribution assembly that includes:
  - overcurrent protective devices
  - protection against indirect contact
  - socket outlets (if required) (604-11-05)
- it is impossible to interconnect different safety and stand-by supplies (604-11-06).

Plugs and socket outlets

Confirm that:

- socket outlets are part of an assembly complying with BS 4363 and BS EN 60439-4 (604-12-01)
- plug and socket outlets comply with BS EN 60309-2 (604-12-01)
- luminaire-supporting couplers have not been used (604-12-03).

Cable couplers

Confirm that cable couplers comply with BS EN 60309-2 (604-13-01).

9.5.22 Highway power supplies

The requirements of this section of the Regulations apply to the installation of highway power supplies such as highway distribution circuits, street furniture and street located equipment. These requirements do **not** apply to distributor's equipment as defined by the Electricity Safety, Quality and Continuity Regulations 2002.

 Note: The requirements of this section of the Regulations also apply to similar equipment located in other areas used by the public but not designated as a highway or part of a building.

Protection against electric shock

For protection against direct contact, check that:

- protection by obstacles has **not** been used
- protection by placing out of reach only applies to low-voltage overhead lines constructed in accordance with the Electricity Safety, Quality and Continuity Regulations 2002
- where items of street furniture (or street located equipment) are within 1.5 m of a low-voltage overhead line, protection against direct contact with the overhead line has been provided by some means **other than** placing out of reach (611-02-01)

 Except, that is, when the maintenance of equipment is restricted to specially trained skilled persons.

- any door used to provide access to electrical equipment in street furniture or street located equipment:
  - has not been used as a barrier or an enclosure
  - is locked with a key or secured by use of a tool if located less than 2.50 m above ground level
  - is protected against direct contact (i.e. when the door is open) by an intermediate barrier providing protection to at least IP2X or IPXXB which can only be removed by the use of a tool (611-02-02)

- access to a light source for luminaries located less than 2.80 m above ground level is only possible after removing a barrier or enclosure that requires the use of a tool (611-02-02)

protection against indirect contact has **not** been provided by:

- – a non-conducting location
- – earth-free equipotential bonding
- – electrical separation (611-02-03)
- circuits feeding fixed equipment that are used in highway power supplies have a maximum disconnection time of 5 s (611-02-04)
- where protection against indirect contact has been provided by the use of earthed equipotential bonding and automatic disconnection, that metallic structures (e.g. fences, grids, not connected to or forming part of the street furniture or street located equipment) are not connected to the main earthing terminal as extraneous conductive parts (611-02-05)
- where protection against indirect contact is provided by Class II equipment (or equivalent insulation):
  - – no protective conductor has been provided and
  - – no conductive parts of the lighting column, street furniture or street located equipment are intentionally connected to the earthing system (611-02-06).

It is permissible to dispense with protective measures against indirect contact **only** if:

- – overhead line insulator brackets (and metal parts connected to them) are not within arm's reach
- – the steel reinforcement of steel reinforced concrete poles is not accessible
- – exposed conductive parts (including small isolated metal parts such as bolts, rivets, nameplates not exceeding 50 mm × 50 mm and cable clips) cannot be gripped or cannot be contacted by a major surface of the human body
- – there is no risk of fixing screws used for non-metallic accessories coming into contact with live parts
- – inaccessible lengths of metal conduit not exceeding 150 mm
- – metal enclosures mechanically protecting against indirect contact
- – unearthed street furniture that is supplied from an overhead line is inaccessible whilst in normal use (471-13-04).

Devices for isolation and switching

Confirm that if:

- the distributor's cut-out is used for isolating a highway power supply and that the approval of the distributor has been obtained (611-03-02)
- it is intended that isolation and switching will only be carried out by instructed persons, that a suitably rated fuse carrier has been provided as the means of isolation (611-03-01).

Identification of cables

Confirm that:

- detailed records have been provided (together with an Electrical Installation Certificate) on completion of installations that include highway distribution circuits and highway power supplies (611-04-01)
- cables buried in the ground are either of an insulated concentric construction or are protected by earthed armour or metal sheath (611-04-02)
- buried cables are marked by cable covers or marking tape (611-04-02)
- all ducting, marker tape and/or cable tiles used with highway power supply cables are suitably colour coded or marked so as to identify them from other services (611-04-03)
- drawings have been prepared showing the position and depth of all underground cables supplying highway power supplies, street furniture and/or street located equipment (611-04-05).

External influences

Confirm that all electrical equipment has a degree of protection not less than IP33 (611-05-02).

Temporary supplies

Confirm that:

- temporary supplies that have been taken from street furniture and street located equipment, do not reduce the safety of the permanent installation (611-06-01)
- temporary supply units have an externally mounted label stating the maximum sustained current that can be supplied from that unit (611-06-02).

9.5.23 Swimming pools

The following requirements apply to basins of swimming pools and paddling pools and their surrounding zones as shown in Figure 9.12.

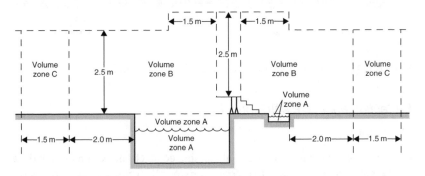

Figure 9.12 Zone dimensions for swimming pools and paddling pools

Where:

Zone A   is the interior of the basin, chute or flume and includes the portions of essential apertures in its walls and floor which are accessible to persons in the basin.

Zone B   is limited:
- by the vertical plane 2 m from the rim of the basin, and
- by the floor or surface expected to be accessible to persons, and
- by the horizontal plane 2.5 m above that floor or surface, except where the basin is above ground, when it shall be 2.5 m above the level of the rim of the basin.

 Note: Where the building containing the swimming pool contains diving boards, spring boards, starting blocks or a chute, zone B will also include the zone limited by the vertical plane (spaced 1.5 m from the periphery of diving boards, spring boards and starting blocks) and within that zone, by the horizontal plane 2.5 m above the highest surface expected to be occupied by persons, or to the ceiling or roof if they exist.

Zone C   is limited:
- by the vertical plane circumscribing zone B, and the parallel vertical plane 1.5 m external to zone B, and
- by the floor or surface expected to be occupied by persons and the horizontal plane 2.5 m above that floor or surface.

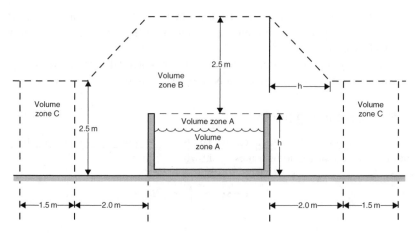

Figure 9.13 Zone dimensions for basin above ground level

Protection against electric shock

Check that:

- local supplementary equipotential bonding connects all extraneous conductive parts (and the protective conductors of all exposed conductive parts) within zones A, B and C (602-03-02)

This requirement is not applicable to equipment supplied by SELV circuits.

- where SELV is used (irrespective of the nominal voltage), protection against direct contact is provided by:
  - barriers or enclosures providing protection to at least IP2X or IPXXB, or
  - insulation capable of withstanding a type-test voltage of 500 V a.c. r.m.s. for 60 seconds (602-03-01)
- if there is a metal grid in a solid floor, it is physically connected to the local supplementary bonding (602-03-02)
- in zones A and B, only protective measure against electric shock by SELV are used
- the safety source has been installed outside zones A, B and C, except:
  - if floodlights are installed (in which case, each floodlight needs to be supplied from its own transformer, or an individual secondary winding of a multi-secondary transformer, having an open circuit voltage not exceeding 18 V)
  - socket outlets are protected by automatic disconnection of supply by means of a residual current device (602-04-01)

 The following protective measures are **not** used in any zone:

- protection by means of obstacles
- protection by means of placing out of reach
- protection by means of a non-conducting location
- protection by means of earth-free local equipotential bonding (602-04-02).

Selection and erection of equipment

Check that the equipment chosen has minimum degrees of protection as shown in Table 9.19 (602-05-01).

Table 9.19 Equipment protection in swimming pools

Zone	Degree of protection
Zone A	IPX8
Zone B	IPX5 IPX4 for swimming pools (where water jets are not likely to be used for cleaning)
Zone C	IPX2 for indoor pools IPX4 for outdoor pools IPX5 for swimming pools (where water jets are likely to be used for cleaning)

Wiring systems

Check that zones A and B:

- contain only wiring necessary to supply equipment situated in those zones (602-06-02)

- the surface wiring system does not use metallic conduits or metallic trunkings, an exposed metallic cable sheath, or an exposed earthing or bonding conductor (602-06-01)
- that accessible metal junction boxes have not been installed (602-06-03).

Switchgear, control gear and accessories

- Check that in zones A and B, switchgear, control gear and accessories have not been installed, unless socket outlets comply with BS EN 60309-2 and they are:
  - installed more than 1.25 m outside of zone A
  - installed at least 0.3 m above the floor
  - protected by either a residual current device or electrical separated with the safety isolating transformer placed outside zones A, B and C (602-07-01).

Check that in zone C, socket outlets, switches or accessories are only permitted if they are:

- protected individually by electrical separation, or
- protected by SELV, or
- protected by a residual current device, or
- a shaver socket complying with BS 3535 (602-07-02).

 This requirement does not apply to the insulating cords of cord operated switches complying with BS 3676.

Other equipment

Check that:

- socket outlets comply with BS EN 60309-2 (602-08-01)
- in zones A and B, only current-using equipment specifically intended for use in swimming pools is installed (602-08-02)
- in zone C, equipment is protected by electrical separation, SELV, or a residual current device (602-08-03)

 This requirement does not apply to instantaneous water heaters.

- if an electric heating unit is embedded in the floor in zone B or C, it is either:
  - connected to the local supplementary equipotential bonding by a metallic sheath or
  - covered by an earthed metallic grid connected to the equipotential bonding (602-08-04).

9.5.24 Hot air saunas

The requirements of this section apply to locations in which hot air sauna heating equipment in accordance with BS EN 60335-2-53 is installed.

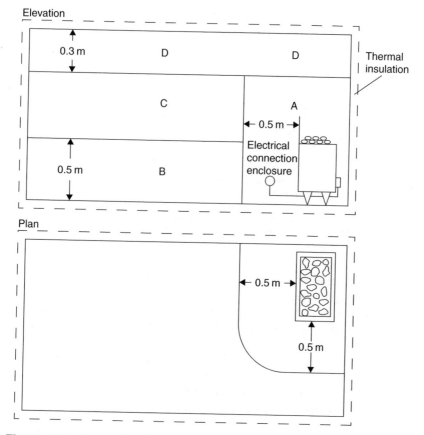

Figure 9.14 Zones of ambient temperature

Protection against electric shock

Confirm that:

- when SELV is used, protection against direct contact is provided either by:
  - insulation that is capable of withstanding a type-test voltage of 500 V a.c. r.m.s. for 60 seconds or
  - barriers (or enclosures) that provide protection to at least IP24 or IPX4B (603-03-01)
- protection by means of obstacles (and by placing out of reach) have not been used as a protective measure against direct contact (603-04-01)
- protection by non-conducting location and by means of earth-free local equipotential bonding, have not been used as a protective measure against indirect contact (603-05-01)
- all equipment is protected to at least IP24 (603-06-01)

- only the equipment shown in Table 9.20 has been installed in each of the four temperature zones (603-06-02)

Table 9.20 Permissible equipment

Temperature zone	Equipment
Zone A	Sauna heaters and equipment directly associated with zone A
Zone B	No special requirement (concerning heat resistance of equipment)
Zone C	Equipment suitable for an ambient temperature of 125°C
Zone D	Luminaries (and their associated wiring) suitable for an ambient temperature of 125°C Control devices for the sauna heater (and their associated wiring) suitable for ambient temperature of 125°C

- wiring systems only use flexible cords with 180°C thermosetting (rubber) insulation protected against mechanical damage (603-07-01)
- all switchgear (other than a thermostat or a thermal cut-out) is installed outside the hot air sauna (as opposed to being built into the sauna heater (603-08-01)
- accessories have not been installed within the hot air sauna unless they:
  - are protected to at least IP24, and
  - comply with the requirements of Table 9.20 (603-08-02)
- luminaries have been mounted so as to prevent overheating (603-09-01).

Heating appliances

Check that the number and type of circuits required for lighting, heating, power, control, signalling, communication and information technology, etc. has taken into consideration:

- the location and points of power demand
- the loads to be expected on the various circuits
- the daily and yearly variation of demand
- any special conditions
- the requirements for control, signalling, communication and information technology, etc. (131-03-01).

9.6 Identification and notices

For safety purposes, the Regulations require a number of notices and labels to be used for electrical installations and these will need to be checked during initial and periodic inspections.

9.6.1 General

Verify that:

- labels (or other means of identification) indicate the purpose of each item of switchgear and control gear (514-01-01)
- wiring is marked and/or arranged so that it can be quickly identified for inspection, testing, repair or alteration of the installation (514-01-02)
- unambiguous marking has been provided at the interface between conductors identified in accordance with these Regulations and conductors identified to previous versions of the Regulations (514-01-03)

Note: Appendix 7 of the Regulations provides guidance on how this can be achieved.

- orange coloured conduits have been used to distinguish an electrical conduit from other services or other pipelines (514-02-01).

9.6.2 Conductors

Verify that:

- conductor cable cores are identified by colour and/or lettering and/or numbering (514-03-01)
- binding and sleeves used for identifying protective conductors comply with BS 3858 (514-03-02)
- neutral or mid-point conductors are coloured blue (514-04-01)
- protective conductor cable cores are identifiable at all terminations (and preferably throughout their length) (514-03-02)
- protective conductors are a bi-colour combination of green and yellow and that neither colour covers more than 70% of the surface being coloured (514-04-02)

This combination of colours shall **not** be used for any other purpose.

- single-core cables used as protective conductors are coloured green-and-yellow throughout their length (514-04-02).
- bare conductors or busbars used as protective conductors are identified by equal green and yellow stripes that are 15 mm to 100 mm wide (514-04-02)

If adhesive tape is used, then it shall be bi-coloured.

- PEN conductors (when insulated) shall either be:
  - green-and-yellow throughout their length, with blue markings at the terminations, or
  - blue throughout its length, with green-and-yellow markings at the terminations (514-04-03)
- bare conductors are painted or identified by a coloured tape, sleeve or disc as per Table 9.21 (514-04-06).

ALL other conductors (including those used to identify conductors switch-board busbars and conductors) shall be coloured as shown in accordance with Table 9.21 (514-03-03 and 514-04-04).

The colour green shall **not** be used on its own (514-04-05).

Table 9.21 Identification of conductors

Function	Alphanumeric	Colour
Protective conductors		Green-and-yellow
Functional earthing conductor		Cream
a.c. power circuit (Note 1)		
Phase of single-phase circuit	L	Brown
Neutral of single- or three-phase circuit	N	Blue
Phase 1 of three-phase a.c. circuit	L1	Brown
Phase 2 of three-phase a.c. circuit	L2	Black
Phase 3 of three-phase a.c. circuit	L3	Grey
Two-wire unearthed d.c. power circuit		
Positive of two-wire circuit	L+	Brown
Negative of two-wire circuit	L−	Grey
Two-wire earthed d.c. power circuit		
Positive (of negative earthed) circuit	L+	Brown
Negative (of negative earthed) circuit (Note 2)	M	Blue
Positive (of positive earthed) circuit (Note 2)	M	Blue
Negative (of positive earthed) circuit	L−	Grey
Three-wire d.c. power circuit		
Outer positive of two-wire circuit derived from three-wire system	L+	Brown
Outer negative of two-wire circuit derived from three-wire system	L−	Grey
Positive of three-wire circuit	L+	Brown
Midwire of three-wire circuit (Notes 2 & 3)	M	Blue
Negative of three-wire circuit	L−	Grey
Control circuits, ELV and other applications		
Phase conductor	L	Brown, black, red, orange, yellow, violet, grey, white, pink or turquoise
Neutral or midwire (Note 4)	N or M	Blue

Notes:
(1) Power circuits include lighting circuits.
(2) M identifies either the midwire of a three-wire d.c. circuit, or the earthed conductor of a two-wire earthed d.c. circuit.
(3) Only the middle wire of three-wire circuits may be earthed.
(4) An earthed PELV conductor is blue.

In accordance with Part P of the Building Regulations:

- these new (harmonised) colour cables may be used on site from 31 March 2004
- new installations or alterations to existing installations may use either new or old colours (but **not** both!) from 31 March 2004 until 31 March 2006
- **only** the new colours may be used after 31 March 2006.

 Note: Part P applies only to fixed electrical installations that are intended to operate at low voltage or extra-low voltage which are not controlled by the Electricity Supply Regulations 1988 as amended, or the Electricity at Work Regulations 1989 as amended.

 Further information concerning cable identification colours for extra-low voltage and d.c. power circuits is available from the IEE website at www.iee.org/cablecolours.

9.6.3 Identification of conductors by letters and/or numbers

Where letters and/or numbers are used to identify conductors, check to confirm that:

- individual conductors and/or groups of conductors are identified by either letters or numbers that are clearly legible (514-05-01)
- all numerals contrast, strongly, with the colour of the insulation (514-05-01)
- numerals 6 and 9 are underlined (514-05-01)
- protective devices are arranged and identified so that the circuit protected is easily recognisable (514-08-01)
- protective conductors coloured green-and-yellow are not numbered other than for the purpose of circuit identification (514-05-02)
- the alphanumeric numbering system is in accordance with Table 9.20 (514-05-03).

9.6.4 Omission of identification by colour or marking

Colour or marking is **not** required for:

- concentric conductors of cables
- metal sheath or armour of cables (when used as a protective conductor)
- bare conductors (where permanent identification is impracticable)
- extraneous conductive parts used as a protective conductor
- exposed conductive parts used as a protective conductor (514-06-01).

9.6.5 Diagrams

Verify that all available diagrams, charts, tables or schedules that have been used, indicate:

- the type and composition of each circuit (points of utilisation served, number and size of conductors, type of wiring)

- the method used for protection against indirect contact
- the data (where appropriate) and characteristics of each protective device used for automatic disconnection
- the identification (and location) of all protection, isolation and switching devices
- the circuits or equipments that are susceptible to a particular test.

Distribution board schedules must be provided within (or adjacent to) each distribution board.

All symbols used shall comply with BS EN 60617 (see Appendix C) (514-09-01).

9.6.6 Warning notices

Check to confirm that the following warning notices are appropriate to the situation, correctly positioned and contain the right information.

Earth-free local equipotential bonding

Where earth-free local equipotential bonding has been used, confirm that an appropriate notice has been fixed in a prominent position adjacent to every point of access to the location concerned (471-12-01).

Earthing and bonding connections

Confirm that a permanent label with the words shown in Figure 9.15 has been permanently fixed at or near:

- the connection point of every earthing conductor to an earth electrode
- the connection point of every bonding conductor to an extraneous conductive part
- the main earth terminal (when separated from the main switchgear) (514-13-01).

Safety Electrical Connection – Do Not Remove

Figure 9.15 Earthing and bonding notice

Electrical installations in caravans, motor caravans and caravan parks

Confirm that:

- a notice has been fixed on (or near) the electrical inlet recess of the caravan or motor caravan installation containing details concerning the:
  - nominal (design) voltage and frequency
  - current rating (608-07-03)

- a notice (worded as shown in Figure 9.16) has been permanently fixed near the main isolating switch inside the caravan or motor caravan (608-07-05).

1.3 INSTRUCTIONS FOR ELECTRICITY SUPPLY

TO CONNECT

1. Before connecting the caravan installation to the mains supply, check that:
   (a) the supply available at the caravan pitch supply point is suitable for the caravan electrical installation and appliances, and
   (b) the caravan main switch is in the OFF position.

2. Open the cover to the appliance inlet provided at the caravan supply point and insert the connector of the supply flexible cable.

3. Raise the cover of the electricity outlet provided on the pitch supply point and insert the plug of the supply cable.

THE CARAVAN SUPPLY FLEXIBLE CABLE MUST BE FULLY UNCOILED TO AVOID DAMAGE BY OVERHEATING.

4. Switch on at the caravan main switch.

5. Check the operation of residual current devices, if any, fitted in the caravan by depressing the test buttons.

IN CASE OF DOUBT OR, IF AFTER CARRYING OUT THE ABOVE PROCEDURE THE SUPPLY DOES NOT BECOME AVAILABLE, OR IF THE SUPPLY FAILS, CONSULT THE CARAVAN PARK OPERATOR OR THE OPERATOR'S AGENT OR A QUALIFIED ELECTRICIAN.

TO DISCONNECT

6. Switch off at the caravan main isolating switch, unplug both ends of the cable.

1.4 PERIODIC INSPECTION

Preferably not less than once every three years and more frequently if the vehicle is used more than normal average mileage for such vehicles, the caravan electrical installation and supply cable should be inspected and tested and a report on their condition obtained as prescribed in BS 7671 Requirements for Electrical Installations published by the Institution of Electrical Engineers.

Figure 9.16 Electrical safety notice

Emergency switching

For any exterior installation for emergency switching, confirm that the switch is placed outside the building, adjacent to the equipment (where this is not possible, a notice showing the position of the switch shall be placed adjacent to the equipment and a notice fixed near the switch shall indicate its use) (476-03-07).

Final circuit distribution boards

Confirm that distribution boards for socket outlet final circuits have a notice that clearly indicates circuits which have a high protective conductor current and that this information is positioned so as to be plainly visible to a person employed in modifying or extending the circuit (607-03-02).

Inspection and testing

Confirm that:

- a notice has been fixed in a prominent position at or near the origin of every installation upon completion of the work (e.g. initial verification, alterations and additions to an installation and periodic inspection and testing)
- the notice has been inscribed in indelible characters and reads as shown on p. 88 of the IEE Wiring Regulations (514-12-01)
- if an installation includes a residual current device then it has a notice (fixed in a prominent position) that reads as shown on p. 89 of the IEE Wiring Regulations (514-12-02)
- following initial verification, an Electrical Installation Certificate (together with a schedule of inspections and a schedule of test results) was given to the person ordering the work (742-01-01) and that:
- the schedule of test results identified every circuit, its related protective device(s) and recorded the results of the appropriate tests and measurements (742-01-02)
- the Certificate took account of the respective responsibilities for the safety of that installation and the relevant schedules (742-01-03)
- defects or omissions revealed during inspection and testing of the installation work covered by the Certificate were rectified before the Certificate is issued (742-01-04).

Confirm that:

- an Electrical Installation Certificate (containing details of the installation together with a record of the inspections made and the test results) or a Minor Electrical Installation Works Certificate (for all minor electrical installations that do not include the provision of a new circuit) was provided for all alterations or additions to electrical circuits (743-01-01) and that
- any defects found in an existing installation were recorded on an Electrical Installation Certificate or a Minor Electrical Installation Works Certificate (743-01-02).

Isolation

Verify that a notice is fixed in each position where there are live parts which are not capable of being isolated by a single device (514-11-01).

Note: The location of each isolator shall be indicated unless there is no possibility of confusion.

If an installation is supplied from more than one source confirm that:

- a main switch is provided for each source of supply
- a notice is placed warning operators that more than one switch needs to be operated (460-01-02).

Unless an interlocking arrangement is provided, confirm that a notice has been provided warning people that they will need to use the appropriate isolating devices if an equipment or enclosure that contains live parts, cannot be isolated by a single device (461-01-03).

Voltage

Verify that:

- items of equipment (or enclosures) within which a nominal voltage exceeding 230 V exists (and where the presence of such a voltage would not normally be expected) are arranged so that before access is gained to a live part, a warning of the maximum voltage present is clearly visible
- where terminals (or other fixed live parts between which a nominal voltage exceeding 230 V exists) are housed in separate enclosures (or items of equipment which, although separated, can be reached simultaneously by a person) a notice has been secured in a position such that anyone, before gaining access to such live parts, is warned of the maximum voltage which exists between those parts
- means of access to all live parts of switchgear and other fixed live parts where different nominal voltages exist are marked to indicate the voltages present (514-10-01).

Warning notice – non-standard colours

If an installation that was wired to a previous version of the Regulations is partially altered or rewired according to the current Regulations (i.e. in accordance with Table 9.20) confirm that a warning notice (for correct wording see p. 89 of the IEE Wiring Regulations) has been placed at (or near) the appropriate distribution board (514-14-01).

9.7 What type of certificates and reports are there?

There are three types of Certificates associated with electrical installations. These are:

The Electrical Installation Certificate (See Figures 9.21 and 9.22)	For the design, construction, inspection and testing of electrical installation work.

The Minor Works Certificate
(See Figures 9.24)

For the inspection and testing of an electrical installation.

The Periodic Inspection Report
(See Figures 9.26 and 9.27)

For the regular inspection and testing of an electrical installation.

All certificates need to be made out and signed (or otherwise authenticated) by a competent person or persons. If the design, construction and inspection and testing is the responsibility of one person, then a certificate (as shown in Figure 9.17) may be used as a replacement for the multiple signatures section of the model form.

FOR DESIGN, CONSTRUCTION, INSPECTION & TESTING

I being the person responsible for the Design, Construction, Inspection & Testing of the electrical installation (as indicated by my signature below), particulars of which are described above, having exercised reasonable skill and care when carrying out the Design, Construction, Inspection & Testing, hereby CERTIFY that the said work for which I have been responsible is to the best of my knowledge and belief in accordance with BS 7671 : . . . . amended to . . . . . (date) except for the departures, if any, detailed as follows.

. . . .

. . . .

. . . .

Name: . . . .   Signature: . . . .   Date: . . . .

Figure 9.17 Single signature declaration form

Schedules of inspections and test results shall be issued along with the associated Electrical Installation Certificate and/or Periodic Inspection Report (see Figures 9.18 and 9.19).

9.7.1 Electrical Installation Certificate

The Electrical Installation Certificate (containing details of the installation together with a record of the inspections made and the test results) is only used for the initial certification of a new installation or for the alteration or addition to an existing installation where new circuits have been introduced (741-01-01).

SCHEDULE OF INSPECTIONS

Methods of protection against electric shock

(a) Protection against both direct and indirect contact:

☐ (i) SELV

☐ (ii) Limitation of discharge of energy

(b) Protection against direct contact:

☐ (i) Insulation of live parts

☐ (ii) Barriers or enclosures

☐ (iii) Obstacles

☐ (iv) Placing out of reach

☐ (v) PELV

☐ (vi) Presence of RCD for supplementary protection

(c) Protection against indirect contact:

☐ (i) EEBADS including:

☐ Presence of earthing conductor

☐ Presence of circuit protective conductors

☐ Presence of main equipotential bonding conductors

☐ Presence of supplementary equipotential bonding conductors

☐ Presence of earthing arrangements for combined protective and functional purposes

☐ Presence of adequate arrangements for alternative source(s), where applicable

☐ Presence of residual current device(s)

☐ (ii) Use of Class II equipment or equivalent insulation

☐ (iii) Non-conducting location: Absence of protective conductors

☐ (iv) Earth-free equipotential bonding: Presence of earth-free equipotential bonding conductors

☐ (v) Electrical separation

Prevention of mutual detrimental influence

☐ (a) Proximity of non-electrical services and other influences

☐ (b) Segregation of band I and band II circuits or band II insulation used

☐ (c) Segregation of safety circuits

Identification

☐ (a) Presence of diagrams, instructions, circuit charts and similar information

☐ (b) Presence of danger notices and other warning notices

☐ (c) Labelling of protective devices, switches and terminals

☐ (d) Identification of conductors

Cables and conductors

☐ (a) Routing of cables in prescribed zones or within mechanical protection

☐ (b) Connection of conductors

☐ (c) Erection methods

☐ (d) Selection of conductors for current-carrying capacity and voltage drop

☐ (e) Presence of fire barriers, suitable seals and protection against thermal effects

General

☐ (a) Presence and correct location of appropriate devices for isolation and switching

☐ (b) Adequacy of access to switchgear and other equipment

☐ (c) Particular protective measures for special installations and locations

☐ (d) Connection of single-pole devices for protection or switching in phase conductors only

☐ (e) Correct connection of accessories and equipment

☐ (f) Presence of undervoltage protective devices

☐ (g) Choice and setting of protective and monitoring devices for protection against indirect contact and/or overcurrent

☐ (h) Selection of equipment and protective measures appropriate to external influences

☐ (i) Selection of appropriate functional switching devices

Inspected by .... Date ....

Notes:
✓ to indicate an inspection has been carried out and the result is satisfactory
✗ to indicate an inspection has been carried out and the result was unsatisfactory
N/A to indicate the inspection is not applicable
LIM to indicate that, exceptionally, a limitation agreed with the person ordering the work prevented the inspection or test being carried out.

Figure 9.18 Schedule of inspections

The Certificate is **only** valid if accompanied by the Schedule of Inspections and the Schedule(s) of Test Results that clearly show when the first periodic inspection must be completed.

Note: The original Certificate shall be given to the person ordering the work (Regulation 742-01-03) and a duplicate retained by the contractor.

SCHEDULE OF TEST RESULTS

Contractor: ....

Test Date: ....

Signature ....

Method of protection against indirect contact: ....

Equipment vulnerable to testing: ....

Address/Location of distribution board: ....

....

Type of Supply: TN-S/TN-C-S/TT

Ze at origin: ....ohms

PFC: ....kA

Instruments

loop impedance: ....

continuity: ....

insulation: ....

RCD tester: ....

Description of Work: ....

Circuit Description	Overcurrent Device			Wiring Conductors		Continuity			Insulation Resistance		Pol ar it y	Earth Loop Imped- ance	Functional Testing		Remarks
	Short-circuit capacity; ......kA	Rating I_n		live	cpc	$(R_1 + R_2)^*$	R_2^*	R i n g	Live/ Live	Live/ Earth		Z_s	RCD time	Other	
	type	A		mm²	mm²	Ω	Ω	Ω	MΩ	MΩ		Ω	ms		
1	2	3		4	5	6	7	8	9	10	11	12	13	14	15

Test Results

Deviations from Wiring Regulations and special notes:

* Complete column 6 or 7.

Figure 9.19 Schedule of test results

The following notice shall be attached to the certificate.

GUIDANCE FOR RECIPIENTS

This safety Certificate has been issued to confirm that the electrical installation work to which it relates has been designed, constructed and inspected and tested in accordance with British Standard 7671 (the IEE Wiring Regulations).

You should have received an original Certificate and the contractor should have retained a duplicate Certificate.

The 'original' Certificate should be retained in a safe place and be shown to any person inspecting or understanding further work on the electrical installation in the future. If you later vacate the property, this Certificate will demonstrate to the new owner that the electrical installation complied with the requirements of British Standard 7671 at the time the Certificate was issued. The Construction (Design and Management) Regulations require that for a project covered by those Regulations, a copy of this Certificate, together with schedules is included in the, project health and safety documentation.

For safety reasons, the electrical installation will need to be inspected at appropriate intervals by a competent person. The maximum time interval recommended before the next inspection is stated on Page 1 under 'Next Inspection'.

This Certificate is intended to be issued only for a new electrical installation or for new work associated with an alteration or addition to an existing installation. It should not have been issued for the inspection of an existing electrical installation. A 'Periodic Inspection Report' should be issued for such a periodic inspection.

Figure 9.20 Guidance notice to accompany the Electrical Installation Certificate

Electrical Installation Certificates:

- may be produced in any durable medium, including written and electronic media (741-01-05)
- shall be compiled and signed by a competent person(s) (741-01-04).

9.7.2 Minor Electrical Installation Works Certificate

A Minor Electrical Installation Works Certificate is used for additions and alterations to an installation such as an extra socket outlet or lighting point to an existing circuit, the relocation of a light switch etc. This Certificate may also be used for the replacement of equipment such as accessories or luminaries, but **not** for the replacement of distribution boards (or similar items) or the provision of a new circuit (741-01-03).

Form 2   Form No. 12

Electrical installation certificate (notes 1 and 2)

(REQUIREMENTS FOR ELECTRICAL INSTALLATIONS - BS 7671 [IEE WIRING REGULATIONS])

DETAILS OF THE CLIENT (note 1) ....
....
....

INSTALLATION ADDRESS
....
....
....Postcode.......................................

DESCRIPTION AND EXTENT OF THE INSTALLATION Tick boxes as appropriate (note 1)	
Description of installation:	New installation ☐
Extent of installation covered by this Certificate:	Addition to an existing installation ☐
....	Alteration to an existing installation ☐

FOR DESIGN

I/We being the person(s) responsible for the design of the electrical installation (as indicated by my/our signatures below), particulars of which are described above, having exercised reasonable skill and care when carrying out the design, hereby CERTIFY that the design work for which I/we have been responsible is to the best of my/our knowledge and belief in accordance with BS 7671: ...., amended to .... (date) except for the departures, if any, detailed as follows:

> Details of departures from BS 7671 (Regulations 120-01-03, 120-02):

The extent of liability of the signatory or the signatories is limited to the work described above as the subject of this Certificate. For the DESIGN of the installation.   **(Where there is mutual responsibility for the design)

Signature: .... Date .... Name (BLOCK LETTERS): .... Designer No. 1

Signature: .... Date .... Name (BLOCK LETTERS): .... Designer No. 2**

FOR CONSTRUCTION

I/We being the person(s) responsible for the construction of the electrical installation (as indicated by my/our signatures below), particulars of which are described above, having exercised reasonable skill and care when carrying out the construction, hereby CERTIFY that the construction work for which I/we have been responsible is to the best of my/our knowledge and belief in accordance with BS 7671: ....amended to ....(date) except for the departures, If any, detailed as follows:

> Details of departures from BS 7671 (Regulations 120-01-03, 120-02):

The extent of liability of the signatory is limited to the work described above as the subject of this Certificate. For CONSTRUCTION of the installation:

Signature: .... Date ....

Name (BLOCK LETTERS): .... Constructor

FOR INSPECTION & TESTING

I/We being the person(s) responsible for the inspection & testing of the electrical installation (as indicated by my/our signatures below), particulars of which are described above, having exercised reasonable skill and care when carrying out the inspection & testing, hereby CERTIFY that the work for which I/we have been responsible is to the best of my knowledge and belief in accordance with BS 7671:........., amended to .... (date) except for the departures, if any, detailed as follows:

> Details of departures from BS 7671 (Regulations 120-01-03, 120-02):

The extent of liability of the signatory is limited to the work described above as the subject of this Certificate. For INSPECTION & TEST of the installation: **(Where there is mutual responsibility for the design)

Signature: .... Date ....

Name (BLOCK LETTERS): .... Inspector

NEXT INSPECTION (notes 4 and 7)

I/We the designer(s) recommend that this installation is further inspected and tested after an interval of not more than............ years/months

Figure 9.21 Electrical Installation Certificate

PARTICULARS OF THE SIGNATORIES TO THE ELECTRICAL INSTALLATION CERTIFICATE (note 3)

Designer (No. 1)
Name: .... Company: ....
Address: ....
.... Postcode: .... Tel No.: ....

Designer (No. 2)
(if applicable) Name: .... Company: ....
Address: ....
.... Postcode: .... Tel No.: ....

Constructor
Name: .... Company: ....
Address: ....
.... Postcode: .... Tel No.: ....

Inspector
Name: .... Company: ....
Address: ....
.... Postcode: .... Tel No.: ....

SUPPLY CHARACTERISTICS AND EARTHING ARRANGEMENTS Tick boxes and enter details, as appropriate

Earthing arrangements	Number and Type of Live Conductors		Nature of Supply Parameters	Supply Protective Device Characteristics
TN-C ☐	a.c. ☐	d.c. ☐	Nominal voltage, $U/U_o^{(1)}$ V	
TN-S ☐	1-phase, 2-wire ☐	2-pole ☐	Nominal frequency, $f^{(1)}$ .... Hz	Type: ....
TN-C-S ☐	1-phase, 3-wire ☐	3-pole ☐	Prospective fault current, $Ipf^{(2)}$ (note 6)...... kA	....
TT ☐	2-phase, 3-wire ☐	other ☐	External loop impedance, $Ze^{(2)}$..........Ω	
IT ☐	3-phase, 3-wire ☐		(Note: (1) by enquiry, (2) by enquiry or by measurement)	Nominal current rating
	3-phase, 4-wire ☐			..........A
Alternative source ☐ of supply (to be detailed on attached schedules)				

PARTICULARS OF INSTALLATION REFERRED TO IN THE CERTIFICATE Tick boxes and enter details, as appropriate

Means of Earthing

Distributor's facility ☐

Maximum Demand
Maximum demand (load) ....Amps per phase

Details of Installation Earth Electrode (where applicable)

Installation earth electrode ☐

Type (e.g. rod(s), tape etc.) ....
Location ....
Electrode resistance to earth ....Ω

Main Protective Conductors

Earthing conductor: material .... csa ....mm² connection verified ☐

Main equipotential bonding conductors: material .... csa ....mm² connection verified ☐

To incoming water and/or gas service ☐   To other elements ....

Main Switch or Circuit-breaker

BS. Type .... No. of poles .... Current rating ....A Voltage rating V

Location .... Fuse rating or setting ....A

Rated residual operating current $I_{\Delta n}$ = ... mA, and operating time of ....ms (at $I_{\Delta n}$)
(applicable only where an RCD is suitable and is used as a main circuit-breaker)

COMMENTS ON EXISTING INSTALLATION: (In the case of an alternation or addition see Section 743)

....
....
....
....

SCHEDULES (note 2)
The attached Schedules are part of this document and this Certificate is valid only when they are attached to it.
.... Schedules of Inspections and .... Schedules of Test Results are attached. (Enter quantities of schedules attached)

Figure 9.22 Electrical Installation Certificate (continued)

The following notice shall be attached to the certificate.

MINOR ELECTRICAL INSTALLATION WORKS CERTIFICATE GUIDANCE FOR RECIPIENTS

This Certificate has been issued to confirm that the electrical installation work to which it relates has been designed, constructed and inspected and tested in accordance with British Standard 767 1, (the IEE Wiring Regulations).

You should have received an 'original' Certificate and the contractor should have retained a duplicate. If you were the person ordering the work, but not the owner of the installation, you should pass this Certificate, or a copy of it, to the owner. A separate Certificate should have been received for each existing circuit on which minor works have been carried out. This Certificate is not appropriate if you requested the contractor to undertake more extensive installation work, for which you should have received an Electrical Installation Certificate.

The Certificate should be retained in a safe place and be shown to any person inspecting or undertaking further work on the electrical installation in the future. If you later vacate the property, this Certificate will demonstrate to the new owner that the minor electrical installation work carried out complied with the requirements of British Standard 7671 at the time the Certificate was issued.

Figure 9.23 Guidance notice to accompany the Minor Electrical Installation Certificate

Minor Electrical Installation Works Certificates:

- may be produced in any durable medium, including written and electronic media (741-01-05)
- shall be compiled and signed by a competent person(s) (741-01-04).

9.7.3 Periodic inspection

A Periodic Inspection Report is used for reporting on the condition of an existing installation and shall include schedules of both the inspection and the test results (741-01-02).

The following notice shall be attached to the certificate on completion of the inspection.

Periodic Inspection Reports:

- may be produced in any durable medium, including written and electronic media (741-01-05)
- shall be compiled and signed by a competent person(s) (741-01-04).

The inspection and test schedule must include a verification that a Periodic Inspection Report (together with a schedule of inspections and a schedule of

MINOR ELECTRICAL INSTALLATION WORKS CERTIFICATE
(REQUIREMENTS FOR ELECTRICAL INSTALLATIONS - BS 7671 [IEE WIRING REGULATIONS])

To be used only for minor electrical work which does not include the provision of a new circuit

---

PART 1: Description of minor works

1. Description of the minor works: ....

2. Location/Address: ....

3. Date minor works completed: ....

4. Details of departures, if any, from BS 7671

....

....

....

PART 2: Installation details

1. System earthing arrangement: TN-C-S ☐   TN-S ☐   TT ☐

2. Method of protection against indirect contact: ....

3. Protective device for the modified circuit: Type BS ....   Rating ....A

4. Comments on existing installation, including adequacy of earthing and bonding arrangements:
(see Regulation 130-07) ....

....

....

....

PART 3: Essential Tests

1. Earth continuity: satisfactory ☐

2. Insulation resistance:

                            Phase/neutral ....$M\Omega$

                            Phase/earth ....$M\Omega$

                            Neutral/earth ....$M\Omega$

3. Earth fault loop impedance: ....Ω

4. Polarity: satisfactory   ☐

5. RCD operation (if applicable): Rated residual operating current $I_{\Delta n}$ ....mA and operating time of ....ms (at $I_{\Delta n}$)

PART 4: Declaration

1. I/We CERTIFY that the said works do not impair the safety of the existing installation, that the said works have been designed, constructed, inspected and tested in accordance with BS 7671: .... (IEE Wiring Regulations), amended to .... and that the said works, to the best of my/our knowledge and belief, at the time of my/our inspection, complied with BS 7671 except as detailed in Part 1.

2. Name: ....   3.   Signature: ....

    For and on behalf of: ....   Position: ....

    Address: ....

    ....   Date: ....

    ....

    ....Postcode: ....

Figure 9.24 Minor Electrical Installation Certificate

test results) was given to the person ordering the inspection (744-01-01) and that this Report included details of:

- any damage, deterioration, defects, dangerous conditions and non-compliance with the requirements of the regulations which may give rise to danger (744-01-02)
- any limitations of the inspection and testing (744-01-02).

PERIODIC INSPECTION REPORT – GUIDANCE FOR RECIPIENTS

This Periodic Inspection Report form is intended for reporting on the condition of an existing electrical installation.

You should have received an original Report and the contractor should have retained a duplicate. If you were the person ordering this Report, but not the owner of the installation, you should pass this Report, or a copy of it, immediately to the owner.

The original Report is to be retained in a safe place and be shown to any person inspecting or undertaking work on the electrical installation in the future. If you later vacate the property, this Report will provide the new owner with details of the condition of the electrical installation at the time the Report was issued.

The 'Extent and Limitations' box should fully identify the extent of the installation covered by this Report and any limitations on the inspection and tests. The contractor should have agreed these aspects with you and with any other interested parties (Licensing Authority, Insurance Company, Building Society etc.) before the inspection was carried out.

The Report will usually contain a list of recommended actions necessary to bring the installation up to the current standard. For items classified as 'requires urgent attention', the safety of those using the installation may be at risk, and it is recommended that a competent person undertakes the necessary remedial work without delay.

For safety reasons, the electrical installation will need to be re-inspected at appropriate intervals by a competent person. The maximum time interval recommended before the next inspection is stated in the Report under 'Next Inspection.'

Figure 9.25 Guidance notice to accompany the Periodic Inspection Report

9.8 Building Regulations

9.8.1 Mandatory requirements

Part P – Electrical safety

Confirm that:

- reasonable provision has been made in the design, installation, inspection and testing of electrical installations to protect persons from fire or injury (Approved Document P1)
- sufficient information has been provided so that persons wishing to operate, maintain or alter an electrical installation can do so with reasonable safety (Approved Document P2).

PERIODIC INSPECTION REPORT FOR AN ELECTRICAL INSTALLATION
(REQUIREMENTS FOR ELECTRICAL INSTALLATIONS - BS 7671 [IEE WIRING REGULATIONS])

DETAILS OF THE CLIENT

Client: ....

Address: ....

Purpose for which this Report is required: ....

DETAILS OF THE INSTALLATION Tick boxes as appropriate

Occupier: ....

Installation: ....

Address: ....

Description of Premises: Domestic ☐   Commercial ☐   Industrial ☐

                             Other ☐

Estimated age of the Electrical   .... years
Installation:

Evidence of Alterations or Additions: Yes ☐   No ☐   Not apparent ☐

If "Yes", estimate age: .... years

Date of last inspection: ....   Records available   Yes ☐   No ☐

EXTENT AND LIMITATIONS OF THE INSPECTION

Extent of electrical installation covered by this report: ....

....

....

Limitations: ....

....

....

This inspection has been carried out in accordance with BS 7671 : 2001 (IEE Wiring Regulations), amended to ....
Cables concealed within trunking and conduits, or cables and conduits concealed under floors, in roof spaces and generally
within the fabric of the building or underground have not been inspected.

NEXT INSPECTION

I/We recommend that this installation is further inspected and tested after an interval of not more than .... months/years,
provided that any observations 'requiring urgent attention' are attended to without delay.

DECLARATION

INSPECTED AND TESTED BY

Name: ....   Signature: ....

For and on behalf of: ....   Position: ....

Address: ....

....   Date: ....

....

Figure 9.26 Periodic Inspection Report

Part M – Access and facilities for disabled people

Confirm that (in addition to the requirements of the Disability Discrimination
Act 1995) precautions have been taken to ensure that:

- new non-domestic buildings and/or dwellings (e.g. houses and flats used
  for student living accommodation etc.)
- extensions to existing non-domestic buildings

SUPPLY CHARACTERISTICS AND EARTHING ARRANGEMENTS	Tick boxes and enter details, as appropriate		
Earthing arrangements	Number and Type of Live Conductors	Nature of Supply Parameters	Supply Protective Device Characteristics
TN-C ☐	a.c. ☐ d.c. ☐	Nominal voltage, $U/U_o^{(1)}$ V	Type: ....
TN-S ☐	1-phase, 2-wire ☐ 2-pole ☐	Nominal frequency, $f^{(1)}$ .... Hz	
TN-C-S ☐	2-phase, 3-wire ☐ 3-pole ☐	Prospective fault current, $I_{pf}^{(2)}$ .... kA	Nominal current rating ....A
TT ☐	3-phase, 3-wire ☐ other ☐	External loop impedance, $Z_e^{(2)}$ ...... Ω	
IT ☐	3-phase, 4-wire ☐	(Note: (1) by enquiry, (2) by enquiry or by measurement)	

PARTICULARS OF INSTALLATION REFERRED TO IN THE REPORT Tick boxes and enter details, as appropriate

Means of Earthing	Details of Installation Earth Electrode (where applicable)		
Distributor's facility ☐	Type	Location	Electrode resistance
Installation earth electrode ☐	(e.g. rod(s), tape etc)	....	to earth .... Ω

Main Protective Conductors

Earthing conductor: material .... csa ....

Main equipotential bonding conductors   material .... csa ....

To incoming water service ☐   To incoming gas service ☐   To incoming oil service ☐   To structural steel ☐
To lightning protection ☐   To other incoming service(s) ☐   (state details...)

Main Switch or Circuit-breaker

BS, Type and number of poles ....   Current rating ....A   Voltage rating V

Location ....   Fuse rating or setting.................A

Rated residual operating current $I_{\Delta n}$ = ... mA, and operating time of .... ms (at $I_{\Delta n}$) (applicable only where an RCD is suitable and is used as a main circuit-breaker)

OBSERVATIONS AND RECOMMENDATIONS Tick boxes as appropriate	Recommendations as detailed below

Referring to the attached Schedule(s) of Inspection and Test Results, and subject to the limitations specified at the Extent and Limitations of the Inspection section
☐ No remedial work is required   ☐ The following observations are made:

....

....

....

....

....

....

One of the following numbers, as appropriate, is to be allocated to each of the observations made above to indicate to the person(s) responsible for the installation the action recommended.

☐1 requires urgent attention   ☐2 requires improvement   ☐3 requires further investigation

☐4 does not comply with BS 7671 : 2001 amended to .... This does not imply that the electrical installation inspected is unsafe.

SUMMARY OF THE INSPECTION

Date(s) of the inspection: ....

General condition of the installation: ....

....

....

Overall assessment: Satisfactory/Unsatisfactory

SCHEDULE(S)

The attached Schedules are part of this document and this Report is valid only when they are attached to it.
.... Schedules of Inspections and .... Schedules of Test Results are attached.
(Enter quantities of schedules attached).

Figure 9.27 Periodic Inspection Report (continued)

- non-domestic buildings that have been subject to a material change of use (e.g. so that they become a hotel, boarding house, institution, public building or shop)

are capable of allowing people, regardless of their disability, age or gender to:

- gain access to buildings
- gain access within buildings

- be able to use the facilities of the buildings (both as visitors and as people who live or work in them)
- use sanitary conveniences in the principal storey of any new dwelling

(Approved Document M).

Part L – Conservation of fuel and power

Confirm that energy efficiency measures have been provided which:

- ensure that lighting systems utilise energy-efficient lamps with:
  - manual switching controls or
  - automatic switching (in the case of external lighting fixed to the building) or
  - both manual and automatic switching controls
- so that the lighting systems can be operated effectively with regards the conservation of fuel and power.

Confirm that building occupiers have been supplied with sufficient information (including results of performance tests carried out during the works) to show how the heating and hot water services can be operated and maintained (Approved Document L1).

9.8.2 Inspection and test

Verify that all electrical installations have been inspected and tested during, at the end of installation and before they are taken into service and that they:

- are reasonably safe and that they comply with BS 7671: 2001 (P1.6)
- meet the relevant equipment and installation standards (P0.1b).

Confirm that all components that are part of an electrical installation have been inspected (during installation as well as on completion) to verify that the components have been:

- selected and installed in accordance with BS 7671 (P1.8a(ii))
- made in compliance with appropriate British Standards or harmonised European Standards (P1.8a(i))
- evaluated against external influences (such as the presence of moisture) (P1.8a(ii))
- checked to see that they have not been visibly damaged (or are defective) so as to be unsafe (P1.8a(iii))
- tested to check satisfactory performance with respect to continuity of conductors, insulation resistance, separation of circuits, polarity, earthing and bonding arrangements, earth fault loop impedance and functionality of all protective devices including residual current devices (P1.8b)

- inspected for conformance with section 712 of BS 7671: 2001 and:
  - been tested as per section 713 of BS 7671: 2001
  - been tested using appropriate and accurate instruments
  - had their test results recorded using the model in Appendix 6 of BS 7671
  - had their test results compared with the relevant performance criteria to confirm compliance (P1.9).

 Note: Inspections and testing of DIY work should **also** meet the above requirements (P1.10).

9.8.3 Extensions, material alterations and material changes of use

Where any electrical installation work is classified as an extension, a material alteration or a material change of use, confirm that:

- the existing fixed electrical installation in the building is capable of supporting the amount of additions and alterations that will be required (P2.1a)
- the earthing and bonding systems are satisfactory and meet the requirements (P2.1a–P2.2c)
- the mains supply equipment is suitable and can carry the additional loads envisaged (P2.1b–P2.2)
- any additions and alterations to the circuits which feed them comply with the requirements of the Regulations (P2.1a)
- the protective measures required to meet the requirements (P2.1a–P2.2b)
- the rating and the condition of existing equipment (belonging to both the consumer and the electricity distributor) is sufficient (P2.2a).

9.8.4 Design

Confirm that electrical installations have been designed and installed so that they:

- are suitably enclosed (and appropriately separated) to provide mechanical and thermal protection (P1.2)
- do not present an electric shock or fire hazard to people (P0.1a)
- meet the requirements of the Building Regulations (P3.1)
- provide adequate protection for persons against the risks of electric shock, burn or fire injuries (P1.2)
- provide adequate protection against mechanical and thermal damage (P0.1a).

9.8.5 Electricity distributor's responsibilities

Prior to starting works, confirm that the electricity distributor has:

- accepted responsibility for ensuring that the supply is mechanically protected and can be safely maintained (P1.3)

- evaluated and agreed the proposal for a new (or significantly altered) installation (P1.4)
- installed the cut-out and meter in a safe location (P1.3).

Confirm that the distributor has:

- provided an earthing facility for new connections (P3.8)
- maintained the supply within defined tolerance limits (P3.8)
- provided certain technical and safety information to the consumer to enable them to design their installation(s) (P3.8)
- ensured that their equipment on consumers' premises:
  - is suitable for its purpose
  - is safe in its particular environment
  - clearly shows the polarity of the conductors (P3.9).

9.8.6 Consumer units

Ensure that accessible consumer units have been fitted with a childproof cover or installed in a lockable cupboard (P1.5).

9.8.7 Earthing

Inspect and confirm that:

- electrical installations have been properly earthed (P App C1)
- lighting circuits include a circuit protective conductor (P App C)
- socket outlets which have a rating of 32 A or less, and which may be used to supply portable equipment for use outdoors, are protected by an RCD (P App C)
- the distributor has provided an earthing facility for all new connections (P3.8)
- new or replacement, non-metallic light fittings, switches or other components do not require earthing (e.g. non-metallic varieties) unless new circuit protective (earthing) conductors have been provided (P App C)
- socket outlets that will accept unearthed (2-pin) plugs do not use supply equipment that needs to be earthed (P App C).

9.8.8 Electrical installations

Verify (by inspection and test) that during installation, at the end of installation and before they are taken into service, all electrical installations:

- have been designed and installed (suitably enclosed and appropriately separated) to provide mechanical and thermal protection (P1.2)
- provide adequate protection for persons against the risks of electric shock, burn or fire injuries (P1.2)
- meet the requirements of the Building Regulations (P3.1).

Electrical installation work

Confirm that all electrical installation work:

- has been carried out professionally (P1.1)
- complies with the Electricity at Work Regulations 1989 as amended (P1.1)
- has been carried out by persons who are competent to prevent danger and injury while doing it, or who are appropriately supervised (P3.4a).

9.8.9 Wiring and wiring systems

Confirm that:

- cables concealed in floors and walls have (if required):
  - an earthed metal covering, or
  - are enclosed in steel conduit, or
  - have some form of additional mechanical protection
- cables to an outside building (e.g. garage or shed) if run underground, have been routed and positioned so as to give protection against electric shock and fire as a result of mechanical damage to a cable
- heat-resisting flexible cables (if required) have been supplied for the final connections to certain equipment (see maker's instructions) (P App A 2d).

Equipotential bonding conductors

Confirm that:

- main equipotential bonding conductors for water service pipes, gas installation pipes, oil supply pipes and certain other 'earthy' metalwork have been provided
- where there is an increased risk of electric shock (such as bathrooms and shower rooms), supplementary equipotential bonding conductors have been installed
- the minimum size of supplementary equipotential bonding conductors (without mechanical protection) is $4\,\text{mm}^2$ (App C).

9.8.10 Socket outlets

Confirm by inspection that:

- older types of socket outlet designed non-fused plugs are not connected to a ring circuit (P App C)
- RCD protection has been provided for all socket outlets which have a rating of 32 A or less and which may be used to supply portable equipment for use outdoors (P App C)
- switched socket outlets indicate whether they are 'ON' (P1.5 and M5.4i)

- socket outlets that will accept unearthed (2-pin) plugs are not used to supply equipment that needs to be earthed (P App C)
- the requirements for wall sockets listed in Table 9.22 have been met.

Table 9.22 Building Regulations requirements for wall sockets

Type of wall	Requirement	Section
Timber framed	Power points: • that have been set in the linings have a similar thickness of cladding behind the socket box • have not been placed back to back across the wall	E (p.14)
Solid masonry	Deep sockets and chases have not been used in separating walls	E 2.32
	The position of sockets has been staggered on opposite sides of the separating wall	E 2.32f
Cavity masonry	The position of sockets has been staggered on opposite sides of the separating wall	E 2.65e
	Deep sockets and chases have not been used in a separating wall	E 2.65d2
	Deep sockets and chases in a separating wall have not been placed back to back	E 2.65d2
Framed walls with absorbent material	Sockets have: • been positioned on opposite sides of a separating wall • not been connected back to back • been staggered a minimum of 150 mm edge to edge	E 2.146b E 2.146b2 E 2.146b2

Confirm (by inspection and testing) that all socket outlets used for lighting:

- are wall-mounted (M-4.30a and b)
- are easily reachable (M2-8.2)
- have been installed between 450 mm and 1200 mm from the finished floor level (M2-8.3)
- are located no nearer than 350 mm from room corners (M-4.30g)

and that:

- switched socket outlets indicate whether they are 'ON' (M-4.30k).

9.8.11 Switches

Ensure that all controls and switches:

- are easy to operate, visible and free from obstruction
- have been located between 750 mm and 1200 mm above the floor
- do not require the simultaneous use of both hands (unless necessary for safety reasons) to operate

and that:

- mains and circuit isolator switches clearly indicate whether they are 'ON' or 'OFF'
- individual switches on panels and on multiple socket outlets have been well separated (M-4.29)
- front plates should contrast visually with their backgrounds (M-5.4i and M-4.30m)
- controls that need close vision (e.g. thermostats) have been located between 1200 mm and 1400 mm above the floor (M-4.30f)
- the operation of switches, outlets and controls does not require the simultaneous use of both hands (unless necessary for safety reasons) (M-4.30j)
- where possible, light switches with large push pads have been used in preference to pull cords (M-5.3)
- the colours red and green have not been used in combination as indicators of 'ON' and 'OFF' for switches and controls (M-5.3 and M-4.28)
- all switches used for lighting:
  - are easily reachable (M2-8.2)
  - have been installed between 450 mm and 1200 mm from the finished floor level (M2-8.3).

Light switches

Confirm that light switches:

- have large push pads (in preference to pull cords)
- align horizontally with door handles
- are within the 900 to 1100 mm from the entrance door opening (M-4.30h and I)
- are located between 750 mm and 1200 mm above the floor (M-4.30c and d)
- are not coloured red and green (i.e. as a combination) as indicators for 'ON' and 'OFF' (M-5.3).

9.8.12 Telephone points and TV sockets

Confirm that all telephone points and TV sockets have been located between 400 mm and 1000 mm above the floor (or 400 mm and 1200 mm above the floor for permanently wired appliances) (M-4.30a and b).

9.8.13 Equipment and components

Emergency alarms

Test and inspect to ensure that:

- emergency assistance alarm systems have:
  - visual and audible indicators to confirm that an emergency call has been received
  - a reset control reachable from a wheelchair, WC, or from a shower/changing seat

- a signal that is distinguishable visually and audibly from the fire alarm (M-5.4h)
- emergency alarm pull cords should be:
  - coloured red
  - located as close to a wall as possible
  - have two red 50 mm diameter bangles (M-4.30e)
- front plates contrast visually with their backgrounds (M-4.30m)
- the colours red and green have not been used (in combination) to indicate 'ON' and 'OFF' for switches and controls (M-4.28).

Fire alarms

Verify (by test and inspection) that fire detection and fire warning systems have been properly designed, installed and maintained and that:

- all buildings have arrangements for detecting fire (B1-1.25)
- all buildings have been fitted with a suitable (electrically operated) fire warning system (in compliance with BS 5839) or have means of raising an alarm in case of fire (e.g. rotary gongs, handbells or shouting 'FIRE') (B1-1.26)
- fire alarms emit an audio and visual signal to warn occupants with hearing or visual impairments (M-5.4g)
- the fire warning signal is distinct from other signals which may be in general use (B1-1.28)
- in premises that are used by the general public (e.g. large shops and places of assembly) a staff alarm system (complying with BS 5839) has been used (B1-1.29).

Heat emitters

Check that heat emitters:

- are either screened or have their exposed surfaces kept at a temperature below 43°C (M-5.4j)
- that are located in toilets and bathrooms do not restrict:
  - the minimum clear wheelchair manoeuvring space
  - the space beside a WC used to transfer from the wheelchair to the WC (M-5.10p).

Portable equipment for use outdoors

Verify that RCDs have been provided for all socket outlets which have a rating of 32 A or less and which may be used to supply portable equipment for use outdoors (P App C).

Power-operated entrance doors

- Confirm that all power-operated doors have been provided with:
  - safety features to prevent injury to people who are struck or trapped (such as a pressure-sensitive door edge which operates the power switch)

- a readily identifiable (and accessible) stop switch
- a manual or automatic opening device in the event of a power failure where and when necessary for health or safety (K5-5.2d).

Confirm that:

- all doors to accessible entrances have been provided with a power-operated door opening and closing system if a force greater than 20 N is required to open or shut a door (M-2.13a)
- once open, all doors to accessible entrances are wide enough to allow unrestricted passage for a variety of users, including wheelchair users, people carrying luggage, people with assistance dogs, and parents with pushchairs and small children (M-2.13b)
- power-operated entrance doors:
  - have a sliding, swinging or folding action controlled manually (by a push pad, card swipe, coded entry, or remote control) or automatically controlled by a motion sensor or proximity sensor such as contact mat
  - open towards people approaching the doors (M-2.21a)
  - provide visual and audible warnings that they are operating (or about to operate) (M-2.21c)
  - incorporate automatic sensors to ensure that they open early enough (and stay open long enough) to permit safe entry and exit (M-2.21c)
  - incorporate a safety stop that is activated if the doors begin to close when a person is passing through (M-2.21b)
  - revert to manual control (or fail safe) in the open position in the event of a power failure (M-2.21d)
  - when open, do not project into any adjacent access route (M-2.21e)
- its manual controls are:
  - located between 750 mm and 1000 mm above floor level
  - operable with a closed fist (M-2.21f)
  - set back 1400 mm from the leading edge of the door when fully open (M-2.21g)
  - clearly distinguishable against the background (M-2.21g)
  - contrast visually with the background (M-2.19 and 2.21g).

9.8.14 Thermostats

Check that all controls that need close vision (e.g. thermostats) are located between 1200 mm and 1400 mm above the floor (M-4.30f).

Smoke alarms – dwellings

Confirm by test and inspection that smoke alarms have been positioned:

- in the circulation space within 7.5 m of the door to every habitable room (B1-1.14a)

- in the circulation spaces between sleeping spaces and places where fires are most likely to start (e.g. kitchens and living rooms) (B1-1.14a)
- on every storey of a house (including bungalows) (B1-1.12).

Confirm by test and inspection that:

- kitchen areas that are not separated from the stairway or circulation space by a door have been equipped with an additional heat detector in the kitchen that is interlinked to the other alarms (B1-1.14b)
- if more than one smoke alarm has been installed in a dwelling then they have been linked so that if a unit detects smoke it will operate the alarm signal of all the smoke detectors (B1-1.13).

Verify by inspection that smoke alarms:

- have ideally been mounted 25 mm and 600 mm below the ceiling (25–150 mm in the case of heat detectors) and at least 300 mm from walls and light fittings (B1-1.14c-d)
- have not been fixed over a stair shaft or any other opening between floors (B1-1.15)
- have not been fitted:
  - in places that get very hot (such as a boiler room)
  - in a very cold area (such as an unheated porch)
  - in bathrooms, showers, cooking areas or garages, or any other place where steam, condensation or fumes could give false alarms
  - next to or directly above heaters or air conditioning outlets
  - on surfaces which are normally much warmer or colder than the rest of the space (B1-1.16).

Test, inspect and confirm that the power supply for a smoke alarm system:

- has been derived from the dwelling's mains electricity supply via a single independent circuit at the dwelling's main distribution board (consumer unit) (B1-1.17)
- includes a stand-by power supply that will operate during mains failure (B1-1.17-18)
- is not (preferably) protected by an RCD (B1-1.20).

9.8.15 Lighting

External lighting fixed to the building

Confirm (by test and inspection) that all external lighting (including lighting in porches, but not lighting in garages and carports):

- automatically extinguishes when there is enough daylight, and when not required at night (L1-1.57a)

- have sockets that can only be used with lamps having an efficacy greater than 40 lumens per circuit watt (such as fluorescent or compact fluorescent lamp types, and **not** GLS tungsten lamps with bayonet cap or Edison screw bases) (L1-1.57b).

Fittings, switches and other components

Confirm that new or replacement, non-metallic light fittings, switches or other components that require earthing (e.g. non-metallic varieties) have been provided with new circuit protective (earthing) conductors (P App C).

Fixed lighting

Ensure that in locations where lighting can be expected to have most use, fixed lighting (e.g. fluorescent tubes and compact fluorescent lamps – but not GLS tungsten lamps with bayonet cap or Edison screw bases) with a luminous efficacy greater than 40 lumens per circuit watt have been made available (L1-1.54).

Lighting circuits

Verify that all lighting circuits include a circuit protective conductor (P App C).

9.8.16 Lecture/conference facilities

In lecture halls and conference facilities, confirm that artificial lighting has been designed to:

- give good colour rendering of all surfaces (M-4.9 and M-4.34)
- be compatible with other electronic and radio-frequency installations (M-4.12.1 and M-4.36f).

9.8.17 Cellars or basements

Ensure that LPG storage vessels and LPG fired appliances that are fitted with automatic ignition devices or pilot lights have not been installed in cellars or basements (J-3.5i).

9.9 What about test equipment?

Although BS 7671 lays great emphasis on the requirements for inspection and testing in Chapter 7 of the Regulations, the only reference (as far as I can tell!) to actual test equipment concerns insulation-monitoring devices which state that:

- an insulation-monitoring device shall be so designed or installed that it can only be possible to modify the setting with the use of a key or a tool (531-06-01)

- in an IT system, an insulation-monitoring device shall be provided so as to indicate the occurrence of a first fault from a live part to an exposed conductive part, or to earth (413-02-24)
- installations forming part of an IT system can make use of an insulation-monitoring device (473-01-05).

Of course, the actual choice of test equipment that the electrician chooses to use will normally be based on personal preference and experience. Nevertheless, it is essential that any piece of test equipment (including software) that is used when installing or inspecting electrical installations for compliance with the Regulations can be relied on to produce accurate results.

ISO 9001: 2000 (i.e. the internationally recognised standard for quality management) specifies the requirements for the control of test equipment (although they actually refer to them as 'measuring and monitoring devices') as follows:

The organisation shall ensure that all measuring and monitoring devices are:

· *calibrated and adjusted periodically prior to use*

· *traceable to international or national standards*

· *safeguarded from adjustments that would invalidate the calibration*

· *protected from damage and deterioration during handling, maintenance and storage.*

Figure 9.28 Mandatory requirements from the Building Regulations

Proof	The controls that an organisation has in place to ensure that equipment (including software) used for conformance to specified requirements is properly maintained
Documentation	Equipment records of maintenance and calibration
	Work instructions

Although the majority of electricians probably work on an individual basis and the requirement to operate as an accredited and registered ISO 9001: 2000 company doesn't really apply, following the recommendations of this standard can only help to improve the quality of any organisation – no matter its size.

In general, therefore:

- all measuring and test equipment that is used by an electrician needs to be well maintained, in good condition and capable of safe and effective operation within a specified tolerance of accuracy
- all measuring and test equipment should be regularly inspected and/or calibrated to ensure that it is capable of accurate operation (and where necessary by comparison with external sources traceable back to national standards)
- any electrostatic protection equipment that is utilised when handling sensitive components is regularly checked to ensure that it remains fully functional
- the control of measuring and test equipment (whether owned by the electrician, on loan, hired or provided by the customer) should always include a check that the equipment:
  - is exactly what is required
  - has been initially calibrated before use
  - operates within the required tolerances
  - is regularly recalibrated, and that
  - facilities exist (either within the organisation or via a third party) to adjust, repair or recalibrate as necessary.

 If the measuring and test equipment is used to verify process outputs against a specified requirement, then the equipment needs to be maintained and calibrated against national and international standards and the results of any calibrations carried out **must** be retained and the validity of previous results reassessed if they are subsequently found to be out of calibration.

9.9.1 Control of inspection, measuring and test equipment

Measuring and test equipment should always be stored correctly and satisfactorily protected between use (to ensure their bias and precision) and should be verified and/or recalibrated at appropriate intervals.

9.9.2 Computers

 Special attention should be paid to computers if they are used in controlling processes and particularly to the maintenance and accreditation of any related software.

9.9.3 Software

Software used for measuring, monitoring and/or testing specified requirements should be validated prior to use.

9.9.4 Calibration

Without exception, all measuring instruments can be subject to damage, deterioration or just general wear and tear when they are in regular use. The electrician should, therefore, take account of this fact and ensure that all of his test equipment is regularly calibrated against a known working standard.

 The accuracy of the instrument will depend very much on what items it is going to be used to test and the frequency of use of the test instrument, and the electrician will have to decide on the maximum tolerance of accuracy for each item of test equipment.

Of course, calibrating against a 'working standard' is pretty pointless if that particular standard cannot be relied upon and so the workshop standard must also be calibrated, on a regular basis, at either a recognised calibration centre or at the UK Physical Laboratory against one of the national standards.

The electrician will, therefore, have to make allowances for:

- the calibration and adjustment of all measuring and test equipment that can affect product quality of their inspection and/or test
- the documentation and maintenance of calibration procedures and records
- the regular inspection of all measuring or test equipment to ensure that they are capable of the accuracy and precision that is required
- the environmental conditions being suitable for the calibrations, inspections, measurements and tests to be completed.

 If the instrument is found to be outside of its tolerance of accuracy, any items previously tested with the instrument must be regarded as suspect. In these circumstances, it would be wise to review the test results obtained from the individual instrument. This could be achieved by compensating for the extent of inaccuracy to decide if the acceptability of the item would be reversed.

9.9.5 Calibration methods

There are various possibilities, such as:

- sending all working equipment to an external calibration laboratory
- sending one of each item (i.e. a 'workshop standard') to a calibration laboratory, then sub-calibrate each working item against the workshop standard
- testing by attributes – i.e. take a known 'faulty' product, and a known 'good' product and then test each one to ensure that the test equipment can identify the faulty and good product correctly.

9.9.6 Calibration frequency

The calibration frequency depends on how much the instrument is used, its ability to retain its accuracy and how critical the items being tested are. Infrequently used instruments are often only calibrated prior to their use whilst frequently used items would normally be checked and re-calibrated at regular intervals depending, again, on product criticality, cost, availability etc. Normally 12 months is considered as about the maximum calibration interval.

9.9.7 Calibration ideals

- Each instrument should be uniquely identified, allowing it to be traced.
- The calibration results should be clearly indicated on the instrument.
- The calibration results should be retained for reference.
- The instrument should be labelled to show the next 'calibration due' date to easily avoid its use outside of the period of confidence.
- Any means of adjusting the calibration should be sealed, allowing easy identification if it has been tampered with (e.g. a label across the joint of the casing).

 Note: Examples of test equipment normally used by electricians is shown in Appendix 9.7.

Appendix 9.1: Examples of test equipment used to test electrical installations

The following are examples of instruments that are required to test electrical installations for compliance with the requirements of BS 7671.

 Quite a lot of test equipment manufacturers now produce dual or multifunctional instruments and so it is quite common to find an instrument that is capable of measuring a number of different types of tests – for example, continuity and insulation resistance, loop impedance and prospective fault current. It is, therefore, wise to carry out a little research before purchasing!

A9.1.1 Continuity tester

All protective and bonding conductors must be tested to ensure that they are electrically safe and correctly connected. Low-resistance ohmmeters and simple multimeters are normally used for continuity testing. Ideally they should have a no load voltage of between 4 V and 24 V, be capable of producing an a.c. or d.c. short-circuit voltage of not less than 200 mA and have a resolution of at least 0.01 milliohms.

| TM INS1600 Digital insulation/ continuity tester by TLC www. tlc-direct. co.uk |  | A compact, easy-to-read, battery-operated insulation and continuity tester covering a wide measuring range up to 2000 Mohm/ 1000 V, a.c. voltage (up to 600 V) and continuity beeper

Capable of testing all requirements of the 16th edition and guidance notes | • Easy to read 0.65 LCD display
• Data hold switch
• Power ON lock facility for hands-free operation
• Rotary switch for easy range selection
• Overload protection
• Dimensions: 165 × 100 × 57 mm
• a.c. voltage range: 600 V
• Batteries: 6 × AA
• Sampling rate: 2.5 times/sec
• Weight: 500 g |

Figure 9.29
Continually Testing

A9.1.2 Insulation Testing

A low resistance between phase and neutral conductors, or from live conductors to earth, will cause a leakage current which will cause weakening of the insulation, as well as involving a waste of energy which would increase the running costs of the installation. To overcome this problem, the resistance between poles or to earth must never be less than 0.5 mΩ for the usual supply voltages.

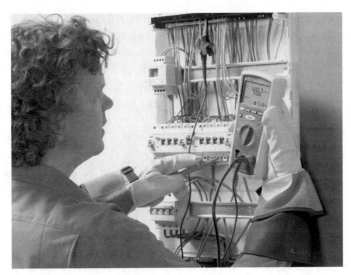

Figure 9.30 Insulation Tester

A9.1.3 Loop impedance tester

Loop testing is a quick, convenient, and highly specific method of testing an electrical circuit for its ability to engage protective devices (circuit breakers and fuses etc.) by simulating a fault from live to earth or from live to neutral (short circuit). The tester first measures the unloaded voltage, then connects a known resistance between the conductors, thereby simulating a fault. The voltage drop is measured across the known resistor, in series with the loop, and the proportion of the supply voltage that appears across the resistor will be dependent on the impedance of the loop.

KMP4120DL
by Robin
Electronics
www.robin
electronics.com

Figure 9.31 Loop Impedance Tester

An earth loop impedance tester with a 0.01 ohm resolution, capable of performing loop tests without tripping most passive RCDs

- 3 pre-selectable loop impedance ranges of 20, 200 & 2000 ohms
- 3 prospective short-circuit (PSC) ranges (200 A, 2000 A, & 20 kA)
- ability to test the majority of passive RCDs without tripping
- 'lock down' test button allowing 'hands free' operation

A9.1.4 RCD tester

The standard method for protecting electrical installations is to make sure that an earth fault results in a fault current that is high enough to operate the protective device quickly so that fatal shock is prevented. However, there are cases where the impedance of the earth-fault loop, or the impedance of the fault itself, are too high to enable enough fault current to flow. In such a case, either:

- the current will continue to flow to earth, perhaps generating enough heat to start a fire, or
- metalwork which can be touched may be at a high potential relative to earth, resulting in severe shock danger.

Either or both of these possibilities can be removed by the installation of a residual current device (RCD).

 Note: RCDs are also, sometimes, referred to as:

RCCD	Residual current operated circuit breaker
SRCD	Socket outlet incorporating an RCD
PRCD	Portable RCD, usually an RCD incorporated into a plug
RCBO	an RCCD which includes overcurrent protection
SRCBO	a socket outlet incorporating an RCBO

An RCD tester allows a selection of out-of-balance currents to flow through the RCD and cause its operation.

The RCD tester should not be operated for longer than 2 s.

| M-RC-70 RCD Tester by TLC www.tlc-direct.co.uk | A compact and simple instrument for monitoring both 10, 30, 100, 300 and 500 mA RCD devices as well as selective 100, 300 and 500 mA RCDs | • Display: LCD
• RCD current: 10, 30, 100, 300 mA
• Power supply: 230 V ± 10%
• Measurement range: 10–300 mA
• Test sequence: 10–300 mA
• Dimensions: 235 × 103 × 70 mm
• Weight: approx. 700 g |

Figure 9.32 RCD Tester

A9.1.5 Prospective fault current tester

A prospective fault current (PFC) tester is used to measure the prospective phase neutral fault current.

 It is usual to find PFC testers are part of a combined PFC/Loop Impendence tester.

| PROFiTEST 0100S-By GMC Instrumentation Ltd techinfo@gmciuk.com | An equipment designed to test prospective fault current as well as other insulation and impedance tests | • Overcurrent protection devices
• Loop and line impedance
• Earth resistance, earth leakage resistance
• Insulation resistance
• Phase sequence indicato |

Figure 9.33

A9.1.6 Test lamp or voltage indicator

These types of tester (often referred to as a *tetrascope* or *neon screwdriver*) are frequently used by electricians and might look like the examples shown in Figure 9.34.

These compact screwdriver multitesters are normally water and impact resistant, with a.c. voltage test, contact test 70–250 V a.c., non-contact 100–1000 V a.c., polarity test 1.5 V d.c. –36 V d.c., continuity check 0–5 ohm and auto power on/off.

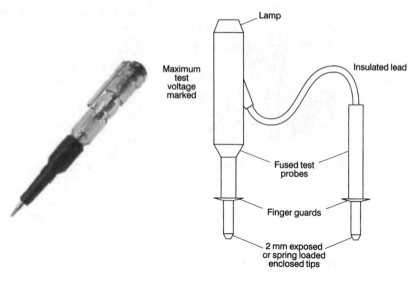

Lamp

Maximum
test
voltage
marked

Insulated lead

Fused test
probes

Finger guards

2 mm exposed
or spring loaded
enclosed tips

Figure 9.34 Typical test lamp and voltage indicators (courtesy TLC)

A9.1.7 Earth electrode resistance

The earth electrode (when used) is the means of making contact with the general mass of earth and should be regularly tested to ensure that good contact is made. In all cases, the aim is to ensure that the electrode resistance is not so high that the voltage from earthed metalwork to earth exceeds 50 V.

 Note: Acceptable electrodes are rods, pipes, mats, tapes, wires, plates and structural steelwork buried or driven into the ground. The pipes of other services such as gas and water must **not** be used as earth electrodes (although they must be bonded to earth).

Fluke 1620 Series GEO Earth Ground Testers By Fluke(UK) Ltd www. fluke.co.uk	The Fluke 1623 and 1625 are earth ground testers that can perform all four types of earth ground measurement. They can measure earth ground loop resistances using only clamps and do not require the use of earth ground stakes or the disconnection of ground rods.	• 3- and 4-pole ground measurement • 4-pole soil resistivity testing • 2-pole resistance measurement AC • 2- and 4-pole resistance measurement DC • Selective testing, no disconnection of ground conductor (1 clamp) • Stakeless testing, quick ground loop testing (2 clamps) • Earth impedance measurement at 55 Hz • Automatic frequency control (AFC) (94, 105, 111, 128 (Hz) • Measuring voltage switchable 20/48V

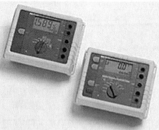

Figure 9.35 Earth electrode resistance tester

10

Installation, maintenance and repair

According to studies recently completed CEN/CENELEC, the installation and maintenance engineer is the primary cause of reliability degradations during the in-service stage of most electrical installations. The problems associated with poorly trained, poorly supported, and/or poorly motivated personnel with respect to reliability and dependability, therefore, requires careful assessment and quantification.

The most important factor, however, that affects the overall reliability of a modern product (system or installation) is the increased number of individual components that are required in that product. Since most system failures are actually caused by the failure of a single component (equipment or subassembly), the reliability of each individual component must be considerably better than the overall system reliability.

Because of this requirement, quality standards for the installation, maintenance, repair and inspection of in-service products have had to be laid down in engineering standards, handbooks and local operating manuals (written for specific items and equipment). These publications are used by maintenance engineers and should always include the most recent amendments. It is essential that assurance personnel also use the same procedures for their inspections as were used for the installation.

 Detailed methods for installing wiring systems (together with guidance for selecting the appropriate size of cable, current ratings and so on) are shown in Appendix 4, Tables 4A1 and 4A2 of the Regulations.

Although this final chapter is a comparatively small chapter (compared with some of the others in this book!), it is nevertheless an extremely important chapter as it provides some guidance on the requirements for installation, maintenance and repair. It also lists the Regulations' requirements for maintenance etc. with respect to electrical installations – but as per the previous chapters which contain similar lists, please remember that this is **only** the author's impression of the most important aspects of the Wiring Regulations and electricians should **always** consult BS 7671 to satisfy compliance. Finally, in Appendix 1 to Chapter 10 there is a complete set of checklists for the quality control of electrical equipment and electrical installations.

The Regulations devote a complete Part to inspection and testing (i.e. Part 7) and emphasise the need for continual improvement by stating:

All electrical installations shall be inspected and tested during erection and on completion before being put into service to verify (so far as is reasonably practicable) that the requirements of BS 7671 have been met.
711-01-01

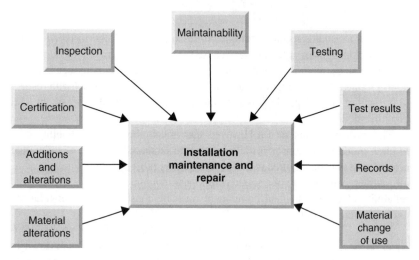

Figure 10.1 Installations, repair and maintenance

10.1 General

Installed equipment and their connections shall be accessible for operational inspection and maintenance purposes.
513-01-01

Wiring systems shall be selected and erected so as to minimise during installation, use and maintenance, damage to the sheath and insulation of cables and insulated conductors and their terminations.
522-08-01

Equipment shall be arranged to allow easy access for periodic inspection, testing and maintenance.
561-01-04

Connections and joints shall be accessible for inspection,   526-04-01
testing and maintenance, unless:

- they are in a compound-filled or encapsulated joint
- the connection is between a cold tail and a heating
  element
- the joint is made by welding, soldering, brazing or a
  compression tool.

Wiring systems shall be capable of: 522-03-01

- withstanding condensation and/or ingress of water
- minimising the ingress of solid foreign bodies

during installation, use and maintenance.   522-04-01

10.2 Inspection and testing

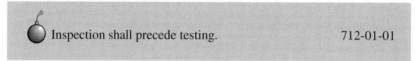

Inspection shall precede testing.   712-01-01

The aim of periodic inspection and testing is to:

- confirm that the installation is in a satisfactory condition for continued service
- confirm that the safety of the installation has not deteriorated or has been
  damaged
- ensure the continued safety of persons and livestock against the effects of
  electric shock and burns
- identify installation defects and non-compliance with the requirements of
  the Regulations which may give rise to danger
- protect property being damaged by fire and heat caused by a defective
  installation.

The Regulations are clear in the requirements for inspection and testing by stating:

'Inspection' shall consist of careful scrutiny of the   731-01-03
installation (dismantled or otherwise) using the appropriate
tests described in Chapter 71 of the Regulations.

The frequency of periodic inspection and testing of   732-01-01
installations will depend on:

- the type of installation, its use and operation
- the frequency and quality of maintenance
- the external influences to which it is subjected.

Precautions shall be taken to ensure that inspection and   731-01-05
testing does not cause:

- danger to persons or livestock
- damage to property and equipment (even if the circuit
  is defective).

Inspections are normally completed with the installation   712-01-01
disconnected from the supply.

The inspection shall be made to verify that the installed   712-01-02
electrical equipment:

- complies with the requirements of the Regulations and
  the appropriate National standards and European
  harmonised Directives
- is correctly selected and erected in accordance with the
  Regulations
- is not visibly damaged or defective so as to impair
  safety.

The inspection shall include the following items, where   712-01-03
relevant:

- access to switchgear and equipment
- cable routing in safe zones
- choice and setting of protective and monitoring devices
- connection of conductors
- connection of single-pole devices for protection or
  switching in phase conductors only
- correct connection of accessories and equipment
- erection methods
- identification of conductors
- labelling of protective devices, switches and terminals
- presence of danger notices and other warning signs
- presence of diagrams, instructions and similar
  information
- presence of fire barriers and suitable seals
- presence of isolation and switching devices (and their
  correct location)
- prevention of mutual (i.e. detrimental) influence
- presence of undervoltage protective devices
- protection against electric shock (direct and indirect
  contact)
- protection against external influences
- protection against direct and indirect contact
- protection against mechanical damage (and compliance
  with Section 522)

- protection against overcurrent
- protection against thermal effects
- selection of conductors for current-carrying capacity and voltage drop.

10.3 Maintenance

Equipment shall **not** be capable of becoming 462-01-03
unintentionally or inadvertently reactivated during
mechanical maintenance.

Generally speaking, the following CEN/CENELEC recommendations are relevant for all installed electrical and/or electronic equipment.

For ease of maintenance, all equipment provided should have:

- easily accessible test points to facilitate fault location
- modules that have been constructed so as to facilitate the connection of test equipment (e.g. logic analysers, emulators and test ROMs etc.)
- fault location provision to allow functional areas within each module or equipment to be isolated.

Under workshop conditions it should be possible to gain access to all circuitry while operating, with a minimum of effort required, to partially dismantle the module concerned with the assumption that there will be a minimum of risk to the components, or the testing maintenance staff.

Special connecting leads, printed wiring extension boards and any other special items required for maintenance purposes, together with the mating half of all necessary connectors, will probably have to be obtained from the manufacturer.

10.3.1 Design life

Equipment is normally expected to have been designed to have a useful life of not less than 20 years. '*Useful life*' normally means '*the period for which the equipment will continue to operate with the specified level of reliability*'.

No components, modules or equipment should, therefore, be used (so far as can be ascertained at the time of manufacture), for which spares cannot be fully guaranteed to be available throughout the life of the equipment.

Prior to commencing work, technicians and users need to be warned if an installation (that was wired in accordance with a previous version of these Regulations) has been altered and the Regulations are very specific on this point as can be seen below.

If an installation wired to a previous version of these   514-14-01
Regulations is partially altered with additions in accordance
with the current Regulations, then a warning notice (see
Figure 10.2) shall be placed at (or near) the appropriate
distribution board.

CAUTION

This installation has wiring colours to two versions of BS 7671.
Great care should be taken before undertaking extension, alteration or repair that all conductors are
correctly identified.

Figure 10.2 Wiring colours – caution sign

Other requirements for safe maintenance include and ensure that:

Care has been taken in reinstating protective measures that have had to be removed in order to carry out maintenance.	529-01-01
Safe access to all parts of the wiring system that may require maintenance has been provided.	529-01-02
If there is a risk of burns or injury from mechanical movement during maintenance, a switching off device has been provided, which is suitably located and identified.	462-01-01 and 462-01-02
Devices for switching off for mechanical maintenance (or a control switch for such a device):	537-03-01
• is either inserted in the main supply circuit or in the control circuit • is manually operated • have an externally visible contact gap or • clearly indicate the OFF or OPEN position.	537-03-02

10.4 Certification

The following certificates shall be completed where appropriate:

Electrical Installation Certificate (See Figures 9.21 and 9.22)	Details of the installation (together with a record of the inspections made and the test results) shall be provided for new installations and changes to existing installations.	741-01-01

| Periodic Inspection Report (See Figures 9.26 and 9.27) | Details of the installation (together with a record of the inspections made and the test results) shall be provided for all installations subject to periodic inspection and testing. | 741-01-02 |
| Minor Electrical Installation Works Certificate (See Figure 9.24) | For all minor electrical installation work that does not include the provision of a new circuit. | 741-01-03 |

> An Electrical Installation Certificate (containing details of   741-01-01
> the installation together with a record of the inspections
> made and the test results) shall be provided for new
> installations and changes to existing installations.

Note: Defects found in an existing installation shall be recorded on an Electrical Installation Certificate or a Minor Electrical Installation Works Certificate.

10.5 Records

> Any damage, deterioration, defects, dangerous   744-01-02
> conditions and non-compliance with the requirements
> of the Regulations which may give rise to danger, shall be
> recorded.

Records are an important part of the management of any electrical installation as they provide objective evidence of activities performed and/or results achieved, and the following requirements come from the Regulations:

> The schedule of test results shall identify every circuit, its   742-01-02
> related protective device(s) and be a record of the results
> of the appropriate tests and measurements.
>
> Defects or omissions revealed during inspection and testing   742-01-04
> of the installation work covered by the Certificate shall be
> rectified before the Certificate is issued.
>
> Any limitations of the inspection and testing shall be   744-01-02
> recorded.

10.6 Additions and alterations to an installation

As mentioned above, clause 711-01-01 of the Regulations requires every electrical installation to be inspected and tested during erection and on completion before being put into service to verify that the requirements of BS 7671 have been met. By definition, this requirement also applies to alterations and/or additions to an existing installation, as well as entirely new installations.

The following is a summary of the relevant requirements for additions and alterations to an installation.

10.6.1 General

Additions and/or alterations (whether temporary or permanent) shall not be made to an existing installation, until it has been confirmed that the rating and condition of existing equipment, earthing and bonding (including that belonging to the distributor) is adequate for the altered circumstances.   130-07-01

All alterations and/or additions to an existing installation: 721-01-02

- shall not impair the safety of an existing installation
- shall be verified for compliance with the Regulations.

All completed installations (including additions and/or alteration to existing installations) shall be inspected and tested for conformance to BS 7621.   133-02-01

Any work carried out on a building needs to ensure that the building not only complies with the requirements of the Regulations but its affect on the implementation of other Statutory Instruments such as: UK Statutory Instruments

- The Electricity at Work Regulations 1989
- The Fire Precautions Act 1971
- The Health and Safety at Work Act 1974
- The Disability Act 2003

and (in particular)

- The Building Act 1984.

Note: Building Regulations also ensure the health and safety of people in and around buildings by providing functional requirements for building design and construction.

If you want to put up a new building, extend or alter an existing one, or provide fixtures and fittings in a building, the Building Regulations will probably apply. They may also apply to certain changes of use of an existing building. (For more details see this book's sister publication *Building Regulations in Brief* (ISBN 0-7506-6703-6) and or Section 2 of this book.

10.6.2 Equipment

Electrical equipment used for alterations and additions:

• shall be capable of withstanding the stresses, environmental conditions and the characteristics of its location or be provided with additional protection	132-01-07
• that is likely to be exposed to weather, corrosive atmospheres or other adverse conditions shall be constructed (or protected) to prevent danger arising from such exposure	131-05-01
• that is vulnerable to risk of fire or explosion, shall be so constructed (or protected) to prevent, so far as is reasonably practicable, danger.	131-05-02

10.6.3 Additional protective methods

The degree of protection of equipment needs to be suitable for its intended use and mode of installation. Where this is not possible, additional protection must be provided as shown below.

Protection against direct and indirect contact

One of the requirements of the Regulations (namely 411-02-05) is that all live parts of a SELV system shall:

- be electrically separated from that of any other higher voltage system
- not be connected to earth
- not be connected to a live part or a protective conductor forming part of another system.

Where this proves impracticable:

SELV circuit conductors shall be enclosed in an insulating sheath additional to their basic insulation.	411-02-06

Protection against overvoltage

Additional protection against overvoltages of atmospheric origin is not necessary for

- installations that are supplied by low-voltage systems   443-02-01
  which do not contain overhead lines
- installations that are supplied by low-voltage networks   443-02-02
  which contain overhead lines and their location is
  subject to less than 25 thunderstorm days per year
- installations that contain overhead lines and their
  location is subject to less than 25 thunderstorm days
  per year

provided that they meet the required minimum equipment impulse withstand voltages shown in Table 44A of the IEE Wiring Regulations.

Protection against electric shock

For installations and locations with increased risk of shock from direct contact (such as those for caravans, swimming pools and saunas etc.) additional measures may be required such as:

- automatic disconnection of supply by means of a   471-08-01
  residual current device with a rated residual operating
  current (I_{sn}) not exceeding 30 mA
- supplementary equipotential bonding
- reduction of maximum fault clearance time.

For installations and locations where the risk of electric shock is increased by a reduction in body resistance and/or by contact with earth potential (and the use of extra-low voltage is impracticable and there is no requirement to use SELV) then a reduced low-voltage system (as specified in Regulations 471-15-02 to 471-15-07) may be used provided that:

the nominal voltage of reduced low-voltage circuits does   471-15-02
not exceed 110 V a.c. r.m.s between phases.

the supply source to a reduced low-voltage circuit shall be   471-15-03
from one of the following:

- a double wound isolating transformer complying with
  BS 3535-2, or

Note: This standard is partially replaced by the BS EN 61558 series of standards, and these standards should be read jointly.

- a motor-generator whose windings provide isolation equivalent to an isolating transformer, or
- a source independent of other supplies (e.g. an engine driven generator).

The neutral (i.e. star) point of the secondary windings of three-phase transformers and generators (or the mid-point of the secondary windings of single-phase transformers and generators) shall be connected to earth.	471-15-04

Protection of live parts against direct contact

In Section 471, the Regulations state that:

- live parts shall be completely covered with insulation which:
  - can only be removed by destruction
  - is capable of durably withstanding electrical, mechanical, thermal and chemical stresses normally encountered during service
- equipment shall be capable of withstanding all mechanical, chemical, electrical and thermal influences stresses normally encountered during service; and
- where insulation is applied during the erection of the installation, the quality of the insulation shall be verified.

Concerning the use of barriers, the Regulations also state that live parts shall be inside enclosures (or behind barriers) protected to at least IP2X or IPXXB and where this is not possible (e.g. because the opening needs to be larger or the enclosure will interfere with the proper functioning of the electrical equipment) then:

- precautions will need to be taken to prevent persons or livestock unintentionally touching a live part
- people are made aware that a live part can be touched through the opening (and that it should not be touched!).

If additional precautions are required, then:

The horizontal top surface of a barrier or an enclosure that is easily accessible shall be protected to at least 1P4X.	471-15-05 412-03-02
All barriers and/or enclosures shall:	471-15-05
• be firmly secured in place	412-03-03

- have sufficient stability and durability to maintain the required degree of protection and appropriate separation from any known live part.

If it is necessary to remove a barrier or open an enclosure or remove a part of an enclosure, the removal or opening shall be possible only: 471-15-05 and 412-03-04

- by using a key or tool
- after disconnecting the supply
- if an1P2X or IPXXB intermediate barrier is used.

 This Regulation does not apply to:

- a ceiling rose complying with BS 67
- a pull cord switch complying with BS 3676
- a bayonet lampholder complying with BS EN 61184
- an Edison screw lampholder complying with BS EN 60238.

Portable equipment

It should always be remembered that supplies for portable equipment used outdoors (where the risk of electric shock is increased by a reduction in body resistance and/or by contact with earth potential) need to ensure that:

circuits that supply portable equipment for use outdoors and which are not connected to a socket by means of flexible cable or cord (having a current-carrying capacity of 32 A or less) should be provided with supplementary protection (i.e. an RCD).   471-16-02

socket outlets rated at 32 A or less are provided with supplementary protection (i.e. an RCD) to reduce the risk associated with direct contact.   471-16-01

 This Regulation does not apply to socket outlets supplied by a circuit protected by:

- SELV
- electrical separation
- automatic disconnection and reduced low-voltage systems.

Self-contained luminaries

Self-contained luminaries (or circuits supplying luminaries) with an open circuit voltage exceeding low voltage require additional protection and this is covered by the following:

> Self-contained luminaries (or circuits supplying luminaries)   476-02-04
> with an open circuit voltage exceeding low voltage shall have:
>
> - an interlock (in addition to the switch normally used
>   for controlling the circuit) that automatically disconnects
>   the live parts of the supply before access to these live
>   parts is permitted
> - some form of isolator (in addition to the switch
>   normally used for controlling the circuit) to isolate the
>   circuit from the supply
> - a switch (with a lock or removable handle) or a
>   distribution board which can be locked.

When this is achieved by a lock or removable handle, the key or handle shall be non-interchangeable with any others used for a similar purpose within the premises.

If an isolating device for a particular circuit is some distance away from the equipment to be isolated, then that isolating device shall be capable of being secured in the open position.

10.6.4 Earth connections

To counteract the effects of thermal, thermomechanical and electromechanical stresses, earthing arrangements need to be sufficiently robust (or have additional mechanical protection) to external influences. The Regulations recognise this fact and state that for PEN conductors:

> The continuity of all joints in the outer conductor of a   546-02-05
> concentric cable (and at a termination of that joint) shall
> be supplemented by an extra conductor additional to any
> means used for sealing and clamping the outer conductor.

And for protective bonding conductors:

> Supplementary bonding not applied to a fixed appliance   547-03-04
> shall be provided by a supplementary conductor or a
> conductive part of a permanent and reliable nature.

10.6.5 Supplies

The following are additional requirements for installations where the generating set provides a switched alternative supply to the distributor's network (stand-by systems).

> When the generator is operating as a switched alternative   551-04-03
> to a TN system, protection through the automatic
> disconnection of the supply shall not rely on the connection
> to the earthed point of the distributor's network.
>
> Precautions shall be taken to ensure that the generator   551-06-01
> cannot operate in parallel with the distributor's network.

 Note: These precautions may include:

- an electrical, mechanical or electromechanical interlock between the operating mechanisms or control circuits of the changeover switching devices
- a system of locks with a single transferable key
- a three-position break-before-make changeover switch
- an automatic changeover switching device with a suitable interlock.

> For TN-S systems, any residual current device shall be   551-06-02
> positioned to avoid incorrect operation due to the existence
> of any parallel neutral earth path.

Additional requirements for installations incorporating static inverters require:

> supplementary equipotential bonding to be provided on the   551-04-04
> load side of the static converter when automatic
> disconnection cannot be completed with earthed equipotential
> bonding and automatic disconnection of the supply.

 Note: Where supplementary equipotential bonding is necessary it shall connect together the exposed conductive parts of equipment in the circuits concerned including the earthing contacts of socket outlets and extraneous conductive parts.

The following additional requirements are for installations where the generating set may operate in parallel with the distributor's network:

When using a generating set in parallel with a distributor's network, care shall be taken to avoid adverse effects in respect of: • power factor • voltage changes • harmonic distortion • unbalance, starting • synchronising or voltage fluctuation effects.	551-07-01
Protection shall be provided to disconnect the generating set from the distributor's network.	551-07-02
If the voltage and frequency of the distributor's network are outside the operational protection, the generating set shall be prevented from being connected to the distributor's network.	551-07-03
The generating shall be capable of being isolated from the distributor's network.	551-07-04

The following are additional requirements for protection by automatic disconnection where the installation and generating set are not permanently fixed.

Installations which are not permanently fixed (e.g. portable generating sets) shall be provided with protective conductors that are part of the cord or cable, between separate items of equipment.	551-04-06
In TN, TT and IT systems a residual current device (with a rated residual operating current not exceeding 30 mA) shall be installed.	551-04-06

 Note: For construction site installations, see Section 604 of the Regulations.

10.6.6 Residual current devices

All alterations that have been made to the setting or calibration of an RCD that can be operated by a non-skilled person shall be designed and installed so that its operating current and/or time delay mechanism cannot be altered without using a key or a tool.	531-02-10

> Where an RCD is used, the product of the rated residual   471-15-06
> operating current (I_{an}) in amperes and the earth fault loop
> impedance in ohms shall not exceed 50.

10.6.7 New materials and inventions

> Where the use of a new material or an invention means (or   120-02-01
> requires) a departure from the Regulations, the resulting
> degree of safety of the installation shall **not** be less than
> that obtained by compliance with the Regulations.

 Note: All departures from the Regulations shall be noted on the Electrical
Installation Certificate (see Figures 9.21 and 9.22 on Page 512 and 513).

10.6.8 Overcurrent protective devices

> Fuses that could be replaced by non-skilled personnel shall   533-01-01
> preferably be of a type that cannot be replaced inadvertently
> by one having a higher nominal current.
>
> Fuse links that could be replaced by non-skilled personnel   533-01-02
> shall:
>
> - clearly indicate the type of fuse link that should be used
> - preferably be of a type that cannot be replaced
>   inadvertently by one having a higher fusing factor.
>
> Any fuse that can be removed or replaced whilst the supply   533-01-03
> is still connected shall be capable of being removed or
> replaced without danger.

10.7 Material alterations

The Regulations make the following requirements for all material alterations
that have been caused by repair, maintenance and/or replacement.

> Wiring shall be marked and/or arranged so that it can be   514-01-02
> quickly identified for inspection, testing, repair or alteration
> of the installation.

Disconnecting devices shall permit disconnection of the   131-10-01
electrical installation, circuits or individual items of apparatus
as required for maintenance, testing, fault detection and repair.

Electrical equipment shall be arranged so that they are fully   131-12-01
accessible (i.e. for operation, inspection, testing, maintenance
and repair) and that there is sufficient space for later
replacement.

Equipment shall not be capable of becoming unintentionally   462-01-03
or inadvertently reactivated during mechanical maintenance.

Installed equipment shall be accessible for operational,   513-01-01
inspection and maintenance purposes.

 This does not apply to inaccessible cable joints.

Wiring systems shall be capable of:

- withstanding condensation and/or ingress of water   522-03-01
- minimising the ingress of solid foreign bodies   522-04-01

during installation, use and maintenance.

Wiring systems shall be selected and erected so as to   522-06-01
minimise mechanical damage (e.g. damage due to impact,
abrasion, penetration, compression or tension) during
installation, use and maintenance.

Wiring systems shall be selected and erected so as to   522-08-01
minimise (during installation, use and maintenance) the
damage to the sheath and insulation of cables and insulated
conductors and their terminations.

Connections and joints shall be accessible for inspection,   526-04-01
testing and maintenance, unless:

- they are in a compound-filled or encapsulated joint
- the connection is between a cold tail and a heating element
- the joint is made by welding, soldering, brazing or a
  compression tool.

Care shall be taken in reinstating protective measures that   529-01-01
have to be removed in order to carry out maintenance.

Safe access to all parts of the wiring system that may   529-01-02
require maintenance shall be provided.

Devices for switching off for mechanical maintenance   537-03-01
shall either be inserted in the main supply circuit (which is
always the preferred option) or in the control circuit.

A device for switching off for mechanical maintenance   537-03-02
(or a control switch for such a device):

- shall be manually operated
- shall have an externally visible contact gap, or
- shall clearly indicate the OFF or OPEN position.

 Note: This indication shall only be permitted to occur when the OFF or OPEN position on each pole has been fully attained.

A device for switching off for mechanical maintenance:

- shall prevent unintentional reclosure, caused by   537-03-03
  mechanical shock or vibration
- shall be capable of cutting off the full load current of   537-03-04
  the relevant part of the installation.

Plug and socket outlets (with a rating not exceeding 16 A)   537-03-05
may be used as a means for switching off for mechanical
maintenance.

Equipment shall be arranged to allow easy access for   561-01-04
periodic inspection, testing and maintenance.

An assessment shall be made of the frequency and quality of   341-01-01
maintenance (e.g. periodic inspection, testing, maintenance
and repair etc.) that an installation can reasonably be
expected to receive during its intended life.

The frequency of periodic inspection and testing of   732-01-01
installations will depend on:

- the type of installation, its use and operation
- the frequency and quality of maintenance, and
- the external influences to which it is subjected.

10.8 Material changes of use

Where there is a material change of use of a building, any work carried out
shall ensure that the building complies with the applicable requirements of the
following paragraphs of Schedule 1 of the Building Act 1984.

(a) In all cases:

- means of warning and escape (B1)
- internal fire spread – linings (B2)
- internal fire spread – structure (B3)

- external fire spread – roofs (B4)(2)
- access and facilities for the fire service (B5)
- resistance to moisture (C1)(2)
- dwelling houses and flats formed by material change of use (E4)
- ventilation (F1)
- sanitary conveniences and washing facilities (G1)
- bathrooms (G2)
- foul water drainage (H1)
- solid waste storage (H6)
- combustion appliances (J1, J2 & J3)
- conservation of fuel and power – dwellings (L1)
- conservation of fuel and power – buildings other than dwellings (L2)
- electrical safety (P1, P2)
- external fire spread – walls (B4 1) – in the case of a building exceeding fifteen metres in height.

(b) In other cases, see Table 10.1 below.

Table 10.1 Building Act Requirements

Material change of use	Requirement	Approved document
The building is used as a dwelling, where previously it was not	Resistance to moisture	C2 E1, E2, E3
The public building consists of a new school	Acoustic conditions in schools	E4
The building contains a flat, where previously it did not	Resistance to the passage of sound	E1, E2 & E3
The building is used as an hotel or a boarding house, where previously it was not	Structure	A1, A2 & A3 E1, E2, E3
The building is used as an institution, where previously it was not	Structure	A1, A2 & A3
The building is used as a public building, where previously it was not		A1, A2 & A3 E1, E2, E3
The building is not a building described in Classes I to VI in Schedule 2, where previously it was not	Structure	A1, A2 & A3
The building, which contains at least one room for residential purposes, contains a greater or lesser number of dwellings than it did previously	Structure	A1, A2 & A3 E1, E2, E3
The building, which contains at least one dwelling, contains a greater or lesser number of dwellings than it did previously	Resistance to the passage of sound	E1, E2 & E3

Note: In some circumstances (particularly when an historic building is under-going a material change of use and where the special characteristics of the building needs to be recognised) it may **not** be practical to improve sound insulation to the standards set out in Part E1 or resistance to contaminants and water as set out in Part C. In these cases, the aim should be to improve the insulation and resistance where it is practically possible – always provided that the work does not prejudice the character of the historic building, or increase the risk of long-term deterioration to the building fabric and/or fittings.

10.8.1 Verification

 All additions and/or alterations to existing installations   133-02-01
shall be inspected and tested for conformance to BS 7621.

Wiring shall be marked and/or arranged so that it can be quickly identified for inspection, testing, repair or alteration of the installation.	514-01-02
Sealing that has been disturbed during alterations work shall be reinstated as soon as practicable.	527-03-02

Note: This requirement is also meant as a fire precaution.

An Electrical Installation Certificate (containing details of the installation together with a record of the inspections made and the test results) or a Minor Electrical Installation Works Certificate (for all minor electrical installations that do not include the provision of a new circuit) shall be provided for all alterations or additions to electrical circuits.	743-01-01

Appendix 10.1: Example stage audit checks

Design stage

Item	Related item	Remark
1 Requirements	1.1 Information	Has the customer fully described his requirement? Has the customer any mandatory requirements? Are the customer's requirements fully understood by all members of the design team? Is there a need to have further discussions with the customer? Are other suppliers or subcontractors involved? If yes, who is the prime contractor?
	1.2 Standards	What international standards need to be observed? Are they available? What national standards need to be observed? Are they available? What other information and procedures are required? Are they available?
	1.3 Procedures	Are there any customer-supplied drawings, sketches or plans? Have they been registered?
2 Quality procedures	2.1 Procedures manual	Is one available? Does it contain detailed procedures and instructions for the control of all drawings within the drawing office?
	2.2 Planning implementation and production	Is the project split into a number of work packages? If so: • are the various work packages listed? • have work package leaders been nominated? • is their task clear? • is their task achievable? Is a time plan available? Is it up to date? Regularly maintained? Relevant to the task?
3 Drawings	3.1 Identification	Are all drawings identified by a unique number? Is the numbering system strictly controlled?
	3.2 Cataloguing	Is a catalogue of drawings maintained? Is this catalogue regularly reviewed and up to date?

(continued)

Design stage (*continued*)

Item	Related item	Remark
	3.3 Amendments and modifications	Is there a procedure for authorising the issue of amendments, changes to drawings? Is there a method for withdrawing and disposing of obsolete drawings?
4 Components	4.1 Availability	Are complete lists of all the relevant components available?
	4.2 Adequacy	Are the selected components currently available and adequate for the task? If not, how long will they take to procure? Is this acceptable?
	4.3 Acceptability	If alternative components have to be used are they acceptable to the task?
5 Records	5.1 Failure reports	Has the design office access to all records, failure reports and other relevant data?
	5.2 Reliability data	Is reliability data correctly stored, maintained and analysed?
	5.3 Graphs, diagrams, plans	In addition to drawings, is there a system for the control of all graphs, tables, plans etc.? Are CAD facilities available? (If so, go to 6.1)
6 Reviews and audits	6.1 Computers	If a processor is being used: • are all the design office personnel trained in its use? • are regular back-ups taken? • is there an anti-virus system in place?
	6.2 Manufacturing division	Is a close relationship being maintained between the design office and the manufacturing division?
	6.3	Is notice being taken of the manufacturing division's exact requirements, their problems and their choices of components etc.?

Installation stage

Item	Related item	Remark
1 Degree of quality	1.1 Quality control procedures	Are quality control procedures available? Are they relevant to the task? Are they understood by all members of the manufacturing team? Are they regularly reviewed and up to date? Are they subject to control procedures?

(*continued*)

Installation stage (*continued*)

Item	Related item	Remark
	1.2 Quality control checks	What quality checks are being observed? Are they relevant? Are there laid down procedures for carrying out these checks? Are they available? Are they regularly updated?
2 Reliability of product design	2.1 Statistical data	Is there a system for predicting the reliability of the product's design? Is sufficient statistical data available to be able to estimate the actual reliability of the design, before a product is manufactured? Is the appropriate engineering data available?
	2.2 Components and parts	Are the reliability ratings of recommended parts and components available? Are probability methods used to examine the reliability of a proposed design? If so, have these checks revealed design deficiencies such as: • Assembly errors? • Operator learning, motivational, or fatigue factors? • Latent defects? • Improper part selection?

(Note: If necessary, use additional sheets to list actions taken)

Acceptance stage

Item	Related item	Remark
1 Product performance		Does the product perform to the required function? If not, what has been done about it?
2 Quality level	2.1 Workmanship	Does the workmanship of the product fully meet the level of quality required or stipulated by the user?
	2.2 Tests	Is the product subjected to environmental tests? If so, which ones? Is the product field-tested as a complete system? If so, what were the results?

(*continued*)

Acceptance stage (*continued*)

Item	Related item	Remark
3 Reliability	3.1 Probability function	Are individual components and modules environmentally tested? If so, how?
	3.2 Failure rate	Is the product's reliability measured in terms of probability function? If so, what were the results? Is the product's reliability measured in terms of failure rate? If so, what were the results?
	3.3 Mean time between failures	Is the product's reliability measured in terms of mean time between failure? If so, what were the results?

In-service stage

Item	Related item	Remark
1 System reliability	1.1 Product basic design	Are statistical methods being used to prove the product's basic design? If so, are they adequate? Are the results recorded and available? What other methods are used to prove the product's basic design? Are these methods appropriate?
2 Equipment reliability	2.1 Personnel	Are there sufficient trained personnel to carry out the task? Are they sufficiently motivated? If not, what is the problem? Have individual job descriptions been developed? Are they readily available?
	2.1.1 Operators	Are all operators capable of completing their duties?
	2.1.2 Training	Do all personnel receive appropriate training? Is a continuous on-the-job training (OJT) programme available to all personnel? If not, why not?
	2.2 Product dependability	What proof is there that the product is dependable? How is product dependability proved? Is this sufficient for the customer?
	2.3 Component reliability	Has the reliability of individual components been considered? Does the reliability of individual components exceed the overall system reliability?

(continued)

In-service stage (*continued*)

Item	Related item	Remark
	2.4 Faulty operating procedures	Are operating procedures available? Are they appropriate to the task? Are they regularly reviewed?
	2.5 Operational abuses	Are there any obvious operational abuses? If so, what are they? How can they be overcome?
	2.5.1 Extended duty cycle	Do the staff have to work shifts? If so, are they allowed regular breaks from their work? Is there a senior shift worker? If so, are his duties and responsibilities clearly defined? Are computers used? If so, are screen filters available? Do the operators have keyboard wrist rests?
	2.5.2 Training	Do the operational staff receive regular on-the-job training? Is there any need for additional in-house or external training?
3 Design capability	3.1 Faulty operating procedures	Are there any obvious faulty operating procedures? Can the existing procedures be improved upon?

Appendix A

Symbols used in electrical installation

Main control or intake point

Main or submain switch

Socket outlet (mains) general symbol

Switched socket outlet

Socket outlet with pilot lamp

Multiple socket outlet
Example: for 3 plugs

Push button

Luminous push button

Electric bell: general symbol

Electric buzzer: general symbol

Time switch

Automatic fire detector

Single-pole, one-way switch

Note: Number of switches at one point may be indicated

Two-pole, one-way switch

Three-pole, one-way switch

Cord-operated single-pole, one-way switch

Two-way switch

Intermediate switch

Lighting point or lamp: general symbol

Note: The number, power and type of the light source should be specified

Example:
Three 40 watt lamps

Lamp or lighting point: wall mounted

Emergency (safety) lighting point

Lighting point with built in switch

Projector or lamp with reflector

Spotlight

Single fluorescent lamp

Appendix B
List of symbols

Symbol	Description
$\beta°$	tube oscillating angle
°C	degrees celsius
Ω	ohm
μg	microgram
$\mu g/m^3$	micrograms per cubic metre
μm	micrometre
μs	microsecond
a	amplitude
A	ampere
A/m	amperes per metre
am	attometre
atm	standard atmosphere
C	coulomb
cd	candela
cd/m^2	candelas per square metre
dB	decibels
dB(A)	decibel amps
dBm	decibel metres
dm^3	cubic decimetre
dm^3/mm	cubic decimetres/millimetre – flow
Em	exametre
eV	electronvolt
f	frequency
F	farad
fm	femtometre
ft	foot
g	gram
G	gauss
G	shock
g^2/Hz	accelerated spectral density
GHz	gigahertz – frequency
Gm	gigametre
g/m^3	grams per cubic metre
g_n	peak acceleration
G_s	setting value of a characteristic quantity
h	hour
H	henry
ha	hectare
hp	horsepower
hr(s)	hour(s) – alternative to h
Hz	hertz
I	amps
I^2R	power
in	inch
J	joule
k	constant of the relay

(*continued*)

List of Symbols (*continued*)

Symbol	Description
K	kelvin
kA	kiloamps
kA/μs	kiloamps per microsecond
kg	kilogram
kg/m^3	kilograms per cubic metre
kgf	kilogram force
kHz	kilohertz
kPa	kilopascal – pressure
ks	kilosecond
kV	kilovolts
kW	kilowatt
kW/m^2	kilowatts per square metre – irradiance
l	litre
lb	pound
lb/in	pounds per square inch
m	metre
m/s	metres per second
m/s^2	metres per second per second – amplitude
m^2	square metres
m^3	cubic metres
mbar	millibar – pressure
MHz	megahertz
min	minute
mm	millimetre
Mm	megametre
mm/h	millimetres per hour
mm/m^2	millimetres/square metre – exposure
mol	mole
ms	millisecond
mV	millivolts
MVA	megavolt amps
N	newton
N/m^2	newtons per square metre
NaCl	sodium chloride
nF	nanofarad
nm	nanometre
pH	alkalinity/acidity value
pm	picometre
Pm	petametre
R	intensity of dropfield in mm/h
R	resistance
rad/s	radians per second
s	second
S	siemens
t	tonne
T	time
T	tesla
Tm	terametre
û	amplitude of voltage surge
U_n	nominal voltage
V	volt
V/μs	volts per microsecond

(*continued*)

List of Symbols (*continued*)

Symbol	Description
V/km	volts per kilometre
Vm	volts per metre
W	watt
Wb	weber
W/m^2	watts per square metre – irradiance
yd	yard
ym	yocotmetre
Ym	yottametre
zm	zeptometre
Zm	zettametre

Appendix C

SI units for existing technology

As Gregor M. Grant explained in his article published in the April/May 1997 issue of *ElectroTechnology*, the *Système International d'Unités* (SI) was a child of the 1960s, a creation of the 11th General Conference on Weights and Measures, *Conférence Générale des Poids et Mesures* (CGPM). This assembly endorsed the Italian physicist Professor Giovanni Giorgi's MKS (i.e. metre-kilogram-second) system of 1901 and decided to base the SI system on it. Seven basic units were adopted, as shown in Table C.1, each of which was harmonised to a standard value.

Of the seven units, only the kilogram (kg) is represented by a physical object, namely a cylinder of platinum-iridium kept at the International Bureau of Weights and Measures at Sèvres, near Paris, with a duplicate at the US Bureau of Standards.

The metre (m), on the other hand, 'is the length of the path travelled by light in a vacuum during a time interval of 1/299,792,458 of a second'.

The second (s) has been defined as 'the duration of 9,192,631,770 periods of radiation corresponding to the energy-level change between the two hyper-fine levels of the ground state of caesium-133 atom'.

The ampere (A) is 'that constant current which, if maintained in two straight parallel conductors of infinite length, of negligible circular cross section and placed 1 m apart in vacuum, would produce between these conductors a force equal to 2×10^{-7} newtons per metre length'.

The unit of temperature is the kelvin (K), which is a thermodynamic measurement as opposed to one based on the properties of real material. Its origin is at absolute zero and there is a fixed point where the pressure and temperature of water, water vapour and ice are in equilibrium, which is defined as 273.16 K.

The mole (mol) is 'that quantity of substance of a system which contains as many elementary entities as there are atoms in 0.012 kg of carbon-12'. For definition purposes the entities *must* be specified (e.g. atoms, electrons, ions or any other particles or groups of such particles).

Finally there is the candela (cd), the unit of light intensity. This is defined as 'the luminous intensity, in the perpendicular direction, of a surface of $1/600,000 \, \text{m}^2$ of a black body at the temperature of freezing platinum under a pressure of $101,325 \, \text{N/m}^2$'.

Two years before the creation of SI units, another international agreement had made the prefixes mega and micro official and introduced some new ones, such as the nano, whose name derives from the Greek 'nanos' meaning dwarf (see C.2). Its symbol is n, and its mathematical representation is 10^{-9}, indicating the number of *digits* to the right of the decimal point, in this case 0.000,000,001.

Even these minute quantities, however, soon became inadequate and, by 1962, it was decided that a thousandth of a picometre be designated a femtometre and one thousandth of this new measurement be termed an attometre. Later on, the zeptometre and yoctometre were introduced.

C.1 Basic SI units

Many SI units are named after people but when these units are written in full, they do not necessarily require initial capital letters, e.g. amperes, coulombs, newtons, siemens.

All the above examples are expressed in the plural, but note that siemens does not drop the final 's' in the singular as this was derived from a person's name (i.e. Siemens). Thus we have one newton, but one siemens.

Table C.1 Basic SI units

SI nomenclature	Abbreviation	Quantity
metre	m	length
kilogram	kg	mass
second	s	time
ampere	A	electrical current
kelvin	K	temperature
mole	mol	amount of substance
candela	cd	luminous intensity

C.2 Small number SI prefixes

Within the SI units there is a distinction between a quantity and a unit. Length is a quantity, but metres (abbreviated to m) is a unit.

Table C.2 Small number SI units

Measurement	Symbol	Equivalent to
millimetre	mm	0.001 m or 10^{-3} m
micrometre	μm	0.000 001 m or 10^{-6} m
nanometre	nm	0.000 000 001 m or 10^{-9} m
picometre	pm	0.000 000 000 001 m or 10^{-12} m
femtometre	fm	0.000 000 000 000 001 m or 10^{-15} m
attometre	am	0.000 000 000 000 000 001 m or 10^{-18} m
zeptometre	zm	0.000 000 000 000 000 000 001 m or 10^{-21} m
yoctometre	ym	0.000 000 000 000 000 000 000 001 m or 10^{-24} m

C.3 Large number SI prefixes

Table C.3 Large number SI prefixes

Measurement	Symbol	Equivalent to
megametre	Mm	1 000 000 m or 10^6 m
gigametre	Gm	1 000 000 000 m or 10^9 m
terametre	Tm	1 000 000 000 000 m or 10^{12} m
petametre	Pm	1 000 000 000 000 000 m or 10^{15} m
exametre	Em	1 000 000 000 000 000 000 m or 10^{18} m
zettametre	Zm	1 000 000 000 000 000 000 000 m or 10^{21} m
yottametre	Ym	1 000 000 000 000 000 000 000 000 m or 10^{24} m

C.4 Deprecated prefixes

Some non-SI fractions and multiples are occasionally used (see below), but they are not encouraged.

Table C.4 Deprecated prefixes

Fractions	Prefix	Abbreviation	Multiple	Prefix	Abbreviation
10^{-1}	deci	d	10	deka	da
10^{-2}	centi	c	10^2	hecto	h

C.5 Derived units

Some units, derived from the basic SI units, have been given special names, many of which originate from a person's name (i.e. Siemens).

Table C.5 Derived units

Quantity	Name of unit	Abbreviation (symbol)	Expression in terms of other SI units
energy	joule	J	Nm
force	newton	N	–
power	watt	W	J/s
electric charge	coulomb	C	As
potential difference (voltage)	volt	V	W/A
electrical resistance (or reactance or impedance)	ohm	Ω	V/A
electrical capacitance	farad	F	C/V
magnetic flux	weber	Wb	Vs
Inductance (note that the plural of henry is henrys)	henry	H	Wb/A
magnetic flux density	tesla	T	Wb/m^2
admittance (electrical conductance)	siemens	S	A/V $(=\Omega^{-1})$
frequency	hertz	Hz	cycles per second (or events per second)

C.6 Units without special names

Other derived units, without special names, are listed below.

Table C.6 Units without special names

Quantity	Unit	Abbreviation
area	square metres	m^2
volume	cubic metres	m^3
density	kilograms per cubic metre	kg/m^3
velocity	metres per second	m/s
angular velocity (angular frequency)	radians per second	rad/s
acceleration	metres per second per second	m/s^2
pressure	newtons per square metre	N/m^2
electric field strength	volts per metre	Vm
magnetic field strength	amperes per metre	A/m
luminance	candelas per square metre	cd/m^2

C.7 Tolerated units

Some non-SI units are tolerated in conjunction with SI units.

Table C.7 Tolerated units

Quantity	Unit	Abbreviation (symbol)	Definition
area	hectare	ha	$10^4 m^2$
volume	litre	l	$10^{-3} m^3$
pressure	standard atmosphere	atm	101,325 Pa
mass	tonne	t	10^3 kg (Mg)
energy	electronvolt	eV	1.6021×10^{19} J
magnetic	gauss	G	10^{-4} T

C.8 Obsolete units

For historical interest (as well as for completeness), the following table gives a list of obsolete units.

Table C.8 Obsolete units

Quantity	Unit	Abbreviation (symbol)	Definition
length	inch	in	0.0254 m
	foot	ft	0.3048 m
	yard	yd	0.9144 m
	mile	mi	1.603 94 km
mass	pound	lb	0.453 923 7 kg
force	dyne	dyn	10^{-5} N
	poundal	pdl	0.138 255 N
	pound force	lbf	4.448 22 N
	kilogram force	kgf	9.806 65 N
pressure	atmosphere	atm	101.325 kN/m^2
	torr	torr	133.322 N/m^2
	pounds per square inch	lb/in	6894.76 N/m^2
energy	erg	erg	10^{-7} J
power	horsepower	hp	745.700 W

Appendix D
Acronyms and abbreviations

a.c.	Alternating Current

ACS	Assemblies For Construction Sites
ADP	Automatic Data Processing
BRE	Building Research Establishment Ltd
BS	British Standard
BSI	British Standards Institution
CAD	Computer-Aided Design
CE	Conformity Europe
CECC	CENELEC Electronic Components Committee
CEN	Comité Européen de Normalisation
CENELEC	Comité Européen de Normalisation Electrotechnique
CNE	Combined Neutral and Earth
CORGI	Council for Registered Gas Installers
CPC	Circuit Protective Conductor
CPS	Control and Protective Switching Device
d.c.	Direct Current
DDA	Disability Discrimination Act
DIY	Do It Youself
DTI	Department of Trade & Industry
EBADS	Equipotential Bonding and Automatic Disconnection of Supplies
EC	European Community
ECA	Electrical Contractors Association
EEC	European Economic Commission
ELECSA	Fenestration Self-Assessment Scheme
ELV	Extra-Low Voltage
EMC	Electromagnetic Compatibility
EMI	Electromagnetic Interference
EN	European Normalisation
EU	European Union
FE	Functional Earth
FELV	Functional Extra-Low Voltage
GLS	As in tungsten lights
HBES	Home and Building Electronic Systems
HD	Harmonised Directive
HELA	Health and Safety Executive/Local Authorities
HEMP	High-Altitude Electromagnetic Pulse
HSE	Health & Safety Executive
HV	High Voltage
I/O	Input/Output
IEC	International Electrotechnical Commission
IEE	Institution of Electrical Engineers
IET	Institution of Engineering and Technology
IIE	Institution of Incorporated Engineers
ILU	Integrated Logistic Unit

(*continued*)

Acronyms and Abbreviations (*continued*)

a.c.	Alternating Current

IPC	Implant Point of Coupling
ISM	Industrial, Scientific and Medical
ISO	International Standards Organisation
IT	Information Technology
ITCZ	International Conveyance Zone
ITE	Information Technology Equipment
LUR	Logical User Requirement
MDD	Medical Devices Directive
MKS	Metre-Kilogram-Second
MMI	Man Machine Interface
MTBF	Mean Time Between Failures
N	Neutral
NAPIT	National Association of Professional Inspectors and Testers
NICEIC	National Inspection Council for Electrical Installation Counciling
NSO	National Standards Organisation
OFTEC	Oil Firing Technical Association
OPSI	Office of Public Sector Information
PCB	Printed Circuit Board
PE	Protective Earth
PELV	Protective Extra-Low Voltage
PEN	Protective and Neutral
PME	Protective Multiple Earthing
prEN	European draft standards
PV	Photovoltaic
PVC	Polyvinyl Chloride
QA	Quality Assurance
QC	Quality Control
QMS	Quality Management System
QP	Quality Procedure
r.m.s.	Root Mean Square
RAH	Relative Air Humidity
RAM	Reliability, Availability and Maintainability
RCBO	Residual Current Operated Circuit Breaker without integral overcurrent protection
RCCB	Residual Current Operated Circuit Breaker with integral overcurrent protection
RCD	Residual Current Device
RF	Radio Frequency
RH	Relative Humidity
S/N	Signal to Noise Ratio
SELV	Safety Extra-Low Voltage
SI	Statutory Instrument
SI	Système International d'Unités
T&E	Time and Expense
TLV	Threshold Limit Values
TQM	Total Quality Management
TVL-C	Threshold Limit Values – Ceiling List
UPS	Uninterruptible Power System
VDU	Visual Display Unit
WAUILF	Workplace Applied Uniform Indicated Low Frequency (application)
WI	Work Instruction
YFR	Yearly Forecast Rationale

Appendix E

British Standards currently used with the Wiring Regulations

E.1 Listed by subject

Title	BS Number
13 A plugs, socket outlets, connection units and adaptors.	BS 1363
13 A plugs, socket outlets, connection units and adaptors. Specification for rewirable and non-rewirable 13 A fused plugs.	BS 1363-1: 1995
13 A plugs, socket outlets, connection units and adaptors. Specification for 13 A switched and unswitched socket outlets.	BS 1363-2: 1995
13 A plugs, socket outlets, connection units and adaptors. Specification for adaptors.	BS 1363-3: 995
13 A plugs, socket outlets, connection units and adaptors. Specification for 13 A fused connection units: switched and unswitched.	BS 1363-4: 1995
300/500 V screened electric cables having low emission of smoke and corrosive gases when affected by fire, for use in thin partitions and building voids.	BS 8436: 2004
Appliance couplers for household and similar general purposes.	BS 4491
Application of equipotential bonding and earthing in buildings with information technology equipment.	BS EN 50310: 2000
Basic and safety principles for man-machine interface, marking and identification. Identification of equipment terminals and of terminations of certain designated conductors, including general rules for an alphanumeric system.	BS EN 60445: 2000
Basic and safety principles for the man-machine interface, marking and identification. Identification of conductors by colours or numerals.	BS EN 60446: 2000
Bayonet lampholders.	BS EN 61184: 1997
Cable trunking.	BS 4678
Cable trunking. Steel surface trunking.	BS 4678-1: 1971
Cable trunking. Steel underfloor (duct) trunking.	BS 4678-2: 1973
Cable trunking. Specification for cable trunking made of insulation material.	BS 4678-4: 1982
Cartridge fuses for voltages up to and including 1000 V a.c. and 1500 V d.c.	BS 88

(*continued*)

E.1 Listed by subject (*continued*)

Title	BS Number
Cartridge fuses for voltages up to and including 1000 V a.c. and 1500 V d.c. Specification for fuses for use by authorised persons (mainly for industrial applications). Also numbered as BS EN 60269-2: 1995.	BS 88-2.1: 1988
Cartridge fuses for voltages up to and including 1000 V a.c. and 1500 V d.c. Specification of supplementary requirements for fuses of compact dimensions for use in 240/415 V a.c. industrial and commercial electrical installations.	BS 88-6: 1988 (1992)
Code of practice for earthing.	BS 7430: 1998
Code of practice for installation of apparatus intended for connection to certain telecommunications systems.	BS 6701: 1994
Code of practice for protection of structures against lightning.	BS 6651: 1999
Common test methods for cables under fire conditions. Test for resistance to vertical flame propagation for a single insulated conductor or cable.	BS EN 50265
Common test methods for cables under fire conditions. Tests for resistance to vertical flame propagation for a single insulated conductor or cable. Apparatus.	BS EN 50265-1: 1999
Common test methods for cables under fire conditions. Tests for resistance to vertical flame propagation for a single insulated conductor or cable. Procedures. 1 kW pre-mixed flame.	BS EN 50265-2-1: 1999
Common test methods for cables under fire conditions. Tests for resistance to vertical flame propagation for a single insulated conductor or cable. Procedures. Diffusion flame.	BS EN 50265-2-2: 1999
Conduits for electrical purposes. Outside diameters of conduits for electrical installations and threads for conduits and fittings.	BS EN 60423: 1995
Elastomer insulated cables for fixed wiring in ships and on mobile and fixed offshore units. Requirements and test methods.	BS 6883: 1999
Electric cables, Calculation of current rating. Parts 1, 2, 3: 1997.	BS 7769
Electric cables. 600\1000 V armoured fire-resistant electric cables having thermosetting insulation and low emission of smoke and corrosive gases when affected by fire.	BS 7846: 2000
Electric cables. Flexible cables rated up to 450n50 V, for use with appliances and equipment intended for industrial and similar environments.	BS 7919: 2001
Electric cables. Flexible cords rated up to 300/00 V, for use with appliances and equipment intended for domestic, office and similar environments.	BS 6500: 2000

(*continued*)

E.1 Listed by subject (*continued*)

Title	BS Number
Electric cables. PVC insulated, non-armoured cables for voltages up to and including 45on50 V, for electric power, lighting and internal wiring.	BS 6004: 2000
Electric fence energisers. Safety requirements for mains-operated electric fence energisers.	BS EN 61011: 1993
Electric fence energisers. Safety requirements for battery-operated electric fence energisers suitable for connection to the supply mains.	BS EN 61011-1: 1993
Electric sauna heating appliances.	BS EN 60335-2-53: 1997
Electric surface heating.	BS 6351
Electric surface heating. Specification for electric surface heating devices.	BS 6351-1: 1983(1993)
Electric surface heating. Guide to the design of electric surface heating systems.	BS 6351-2: 1983 (1993)
Electric surface heating. Code of practice for the installation, testing and maintenance of electric surface heating systems.	BS 6351-3: 1983 (1993)
Electrical apparatus for potentially explosive atmospheres. General requirements.	BS EN 50014: 1998
Electrical apparatus for potentially explosive gas atmospheres.	BS EN 60079
Electrical apparatus for potentially explosive gas atmospheres. Classification of hazardous areas.	BS EN 60079-10: 1996
Electrical apparatus for potentially explosive gas atmospheres. Electrical installations in hazardous areas (other than mines).	BS EN 60079-14: 1997
Electrical apparatus for potentially explosive gas atmospheres. Inspection and maintenance of electrical installations in hazardous areas (other than mines).	BS EN 60079-17: 1997
Electrical apparatus for use in the presence of combustible dust.	BS EN 50281
Electrical installations for open cast mines and quarries.	BS 6907
Electrical supply track systems for luminaires.	BS EN 60570: 1997
Electromagnetic compatibility. Generic emission standard.	BS EN 50081
Electromagnetic compatibility. Generic immunity standard.	BS EN 50082
Emergency lighting.	BS 5266
Fire detection and alarm systems for buildings.	BS 5839
Note: BS 5839 is a multiple part standard.	
Fire detection and alarm systems for buildings. Specification for control and indicating equipment.	BS 5839-4: 1988
Fire hazard testing for electrotechnical products.	BS 6458

(*continued*)

E.1 Listed by subject (*continued*)

Title	BS Number
Fire hazard testing for electrotechnical products. Glow-wire test.	BS 6458-2.1: 1984
Fire tests on building materials and structure.	BS 476
Fire tests on building materials and structure. Non-combustibility test for materials.	BS 476-4: 970 (1984)
Fire tests on building materials and structure. Method of test for ignitability, now withdrawn.	BS 476 Part 5: 1979
Fire tests on building materials and structure. Method of test for ignitability also refers but methods of test are not identical.	BS 476-12: 991
Fire tests on building materials and structure. Methods for determination of the contribution of components to the fire resistance of a structure.	BS 476-23: 1987
Flexible steel conduit and adaptors for the protection of electric cables.	BS 731-1: 1952 (1993)
Flexible steel conduit for cable protection and flexible steel tubing to enclose flexible drives.	BS 731
Glossary of electrotechnical, power, telecommunications, electronics, lighting and colour terms.	BS 4727
Graphical symbols for diagrams.	BS EN 60617
Guide to electrical earth monitoring and protective conductor proving.	BS 4444: 1989 (1995)
Insulation coordination for equipment within low-voltage systems.	BS 7822
Insulation coordination for equipment within low-voltage systems. Principles, requirements and tests.	BS 7822-1: 1995
Isolating transformers and safety isolating transformers.	BS 3535

 Note: This standard is partially replaced by the BS EN 61558 series of standards, and these standards should be read jointly.

Isolating transformers and safety isolating transformers. Specification for transformers for reduced system voltage.	BS 3535-2: 1990

 Note: To be read in conjunction with BS 3535-1.

Isolating transformers and safety isolating transformers. Requirements.	BS EN 60742: 1996
Leisure accommodation vehicles. 12 V direct current extra-low-voltage electrical installations.	BS EN 1648
Leisure accommodation vehicles. 12 V direct current extra-low-voltage electrical installations. Caravans.	BS EN 1648-1: 1998
Leisure accommodation vehicles. 12 V direct current extra-low-voltage electrical installations. Motor caravans.	BS EN 1648-2: 1998
Lifts and service lifts.	BS 5655

 Note: BS 5655 is a multiple set of standards.

(*continued*)

E.1 Listed by subject (*continued*)

Title	BS Number
Low-voltage fuses.	BS EN 60269
Low-voltage fuses. General requirements.	BS EN 60269-1: 1999
Low-voltage fuses. Supplementary requirements for fuses for use by authorised persons.	BS EN 60269-2: 1995
Luminaries.	BS EN 60598
Luminaries. Luminaries with limited surface temperature.	BS EN 60598-2-24: 1999
Memorandum. Construction of electrical equipment for protection against electric shock.	BS 2754: 1976 (1999)
Method for calculation of thermally permissible short-circuit currents taking into account non-adiabatic heating effects.	BS 7454: 1991
Mineral insulated cables and their terminations with a rated voltage not exceeding 750 V. Cables.	BS EN 60702-1: 2002
Mineral insulated cables with a rated voltage not exceeding 750 V.	BS 6207
Mineral insulated cables with a rated voltage not exceeding 750 V. Cables.	BS 6207-1: 1995
Mineral insulated cables with a rated voltage not exceeding 750 V. Terminations.	BS 6207-2: 1995
Nominal voltages for low-voltage public electricity supply systems.	BS 7697
Non-metallic conduits and fittings for electrical installations.	BS 4607
Plugs, socket outlets and couplers for industrial purposes. General requirements.	BS EN 60309-1: 1999
Plugs, socket outlets and couplers for industrial purposes.	BS EN 60309
Plugs, socket outlets and couplers for industrial purposes. Dimensional interchangeability requirements for pin and contact-tube accessories of harmonised configurations.	BS EN 60309-2: 1998
Residual current operated circuit breakers with integral overcurrent protection for household and similar uses (RCBOs).	BS EN 61009
Residual current operated circuit breakers with integral over-current protection for household and similar uses (RCBOs). General rules.	BS EN 61009-1: 1995
Residual current operated circuit breakers without integral overcurrent protection for household and similar uses (RCCBs).	BS EN 61008
Residual current operated circuit breakers without integral overcurrent protection for household and similar uses (RCCBs). General rules.	BS EN 61008-1: 1995

(*continued*)

E.1 Listed by subject (*continued*)

Title	BS Number
Safety of power transformers, power supply units and similar.	BS EN 61558
Safety of power transformers, power supply units and similar. General requirements and tests.	BS EN 61558-2-1: 1998
Safety of power transformers, power supply units and similar. Particular requirements for separating transformers for general use.	BS EN 61558-2-2: 1998
Safety of power transformers, power supply units and similar. Particular requirements for isolating transformers for general use.	BS EN 61558-2-4: 1998
Safety of power transformers, power supply units and similar. Particular requirements for shaver transformers and shaver units.	BS EN 61558-2-5: 1998
Safety of power transformers, power supply units and similar. Particular requirements for safety isolating transformers for general use.	BS EN 61558-2-6
Safety of power transformers, power supply units and similar. Particular requirements for bell and chime transformers.	BS EN 61558-2-8: 1999
Safety of power transformers, power supply units and similar. Particular requirements for switch mode power supplies.	BS EN 61558-2-17: 1998
Specification for 2-pin reversible plugs and shaver socket outlets.	BS 4573: 1970 (1979)
Specification for 3001500 V fire-resistant electric cables having low emission of smoke and corrosive gases when affected by fire. Multicore cables.	BS 7629-1: 1997
Specification for 600/1000 V and 1900/3300 V armoured electric cables having thermosetting insulation.	BS 5467: 1997
Specification for 600/1000 V single-phase split concentric electric cables.	BS 4553
Specification for 600/1000 V single-phase split concentric electric cables. Cables having PVC insulation.	BS 4553-1: 1998
Specification for 600/1000 V single-phase split concentric electric cables. Cables having thermosetting insulation.	BS 4553-2: 1998
Specification for 600/1000 V single-phase split concentric electric cables. Cables having thermosetting insulation and low emission of smoke and corrosive gases when affected by fire.	BS 4553-3: 1998
Specification for 60011000 V and 190013000 V armoured cables having PVC insulation.	BS 6346: 1997
Specification for 60011000 V and 190013300 V armoured cables having thermosetting insulation and low emission of smoke and corrosive gases when affected by fire.	BS 6724: 1997
Specification for 60011000 V single-core unarmoured electric cables having thermosetting insulation.	BS 7889: 1997

(*continued*)

E.1 Listed by subject (*continued*)

Title	BS Number
Specification for 6110 amp two-pole weather-resistant couplers for household, commercial and light industrial equipment.	BS 6991: 1990
Specification for bayonet lampholders with enhanced safety	BS 7895: 1997
Specification for bayonet lampholders.	BS 5042: 1987

Note: Replaced by BS EN 61184: 1997.

Title	BS Number
Specification for binding and identification sleeves for use on electric cables and wires.	BS 3858: 1992
Specification for cable trunking and ducting systems for electrical installations.	BS EN 50085
Specification for cable trunking and ducting systems for electrical installations. General requirements.	BS EN 50085-1: 1999
Specification for cartridge fuses for a.c. circuits in domestic and similar premises.	BS 1361: 1971 (1986)
Specification for ceiling roses.	BS 67: 1987 (1999)
Specification for circuit breakers for overcurrent protection for household and similar installations.	BS EN 60898: 1991
Specification for clamps for earthing and bonding purposes.	BS 951: 1999
Specification for conduit systems for electrical installations.	BS EN 50086
Specification for conduit systems for electrical installations. General requirements.	BS EN 50086-1: 1994
Specification for conduit systems for electrical installations. Rigid conduit systems.	BS EN 50086-2-1: 1996
Specification for distribution assemblies for electricity supplies for construction and building sites.	BS 4363: 1998
Specification for Edison screw lampholders.	BS EN 60238: 1999
Specification for electric signs and high-voltage luminous discharge tube installations.	BS 559: 998
Specification for electronic variable control switches (dimmer switches) for tungsten filament lighting.	BS 5518: 1977 (1999)
Specification for flexible insulating sleeving for electrical purposes.	BS 2848: 1973
Specification for general-purpose fuse links for domestic and similar purposes (primarily for use in plugs).	BS 1362: 1973 (1992)
Specification for general requirements for electrical accessories.	BS 5733: 995
Specification for general requirements for luminaire supporting couplers for domestic, light industrial and commercial use.	BS 6972: 1988
Specification for insulated cables and flexible cords for use in high-temperature zones.	BS 6141: 1991
Specification for interchangeability and safety of a standardised luminaire supporting coupler.	BS 7001: 1988

Note: To be read in conjunction with BS 6972: 1988.

(*continued*)

E.1 Listed by subject (*continued*)

Title	BS Number
Specification for low-voltage switchgear and control gear assemblies.	BS EN 60439
Specification for low-voltage switchgear and control gear assemblies. Specification for type-tested and partially type-tested assemblies.	BS EN 60439-1: 1999
Specification for low-voltage switchgear and control gear assemblies. Particular requirements for busbar trunking systems (busways).	BS EN 60439-2: 1993
Specification for low-voltage switchgear and control gear assemblies. Particular requirements for Assemblies For Construction Sites (ACS).	BS EN 60439-4: 1991
Specification for low-voltage switchgear and control gear.	BS EN 60947
Specification for low-voltage switchgear and control gear. General rules.	BS EN 60947-1: 1999
Specification for low-voltage switchgear and control gear. Circuit breakers.	BS EN 60947-2: 1996
Specification for low-voltage switchgear and control gear. Switches, disconnecters, switch-disconnectors and fuse-combination units.	BS EN 60947-3: 999
Specification for low-voltage switchgear and control gear. Contactors and motor starters.	BS EN 60947-4
Specification for low-voltage switchgear and control gear. Electromechanical contactors and motor starters.	BS EN 60947-4-1: 1992
Specification for low-voltage switchgear and control gear assemblies. Particular requirements for low-voltage switchgear and control gear assemblies intended to be installed in places where unskilled persons have access to their use.	BS EN 60439-3: 1991
Specification for portable residual current devices.	BS 7071: 1992 (1998)
Specification for protected-type non-reversible plugs, socket outlets, cable couplers and appliance couplers with earthing contacts for single-phase a.c. circuits up to 250 volts.	BS 196: 1961
Specification for safety of information technology equipment including electrical business equipment.	BS EN 60950: 1992
Specification for socket outlets incorporating residual current devices (SRCDs).	BS 7288: 1990 (1998)
Specification for steel conduit and fittings with metric threads of ISO form for electrical installations.	BS 4568
Specification for thermosetting insulated cables (non-armoured) for electric power and lighting with low emission of smoke and corrosive gases when affected by fire.	BS 7211: 1998

(*continued*)

E.1 Listed by subject (*continued*)

Title	BS Number
Specification. Cartridge fuse links (rated up to 5 amperes) for a.c. and d.c. service.	BS 646: 1958 (1991)
Specification. Semi-enclosed electric fuses (rating up to 100 amperes and 240 volts to earth).	BS 3036: 1958 (1992)
Specification. Two-pole and earthing pin plugs, socket outlets and socket outlet adaptors.	BS 546: 1950 (1988)
Specification. Steel conduit and fittings for electrical wiring.	BS 31: 1940 (1988)
Steel conduit, bends and couplers.	BS 4568 Part 1: 1970 (1973)
Switches for household and similar fixed electrical equipment.	BS EN 60669
Switches for household and similar fixed electrical equipment. Particular requirements – isolating switches.	BS EN 60669-2-4
Switches for household and similar fixed electrical installations.	BS 3676
Tests on electric cables under fire conditions.	BS 4066-3: 1994
Tests on bunched wires or cables.	
Tests on electric cables under fire conditions.	BS 4066

E.2 Listed by standard

BS Number	Title
BS 31: 1940 (1988)	Specification. Steel conduit and fittings for electrical wiring.
BS 67: 1987 (1999)	Specification for ceiling roses.
BS 88	Cartridge fuses for voltages up to and including 1000 V a.c. and 1500 V d.c.
BS 88-2.1: 1988	Cartridge fuses for voltages up to and including 1000 V a.c. and 1500 V d.c. Specification for fuses for use by authorised persons (mainly for industrial applications). Also numbered as BS EN 60269-2: 1995.
BS 88-6: 1988 (1992)	Cartridge fuses for voltages up to and including 1000 V a.c. and 1500 V d.c. Specification of supplementary requirements for fuses of compact dimensions for use in 240/415 V a.c. industrial and commercial electrical installations.
BS 196: 1961	Specification for protected-type non-reversible plugs, socket outlets, cable couplers and appliance couplers with earthing contacts for single-phase a.c. circuits up to 250 volts.
BS 476	Fire tests on building materials and structure.
BS 476 Part 5: 1979	Fire tests on building materials and structure. Method of test for ignitability, now withdrawn.

(*continued*)

E.2 Listed by standard (*continued*)

BS Number	Title
BS 476-12: 1991	Fire tests on building materials and structure. Method of test for ignitability also refers but methods of test are not identical.
BS 476-23: 1987	Fire tests on building materials and structure. Methods for determination of the contribution of components to the fire resistance of a structure.
BS 476-4: 970 (1984)	Fire tests on building materials and structure. Non-combustibility test for materials.
BS 546: 1950 (1988)	Specification. Two-pole and earthing pin plugs, socket outlets and socket outlet adaptors.
BS 559: 1998	Specification for electric signs and high-voltage luminous discharge tube installations.
BS 646: 1958 (1991)	Specification. Cartridge fuse links (rated up to 5 amperes) for a.c. and d.c. service.
BS 731	Flexible steel conduit for cable protection and flexible steel tubing to enclose flexible drives.
BS 731-1: 1952 (1993)	Flexible steel conduit and adaptors for the protection of electric cables.
BS 951: 1999	Specification for clamps for earthing and bonding purposes.
BS 1361: 1971 (1986)	Specification for cartridge fuses for a.c. circuits in domestic and similar premises.
BS 1362: 1973 (1992)	Specification for general purpose fuse links for domestic and similar purposes (primarily for use in plugs).
BS 1363	13 A plugs, socket outlets, connection units and adaptors.
BS 1363-1: 1995	13 A plugs, socket outlets, connection units and adaptors. Specification for rewirable and non-rewirable 13 A fused plugs.
BS 1363-2: 1995	13 A plugs, socket outlets, connection units and adaptors. Specification for 13 A switched and unswitched socket outlets.
BS 1363-3: 995	13 A plugs, socket outlets, connection units and adaptors. Specification for adaptors.
BS 1363-4: 1995	13 A plugs, socket outlets, connection units and adaptors. Specification for 13 A fused connection units: switched and unswitched.
BS 2754: 1976 (1999)	Memorandum. Construction of electrical equipment for protection against electric shock.
BS 2848: 1973	Specification for flexible insulating sleeving for electrical purposes.
BS 3036: 1958 (1992)	Specification. Semi-enclosed electric fuses (rating up to 100 amperes and 240 volts to earth).
BS 3535	Isolating transformers and safety isolating transformers. **Note:** This standard is partially replaced by the BS EN 61558 series of standards, and these standards should be read jointly.
BS 3535-2: 1990	Isolating transformers and safety isolating transformers. Specification for transformers for reduced system voltage. **Note:** To be read in conjunction with BS 3535-1.

(*continued*)

E.2 Listed by standard (*continued*)

BS Number	Title
BS 3676	Switches for household and similar fixed electrical installations.
BS 3858: 1992	Specification for binding and identification sleeves for use on electric cables and wires.
BS 4066	Tests on electric cables under fire conditions.
BS 4066-3: 1994	Tests on electric cables under fire conditions. Tests on bunched wires or cables.
BS 4363: 1998	Specification for distribution assemblies for electricity supplies for construction and building sites.
BS 4444: 1989 (1995)	Guide to electrical earth monitoring and protective conductor proving.
BS 4491	Appliance couplers for household and similar general purposes.
BS 4553	Specification for 600/1000 V single-phase split concentric electric cables.
BS 4553-1: 1998	Specification for 600/1000 V single-phase split concentric electric cables. Cables having PVC insulation.
BS 4553-2: 1998	Specification for 600/1000 V single-phase split concentric electric cables. Cables having thermosetting insulation.
BS 4553-3: 1998	Specification for 600/1000 V single-phase split concentric electric cables. Cables having thermosetting insulation and low emission of smoke and corrosive gases when affected by fire.
BS 4568	Specification for steel conduit and fittings with metric threads of ISO form for electrical installations.
BS 4568 Part 1: 1970 (1973)	Steel conduit, bends and couplers.
BS 4573: 1970 (1979)	Specification for 2-pin reversible plugs and shaver socket outlets.
BS 4607	Non-metallic conduits and fittings for electrical installations.
BS 4678	Cable trunking.
BS 4678-1: 1971	Cable trunking. Steel surface trunking.
BS 4678-2: 1973	Cable trunking. Steel underfloor (duct) trunking.
BS 4678-4: 1982	Cable trunking. Specification for cable trunking made of insulation material.
BS 4727	Glossary of electrotechnical, power, telecommunications, electronics, lighting and colour terms.
BS 5042: 1987	Specification for bayonet lampholders. **Note:** Replaced by BS EN 61184: 1997.
BS 5266	Emergency lighting.

(*continued*)

E.2 Listed by standard (*continued*)

BS Number	Title
BS 5467: 1997	Specification for 600/1000 V and 1900/3300 V armoured electric cables having thermosetting insulation.
BS 5518: 1977 (1999)	Specification for electronic variable control switches (dimmer switches) for tungsten filament lighting.
BS 5655	Lifts and service lifts.
	Note: BS 5655 is a multiple set of standards.
BS 5733: 995	Specification for general requirements for electrical accessories.
BS 5839	Fire detection and alarm systems for buildings.
	Note: BS 5839 is a multiple part standard.
BS 5839-4: 1988	Fire detection and alarm systems for buildings. Specification for control and indicating equipment.
BS 6004: 2000	Electric cables. PVC insulated, non-armoured cables for voltages up to and including 45on50 V, for electric power, lighting and internal wiring.
BS 6141: 1991	Specification for insulated cables and flexible cords for use in high-temperature zones.
BS 6207	Mineral insulated cables with a rated voltage not exceeding 750 V.
BS 6207-1: 1995	Mineral insulated cables with a rated voltage not exceeding 750 V. Cables.
BS 6207-2: 1995	Mineral insulated cables with a rated voltage not exceeding 750 V. Terminations.
BS 6346: 1997	Specification for 60011000 V and 190013000 V armoured cables having PVC insulation.
BS 6351-1: 1983(1993)	Electric surface heating. Specification for electric surface heating devices.
BS 6351	Electric surface heating.
BS 6351-2: 1983 (1993)	Electric surface heating. Guide to the design of electric surface heating systems.
BS 6351-3: 1983 (1993)	Electric surface heating. Code of practice for the installation, testing and maintenance of electric surface heating systems.
BS 6458	Fire hazard testing for electrotechnical products.
BS 6458-2.1: 1984	Fire hazard testing for electrotechnical products. Glow-wire test.
BS 6500: 2000	Electric cables. Flexible cords rated up to 300/00 V, for use with appliances and equipment intended for domestic, office and similar environments.
BS 6651: 1999	Code of practice for protection of structures against lightning.

(continued)

E.2 Listed by standard (*continued*)

BS Number	Title
BS 6701: 1994	Code of practice for installation of apparatus intended for connection to certain telecommunications systems.
BS 6724: 1997	Specification for 600 1000 V and 1900 3300 V armoured cables having thermosetting insulation and low emission of smoke and corrosive gases when affected by fire.
BS 6883: 1999	Elastomer insulated cables for fixed wiring in ships and on mobile and fixed offshore units. Requirements and test methods.
BS 6907	Electrical installations for open cast mines and quarries.
BS 6972: 1988	Specification for general requirements for luminaire supporting couplers for domestic, light industrial and commercial use.
BS 6991: 1990	Specification for 6 110 amp two pole weather-resistant couplers for household, commercial and light industrial equipment.
BS 7001: 1988	Specification for interchangeability and safety of a standardised luminaire supporting coupler.
	Note: To be read in conjunction with BS 6972: 1988.
BS 7071: 1992 (1998)	Specification for portable residual current devices.
BS 7211: 1998	Specification for thermosetting insulated cables (non-armoured) for electric power and lighting with low emission of smoke and corrosive gases when affected by fire.
BS 7288: 1990 (1998)	Specification for socket outlets incorporating residual current devices (SRCDs).
BS 7430: 1998	Code of practice for earthing.
BS 7454: 1991	Method for calculation of thermally permissible short-circuit currents taking into account non-adiabatic heating effects.
BS 7629-1: 1997	Specification for 300 1500 V fire-resistant electric cables having low emission of smoke and corrosive gases when affected by fire. Multicore cables.
BS 7697	Nominal voltages for low-voltage public electricity supply systems.
BS 7769	Electric cables, Calculation of current rating. Parts 1, 2, 3: 1997.
BS 7822	Insulation coordination for equipment within low-voltage systems.
BS 7822-1: 1995	Insulation coordination for equipment within low-voltage systems.
	Principles, requirements and tests.
BS 7846: 2000	Electric cables. 600 1000 V armoured fire-resistant electric cables having thermosetting insulation and low emission of smoke and corrosive gases when affected by fire.
BS 7889: 1997	Specification for 600 1000 V single-core unarmoured electric cables having thermosetting insulation.

(*continued*)

E.2 Listed by standard (*continued*)

BS Number	Title
BS 7895: 1997	Specification for bayonet lampholders with enhanced safety.
BS 7919: 2001	Electric cables. Flexible cables rated up to *45on50* V, for use with appliances and equipment intended for industrial and similar environments.
BS 8436: 2004	300/500 V screened electric cables having low emission of smoke and corrosive gases when affected by fire, for use in thin partitions and building voids.
BS EN 1648	Leisure accommodation vehicles. 12 V direct current extra-low-voltage electrical installations.
BS EN 1648-1: 1998	Leisure accommodation vehicles. 12 V direct current extra-low-voltage electrical installations. Caravans.
BS EN 1648-2: 1998	Leisure accommodation vehicles. 12 V direct current extra-low-voltage electrical installations. Motor caravans.
BS EN 50014: 1998	Electrical apparatus for potentially explosive atmospheres. General requirements.
BS EN 50081	Electromagnetic compatibility. Generic emission standard.
BS EN 50082	Electromagnetic compatibility. Generic immunity standard.
BS EN 50085	Specification for cable trunking and ducting systems for electrical installations.
BS EN 50085-1: 1999	Specification for cable trunking and ducting systems for electrical installations. General requirements.
BS EN 50086	Specification for conduit systems for electrical installations.
BS EN 50086-1: 1994	Specification for conduit systems for electrical installations. General requirements.
BS EN 50086-2-1: 1996	Specification for conduit systems for electrical installations. Rigid conduit systems.
BS EN 50265	Common test methods for cables under fire conditions. Test for resistance to vertical flame propagation for a single insulated conductor or cable.
BS EN 50265-1: 1999	Common test methods for cables under fire conditions. Tests for resistance to vertical flame propagation for a single insulated conductor or cable. Apparatus.
BS EN 50265-2-1: 1999	Common test methods for cables under fire conditions. Tests for resistance to vertical flame propagation for a single insulated conductor or cable. Procedures. 1 kW pre-mixed flame.
BS EN 50265-2-2: 1999	Common test methods for cables under fire conditions. Tests for resistance to vertical flame propagation for a single insulated conductor or cable. Procedures. Diffusion flame.

(*continued*)

E.2 Listed by standard (*continued*)

BS Number	Title
BS EN 50281	Electrical apparatus for use in the presence of combustible dust.
BS EN 50310: 2000	Application of equipotential bonding and earthing in buildings with information technology equipment.
BS EN 60079	Electrical apparatus for potentially explosive gas atmospheres.
BS EN 60079-10: 1996	Electrical apparatus for potentially explosive gas atmospheres. Classification of hazardous areas.
BS EN 60079-14: 1997	Electrical apparatus for potentially explosive gas atmospheres. Electrical installations in hazardous areas (other than mines).
BS EN 60079-17: 1997	Electrical apparatus for potentially explosive gas atmospheres. Inspection and maintenance of electrical installations in hazardous areas (other than mines).
BS EN 60238: 1999	Specification for Edison screw lampholders.
BS EN 60269	Low-voltage fuses.
BS EN 60269-1: 1999	Low-voltage fuses. General requirements.
BS EN 60269-2: 1995	Low-voltage fuses. Supplementary requirements for fuses for use by authorised persons.
BS EN 60309	Plugs, socket outlets and couplers for industrial purposes.
BS EN 60309-1: 1999	Plugs, socket outlets and couplers for industrial purposes. General requirements.
BS EN 60309-2: 1998	Plugs, socket outlets and couplers for industrial purposes. Dimensional interchangeability requirements for pin and contact-tube accessories of harmonised configurations.
BS EN 60335-2-53: 1997	Electric sauna heating appliances.
BS EN 60423: 1995	Conduits for electrical purposes. Outside diameters of conduits for electrical installations and threads for conduits and fittings.
BS EN 60439	Specification for low-voltage switchgear and control gear assemblies.
BS EN 60439-1: 1999	Specification for low-voltage switchgear and control gear assemblies. Specification for type-tested and partially type-tested assemblies.
BS EN 60439-2: 1993	Specification for low-voltage switchgear and control gear assemblies. Particular requirements for busbar trunking systems (busways).
BS EN 60439-3: 1991	Specification for low-voltage switchgear and control gear assemblies.

(*continued*)

E.2 Listed by standard (*continued*)

BS Number	Title
	Particular requirements for low-voltage switchgear and control gear assemblies intended to be installed in places where unskilled persons have access to their use.
BS EN 60439-4: 1991	Specification for low-voltage switchgear and control gear assemblies. Particular requirements for Assemblies For Construction Sites (ACS).
BS EN 60445: 2000	Basic and safety principles for man-machine interface, marking and identification. Identification of equipment terminals and of terminations of certain designated conductors, including general rules for an alphanumeric system.
BS EN 60446: 2000	Basic and safety principles for the man-machine interface, marking and identification. Identification of conductors by colours or numerals.
BS EN 60570: 1997	Electrical supply track systems for luminaires.
BS EN 60598	Luminaries.
BS EN 60598-2-24: 1999	Luminaries. Luminaries with limited surface temperature.
BS EN 60617	Graphical symbols for diagrams.
BS EN 60669	Switches for household and similar fixed electrical equipment.
BS EN 60669-2-4	Switches for household and similar fixed electrical equipment. Particular requirements – isolating switches.
BS EN 60702-1: 2002	Mineral insulated cables and their terminations with a rated voltage not exceeding 750 V. Cables.
BS EN 60742: 1996	Isolating transformers and safety isolating transformers. Requirements.
BS EN 60898: 1991	Specification for circuit breakers for overcurrent protection for household and similar installations.
BS EN 60947	Specification for low-voltage switchgear and control gear.
BS EN 60947-1: 1999	Specification for low-voltage switchgear and control gear. General rules.
BS EN 60947-2: 1996	Specification for low-voltage switchgear and control gear. Circuit breakers.
BS EN 60947-3: 999	Specification for low-voltage switchgear and control gear. Switches, disconnecters, switch-disconnectors and fuse-combination units.
BS EN 60947-4	Specification for low-voltage switchgear and control gear. Contactors and motor starters.
BS EN 60947-4-1: 1992	Specification for low-voltage switchgear and control gear. Electromechanical contactors and motor starters.
BS EN 60950: 1992	Specification for safety of information technology equipment including electrical business equipment.
BS EN 61008	Residual current operated circuit breakers without integral overcurrent protection for household and similar uses (RCCBs).

(*continued*)

E.2 Listed by standard (*continued*)

BS Number	Title
BS EN 61008-1: 1995	Residual current operated circuit breakers without integral overcurrent protection for household and similar uses. General rules.
BS EN 61009	Residual current operated circuit breakers with integral overcurrent protection for household and similar uses (RCBOs).
BS EN 61009-1: 1995	Residual current operated circuit breakers with integral overcurrent protection for household and similar uses. General rules.
BS EN 61011: 1993	Electric fence energisers. Safety requirements for mains operated electric fence energisers.
BS EN 61011-1: 1993	Electric fence energisers. Safety requirements for battery operated electric fence energisers suitable for connection to the supply mains.
BS EN 61184: 1997	Bayonet lampholders.
BS EN 61558	Safety of power transformers, power supply units and similar.
BS EN 61558-2-1: 1998	Safety of power transformers, power supply units and similar. General requirements and tests.
BS EN 61558-2-17: 1998	Safety of power transformers, power supply units and similar. Particular requirements for switch mode power supplies.
BS EN 61558-2-2: 1998	Safety of power transformers, power supply units and similar. Particular requirements for separating transformers for general use.
BS EN 61558-2-4: 1998	Safety of power transformers, power supply units and similar. Particular requirements for isolating transformers for general use.
BS EN 61558-2-5: 1998	Safety of power transformers, power supply units and similar. Particular requirements for shaver transformers and shaver units.
BS EN 61558-2-6	Safety of power transformers, power supply units and similar. Particular requirements for safety isolating transformers for general use.
BS EN 61558-2-8: 1999	Safety of power transformers, power supply units and similar. Particular requirements for bell and chime transformers.

Appendix F
Useful contacts

BSI
389 Chiswick High Road
London W4 4AL
Telephone: +44 (0)20 8996 9000
Fax: +44 (0)20 8996 7001
email: cservices@bsi-global.com
Web: www.bsi-global.com

As the National Standards Body for the UK, BSI are responsible for developing standards and standardising solutions to meet the needs of business and society.

ECA
ESCA House
34 Palace Court
London W2 4HY
Tel: 020 7313 4800
Fax: 020 7221 7344
email:electricalcontractors@eca.co.uk
Web: www.eca.co.uk

The aims of the Electrical Contractors Association's are to:
- promote quality and safety through:
  - the qualification of companies, the training, qualification and reward of individuals
- promote the compliance of all electrical and related installation work to relevant standards, encourage the adoption of beneficial new technologies and installation practices
- influence the market to ensure that there is an equitable commercial environment.

ELECSA Limited
44–48 Borough High Street
London SE1 1XB
Tel: +44 (0) 870 749 0080
Fax: +44 (0) 870 749 0085
email: enquiries@elecsa.org.uk
Web: www.elecsa.org.uk

The Fenestration Self-Assessment Scheme is the recognised competent person's scheme for Approved Document P of the Building Regulations.

IET
Savoy Place
London WC2R 0BL
Tel: + 44 (0)20 7240 1871
Fax: + 44 (0)20 7240 7735
email: postmaster@theiet.org.
Web: www.theiet.co.uk

The IET was formed in spring 2006 by bringing together the IEE (Institution of Electrical Engineers) and the IIE (Institution of Incorporated Engineers).
The IET is Europe's largest professional society for engineers.

NAPIT
Suite L4A, Mill 3
Pleasley Vale Business Park
Mansfield
Nottinghamshire
NG19 8RL
Tel: 0870 4441392
Fax: 0870 4441427
email: info@napit.org.uk
Web: www.napit.org.uk

The National Association of Professional Inspectors and Testers provides an independent professional trade body for electrical inspectors, electrical contractors, electricians and allied trades throughout the UK.

(continued)

Useful contacts (*continued*)

NICEIC Warwick House Houghton Hall Park Houghton Regis, Dunstable Bedfordshire LU5 5ZX Tel: 01582 531000 Fax: 01582 531010 email: enquiries@niceic.com Web: niceic.org.uk	The National Inspection Council for Electrical Installation Contracting is the industry's independent, non profit-making, voluntary regulatory body covering the whole of the United Kingdom. The NICEIC's sole purpose is to protect consumers from unsafe and unsound electrical work. They are not a trade association and do not represent the interests of electrical contractors.
Department for Communities and Local Government (DCGL) Eland House Bressenden Place London SW1E 5DU Tel: 020 7944 4400 Fax: 020 7944 9645 email: contactus@communities.gsi.gov.uk Web: www.communities.gov.uk	The job of the DCLG is to help create sustainable communities, working with other government departments, local councils, businesses, the voluntary sector, and communities themselves.

Appendix G

Books by the same author

Title	Details	Publisher
Building Regulations in Brief (fourth edition) 	Handy reference guide to the requirements of the Building Act and its associated Approved Documents. Aimed at experts as well as DIY enthusiasts and those undertaking building projects	Butterworth-Heinemann ISBN: 0 7506 8058 X
Environmental Requirements for Electromechanical and Electronic Equipment	Definitive reference containing all the background guidance, ranges, test specifications, case studies and regulations worldwide.	Butterworth-Heinemann ISBN: 0 7506 3902 4
CE Conformity Marking	Essential information for any manufacturer or distributor wishing to trade in the European Union. Practical and easy to understand.	Butterworth-Heinemann ISBN: 0 7506 4813 9

Books by the same author (*continued*)

Title	Details	Publisher
ISO 9001:2000 for Small Businesses (third edition) 	New edition of this top-selling quality management handbook. Contains a full description of the ISO 9001: 2000 standard plus detailed information on quality control and quality assurance. Fully updated following 4 years practical field experience of the standard. Includes a sample quality manual (that can be customised to suit individual requirements) and assistance on self-certification etc.	Butterworth-Heinemann ISBN: 0 7506 6617 X
ISO 9001:2000 Audit Procedures (second edition) 	A complete set of audit check sheets and explanations to assist quality managers and auditors in completing internal, external and third part audits of ISO 9001:2000 Quality Management Systems.	Butterworth-Heinemann ISBN: 0 7506 6615 3
ISO 9001:2000 in Brief (second edition) 	Revised and expanded, this new edition of an easy-to-understand guide provides practical information on how to set up a cost-effective ISO 9001:2000 compliant Quality Management System	Butterworth-Heinemann ISBN: 0 7506 6616 1
ISO 9001:2000 Quality Manual & Audit Checksheets	A CD containing a soft copy of the generic quality management system featured in ISO 9001:2000 for Small Businesses (3rd edition) plus a soft copy of all the check sheets and example audit forms contained in ISO 9001:2000 Audit Procedures (2nd edition).	ISBN 0-9548647-2-7

(*continued*)

Books by the same author (*continued*)

Title	Details	Publisher
ISO 9001:2000 Audit Checklists (edition two)	A CD containing all of the major audit check sheets and forms required to conduct either: • a simple internal review of a particular process • an external (e.g. third party) assessment of an organisation against the formal requirements of ISO 9001:2000 • an assessment of an organisation against the formal requirements of other associated management standards such as ISO 140001 and OHSAS etc.	ISBN 0-9548647-1-9
Aide Mémoire for ISO 9001: 2000 (edition two)	An electronic book (in CD format) that cuts out all the lengthy explanations, hypothesis and discussions about ISO 9001:2000 and just provides the reader with the bare details of the standard (e.g. What does the standard actually say? What needs to be checked? Where is a particular requirement covered in the standard? What will an auditor be looking for? etc.). Also includes additional auditors checksheets and example forms etc. from *ISO 9001:2000 Audit Procedures* (edition 1) and *ISO 9001:2000 for Small Businesses* (edition 2).	ISBN 0-9548647-0-0
Quality and Standards in Electronics 	Ensures that manufacturers are aware of all the UK, European and international necessities, knows the current status of these regulations and standards, and where to obtain them.	Butterworth-Heinemann ISBN: 0 7506 2531 7

Books by the same author (*continued*)

Title	Details	Publisher
Optoelectronic and Fiber Optic Technology	An introduction to the fascinating technology of fiber optics	Butterworth-Heinemann ISBN 0 7506 5370 1
MDD Compliance using Quality Management Techniques	Easy-to-follow guide to MDD, enabling the purchaser to customise the quality management system to suit his own business.	Butterworth-Heinemann ISBN: 0 7506 4441 9

Index